George Gheverghese Joseph

◊◊◊

The Crest of the Peacock

The Non-European Roots of Mathematics

Princeton University Press
Princeton and Oxford

Published in the United States and Canada by
Princeton University Press, 41 William Street, Princeton, N.J. 08540

First published by I. B. Tauris 1991
Published in Penguin Books 1992
Reprinted with minor revisions 1993
Reprinted with revisions 1994
Reprinted with additional material 2000

Library of Congress Catalog Card Number: 00-102425

ISBN 0-691-00659-8

Printed in England by Clays Ltd, St Ives plc

The paper used in this publication meets the minimum requirements of
ANSI NISO Z39.48-1992 (R1997) (*Permanence of Paper*)

www.pup.princeton.edu

1 3 5 7 9 10 8 6 4 2

Like the crest of a peacock, like the gem on the head of
a snake, so is mathematics at the head of all knowledge.

Vedanga Jyotisa (*c.* 500 BC)

Contents

◇◇

10 Prelude to Modern Mathematics:

Acknowledgements

◇◇

I am grateful to authors and publishers for the use of copyright material
reproduced or adapted with permission in the following:

Figures 2.2, 2.7, 2.9, 3.3, 3.4(a), 4.2(a), 4.3, 4.7(a), 6.6, 7.1, 7.6(a), 7.9(a), 8.9
and 8.10; Tables 9.1 and 9.2.

Figure 2.8 is reprinted from *Native American Mathematics*, M.D. Closs (ed.),
by permission of the author and the University of Texas Press.

Every effort has been made to obtain permission from relevant copyright
holders. The publishers would like to hear from copyright holders of any
material reproduced in this book which is not properly acknowledged.

Note on foreign words and names

In this book I have decided, for simplicity, to omit diacritical marks from words
and names transliterated from languages such as Sanskrit and Arabic. For
Chinese words and names, the historically familiar forms of the older Wade–
Giles system have, in some cases, been used in preference to the modern Pinyin
transliterations.

G.G.J.

Preface to the First Edition

◇◇

In 1987 I visited the birthplace of the Indian mathematician Srinivasa Ramanujan. Exactly a hundred years had passed since Ramanujan was born in a small town called Erode in southern India. At his death, aged 32, he was recognized by some as a natural genius, the like of whom could be found only by going back two centuries to Euler and Gauss. Among his contemporaries, particularly his close collaborator G.H. Hardy, there was a sense of disappointment – the feeling that Ramanujan's ignorance of modern mathematics, his strange ways of 'doing' mathematics and his premature death had diminished his achievements and therefore his influence on the future of the subject. Yet today few mathematicians would accept this assessment. In 1976, George Andrews, an American mathematician, was rummaging through some of Ramanujan's papers in a library at Cambridge University and came across a hundred and thirty pages of scrap paper filled with notes representing Ramanujan's work during the last year of his life in Madras. This is what Richard Askey, a collaborator of Andrews, had to say about what has come to be known as Ramanujan's 'Lost Notebook':

> The work of that one year, while he was dying [and obviously in considerable pain a lot of the time, according to his wife], was the equivalent of a lifetime of work of a very great mathematician. What he accomplished was unbelievable. If it were in a novel, nobody would believe it.

(My comment in brackets.) The riches contained in the 'Lost Notebook' and his earlier work are being mined with increasing success and excitement by mathematicians today. They have contributed to the creation of one of the most revolutionary concepts of recent theoretical physics – superstring theory in cosmology. A 1914 paper of Ramanujan on 'Modular Equations and Approximations to π' was used to program a computer a few years ago to evaluate π to a level of accuracy (to millions of digits) never attained previously. But one should not make excessive claims for the practical applications of Ramanujan's work. As Kanigel (1991, pp. 349–50) states

> What makes Ramanujan's work so seductive is not the prospect of its use in the solution of real-world problems, but its richness, beauty and mystery – its sheer mathematical loveliness.

However, for me the most intriguing aspect of Ramanujan's mathematical work remains his method. Here was someone poorly educated in modern mathematics and isolated for most of his life from work going on at the frontiers of the subject, yet who produced work of a quality and durability that is increasingly tending to overshadow some of his more prominent contemporaries, including Hardy. Ramanujan's style of doing mathematics was very different from that of the conventional mathematician trained in the deductive axiomatic method of proof. From the accounts of his wife and close associates, he made extensive use of a slate on which he was always jotting down and erasing what his wife described as 'sums', and then transferring some of the final results into his Notebook when he was satisfied with his conclusions. He felt no strong compulsion to prove that the results were true – what mattered were the results themselves. This has provided a growing number of mathematicians with a singular task: to prove the results that Ramanujan simply stated. And from the endeavours of these mathematicians have emerged a number of sub-disciplines, promoting gatherings and collaborations among their practitioners, whose approach stands in stark contrast to that taken by the original inspirer.

In writing this book, I found the life and work of Ramanujan instructive because it raises a number of interesting questions. First (and this is a question that is rarely addressed by historians of mathematics, one for which there can in any case never be a fully satisfactory answer), how far did cultural influences determine Ramanujan's choice of subjects or his methods? It is interesting in this context that Ramanujan came from the Ayyangar Brahmans of Tamil Nadu in southern India, a group that enjoyed a high social status for their traditional learning and religious observances. Given this background, Ramanujan's tendency to credit his discoveries to the intervention of the family goddess, Namagiri, is understandable, though it must have been a source of embarrassment to some of his admirers, both in India and in the West. But it is perfectly consistent with a culture which saw mathematics in part as an instrument of divine intervention and astrological prediction. The Western mathematical temperament finds it difficult to come to terms with the speculative, extra-rational and intuitive elements in Ramanujan's make-up.

At another level, the example of Ramanujan is a sure indication that the highest level of mathematical achievement is well within the scope of those educated and brought up in traditions and environments far removed from Western society. However, a second question, an interesting and indeed a central one, is raised by Ramanujan's work: is it possible to identify any features in his own culture which were conducive to creative work in mathematics? Any attempt to answer this question should delve deeply into the role of Ayyangars as custodians of traditional knowledge of astronomy and mathematics. Ramanujan's mother was a well-known local astrologer, and it is likely that his first exposure to mathematics, and in particular his special interest in the theory of numbers, came about through his mother's astrology. Mathematics and numbers had a special significance within the Brahmanical tradition as extra-rational instruments for controlling fate and nature.

Ramanujan's work also raises questions about what constitutes mathematics. Is there a need to conform to a particular method of presentation before something is recognizable as mathematics? His notebooks contain many jottings which do not conform to a conventional view of mathematical results, since there is no attempt at any demonstration or examination of the theory behind these results. Yet a number of mathematicians not only have found these jottings sufficiently worth while to devote years of their time to proving theorems Ramanujan knew to be true, but may even have gained more from the very act of deriving the formulae than the knowledge of the formulae themselves. This is quite consistent with both the Indian and Chinese traditions, where great mathematicians merely state the results, leaving their students to provide oral demonstrations or written commentaries. The students are thus encouraged to allow both their critical and their creative faculties to develop at the same time.

An author is not expected to explain why he writes a book. But the motives are often quite revealing. If I am to explain why I have spent the last three years working on this book, I would think my being a product of four different heritages is highly relevant. I was born in Kerala, southern India, and spent the first nine years of my life there and in the town of Madurai, the cultural capital of the neighbouring state of Tamil Nadu. My early awareness of the sheer diversity of Indian culture was helped by living close to the famous Meenakshi Temple at Madurai, a great centre of pilgrimage, dance, music and religious festivals. This exposure during my formative years contributed to my Indian heritage. I come from a family of Syrian Orthodox Christians which traces its descent directly to one of the families (the Sankarapuri) who were converted by Christ's disciple St Thomas in about AD 50. This is my Middle Eastern Christian heritage. My family moved to Mombasa in Kenya, where I was brought up in that rich mixture of African and Arab influences that makes the distinctive Swahili culture. My African heritage is a result of the time I spent there, first at school and then at work, both in Mombasa and in the neighbouring country of Tanzania. The period I have spent in Britain, both at the University of Leicester, where I did my first degree, and at the University of Manchester, where I continued my postgraduate studies and subsequently worked, now accounts for more than half my life. This is my Western heritage. To keep a balance between my four heritages and not allow any one of them to take over permanently is important to me. Hence my travels and jobs abroad, which have taken me to East and Central Africa, to India, to Papua New Guinea, and to South and South East Asia. And hence, in a different way, the driving passion behind this book, which emphasizes the global nature of mathematical pursuits and creations.

In writing this book, I am indebted to many who have over the years patiently and skilfully translated and interpreted the original sources of mathematics from different cultures so that they are now more accessible and comprehensible. I owe them more than I can acknowledge merely through entries in the Bibliography at the back of this book. They have often had to work in environments which are not particularly sympathetic to their efforts, and have rarely received sufficient academic recognition. In a number of cases, their attempts at collecting and transliterating ancient manuscripts show a desperate sense of urgency, as the storage and preservation of these documents often leave much to be desired.

During the time I have been working on this book, several people have given me advice, constructive criticism and encouragement. Burjor Avari and I have shared a long and close association which has taken the form, among other things, of a study of the nature and consequences of a Eurocentric view of the history of knowledge. Our collaboration in this area is clearly reflected in some of the ideas found in the first chapter. In particular, the historical backgrounds to a number of chapters have benefited from his criticisms. David Nelson read the whole manuscript carefully and suggested a number of changes to improve the clarity and balance of the book. I found his comments so useful and persuasive that I have tried in almost all cases to respond to them. Even so, I am conscious of having fallen short of the thoroughness that his detailed comments deserve. To Bill Farebrother I am grateful for having gone through the manuscript at various stages, making useful criticisms of the style and mathematical presentation. I should also like to acknowledge my debt to other colleagues in the Department of Econometrics and Social Statistics, University of Manchester, who not only tolerated my project (as removed as it was from the usual concerns of the Department) but in some cases went through individual chapters and provided constructive responses. Finally, at the stage of preparing the manuscript for publication, John Woodruff's role has been invaluable. Not only has he provided meticulous editorial assistance in spotting ambiguities, omissions and inconsistencies in the text and bibliography, but at various places he has suggested changes which have significantly improved the presentation. It is appropriate, given my insufficient response to some of the advice offered me, that I exclude all those mentioned above from responsibility for any errors of fact and interpretation present in this book.

GEORGE GHEVERGHESE JOSEPH
University of Manchester
1991

Preface to the Second Edition

◇◇

In April 1991, the first edition of *The Crest of the Peacock* came out in print. Though commissioned by Penguin Books, they had decided in their wisdom to sell the hardback rights to a small and respectable publisher, I. B. Tauris, who had established a reputation for publishing scholarly books on the Middle East. The thinking behind this action was to have the book noticed, and widely reviewed, before Penguin brought out a paperback edition a year later. To my surprise, the book that was intended to be a history of mathematics received notice in a number of unlikely quarters. An important purpose of writing the book was to set some accounts straight – to correct a widely held view that mathematics was essentially a European product. This has gratifyingly found resonance among many all over the world.

In this second edition of the book I have retained the basic format, content and structure of the first. The additions, updates and second thoughts are incorporated into notes on each chapter appended at the end of the book in a new section entitled 'Reflections'. These notes are fairly substantial and will, hopefully, bring some of the more recent research, evidence and thinking to the attention of the reader. The topics are covered in the notes in the same order as in the chapters. The bibliography has been expanded at the end of the book to enable the reader to follow up new references. However, I would encourage any reader coming to this book for the first time to read the main text before examining the notes. For those who have read the first edition, it is hoped that the material in the notes would prove to be of sufficient interest in themselves so as to promote a greater awareness of the global heritage of mathematics.

I owe a debt to literally hundreds of people (including the reviewers) from all over the world who have taken the trouble to speak or write about the first edition. I greatly value the friendship and the warmth that have resulted from these encounters. They will notice, I hope, the improvements and additions due to them, both in the main text and in the notes. There are, however, several people that I would like to single out for their wise contributions during the preparation of this edition: Milo Gardner, Takao Hayashi, Ma Li, Beatrice Lumpkin and Bobby Sayyid have devoted so much of their valuable time in commenting on various sections of the book. I found their comments so persuasive that I have tried in almost all cases to respond to them. I owe them a great debt of gratitude but, at the same time, I absolve them from responsibility for any errors of fact and interpretation present in this book. Finally, there is one person who has consistently encouraged me in all my endeavours and provided the necessary confidence and fortitude to complete the task, and that has been my wife, Leela. She has also been a wonderful travelling companion on the trips we have made together to different parts of the world to promote the ideas in this book. To her, my deepest gratitude and love.

GEORGE GHEVERGHESE JOSEPH
University of Manchester
November 1999

1 The History of Mathematics: Alternative Perspectives

◇◇◇

A JUSTIFICATION FOR THIS BOOK

An interest in history marks us for life. How we see ourselves and others is shaped by the history we absorb, not only in the classroom but from films, newspapers, television programmes, novels and even strip cartoons. From the time we first become aware of the past, it can fire our imagination and excite our curiosity: we ask questions and then seek answers from history. As our knowledge develops, differences in historical perspectives emerge. And, to the extent that different views of the past affect our perception of ourselves and of the outside world, history becomes an important point of reference in understanding the clash of cultures and of ideas. Not surprisingly, rulers throughout history have recognized that to control the past is to master the present and thereby consolidate their power.

During the last four hundred years, Europe and her cultural dependencies* have played a dominant role in world affairs. This is all too often reflected in the character of some of the historical writing produced by Europeans. Where other people appear, they do so in a transitory fashion whenever Europe has chanced in their direction. Thus the history of the Africans or the indigenous peoples of the Americas often appears to begin only after their encounter with Europe.

An important aspect of this Eurocentric approach to history is the manner in which the history and potentialities of non-European societies are represented, particularly with respect to their creation and development of science and technology. The progress of Europe during the last four hundred years is often inextricably – or even causally – linked with the rapid growth

---

*The term 'cultural dependencies' is used here to describe those countries – notably the United States, Canada, Australia and New Zealand – which are inhabited mainly by populations of European origin or with similar historical and cultural roots. For the sake of brevity, the term 'Europe' is used hereafter to include these areas as well.

of science and technology during that period. In the minds of some, scientific progress becomes a uniquely European phenomenon which can be emulated by other nations only if they follow a specifically European path of scientific and social development.

Such a representation of societies outside the European cultural milieu raises a number of issues which are worth exploring, however briefly. First, these societies, many of them still in the grip of an intellectual dependence that is the legacy of European political domination, should ask themselves some questions. Was their indigenous scientific and technological base innovative and self-sufficient during their pre-colonial period? Recent case studies of India, China and parts of Africa, contained, for example, in the work of Dharampal (1971), Needham (1954) and Van Sertima (1983), seem to indicate the existence of scientific creativity and technological achievements long before the incursions of Europe into these areas. If this is so, we need to understand the dynamics of pre-colonial science and technology in these and other societies and to identify the material conditions that gave rise to these developments. This is essential if we are to see why modern science did not develop in these societies, only in Europe, and to find meaningful ways of adapting to present-day requirements the indigenous and technological forms that still remain.

Second, there is the wider issue of who 'makes' science and technology. In a material and non-elitist sense, each society, impelled by the pressures and demands of its environment, has found it necessary to create a scientific base to cater for its material requirements. The perceptions of what constitute the particular requirements of a society would vary according to time and place, but it would be wrong to argue that the capacity to 'make' science and technology is a prerogative of one culture alone.

Third, if one attributes all significant historical developments in science and technology to Europe, then the rest of the world can impinge only marginally, either as an unchanging residual experience to be contrasted with the dynamism and creativity of Europe, or as a rationale for the creation of academic disciplines congealed in subjects such as Development Studies, Anthropology and Oriental Studies. These subjects in turn serve as the basis from which more elaborate Eurocentric theories of social development and history are developed and tested.

One of the more heartening aspects of academic research in recent years is that the shaky foundations of these 'adjunct' disciplines are being increasingly exposed by scholars, a number of whom originate from countries which provide the subject-matter of these disciplines. 'Subversive' analyses aimed at nothing less than the modification or destruction of prevailing Eurocentric paradigms have become the major preoccupation of many of these scholars. Syed Husain Alatas (1976) has studied intellectual dependence and imitative thinking among social scientists in developing countries. The growing movement towards promoting a form of indigenous anthropology which sees its primary task as questioning, redefining and, if necessary, rejecting particular concepts which grew out of colonial experience in Western anthropology is thoroughly examined by Fahim (1982). Edward Said (1978) has brilliantly described the motives and methods of the so-called orientalists who set out to construct a fictitious entity called 'the Orient' and then ascribe to it qualities which are a mixture of the exotic, the mysterious and the other-worldly. The rationale for such constructs needs to be examined in terms of the recent history of Europe's relations with the rest of the world.

In a similar vein, I propose to show in this chapter that the standard treatment of the history of non-European mathematics exhibits a deep-rooted historiographical bias in the selection and interpretation of facts, and that mathematical activity outside Europe has as a consequence been ignored, devalued or distorted. Both as a foretaste of subsequent chapters and as a means of presenting the principal issues relating to Eurocentrism in the history of mathematics, I begin by examining different views of the origins and growth of mathematical knowledge.

THE DEVELOPMENT OF MATHEMATICAL KNOWLEDGE

A concise and meaningful definition of mathematics is virtually impossible. In the context of this book, the following aspects of the subject are highlighted. Modern mathematics has developed into a worldwide language with a particular kind of logical structure. It contains a body of knowledge relating to number and space, and prescribes a set of methods for reaching conclusions about the physical world. And it is an intellectual activity which calls for both intuition and imagination in deriving 'proofs' and reaching conclusions. Often it rewards the creator with a strong sense of aesthetic satisfaction.

The 'classical' Eurocentric trajectory

Most histories of mathematics that were to have a great influence on later work were written in the late nineteenth or early twentieth century. During that period, two contrasting developments were taking place which had an impact on both the content and the balance of these books, especially those produced in Britain and the United States. Exciting discoveries of ancient mathematics on papyri in Egypt and clay tablets in Mesopotamia pushed back the origins of written mathematical records by at least 1500 years. But a far stronger and adverse influence was the culmination of European domination in the shape of political control of vast tracts of Africa and Asia. Out of this domination arose the ideology of European superiority which permeated a wide range of social and economic activities, with traces to be found in histories of science that emphasized the unique role of Europe in providing the soil and spirit for scientific discovery. The contributions of the colonized peoples were ignored or devalued as part of the rationale for subjugation and dominance. And the development of mathematics before the Greeks − notably in Egypt and Mesopotamia − suffered a similar fate, dismissed as of little importance to the later history of the subject. In his book *Black Athena* (1987), Martin Bernal has shown how the respect for ancient Egyptian science and civilization, shared by ancient Greece and pre-nineteenth-century Europe alike, was gradually eroded, leading eventually to a Eurocentric model with Greece as the source and Europe as the inheritor and guardian of the Greek heritage.

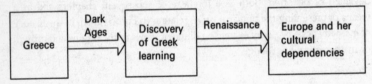

Figure 1.1 The 'classical' Eurocentric trajectory.

Figure 1.1 presents the 'classical' Eurocentric view of how mathematics developed over the ages. This development is seen as taking place in two sections, separated by a period of stagnation lasting for over a thousand years: Greece (from about 600 BC to AD 400), and post-Renaissance Europe from the sixteenth century to the present day. The intervening period of inactivity was the 'Dark Ages' − a convenient label that expressed both post-Renaissance Europe's prejudices about its immediate past and the intellectual self-confidence of those who saw themselves as the true inheritors of the 'Greek miracle' of two thousand years earlier.

Two passages, one by a well-known historian of mathematics writing at the turn of the century, and the other by a contemporary writer whose books are still widely referred to on both sides of the Atlantic, show the durability of this Eurocentric view and its imperviousness to new evidence and sources:

> The history of mathematics cannot with certainty be traced back to any school or period before that of the Ionian Greeks. (Rouse Ball, 1908, p. 1)

[Mathematics] finally secured a new grip on life in the highly congenial soil of Greece and waxed strongly for a short period . . . With the decline of Greek civilization the plant remained dormant for a thousand years . . . when the plant was transported to Europe proper and once more imbedded in fertile soil. (Kline, 1953, pp. 9–10)

The first statement is a reasonable summary of what was popularly known and accepted as the origins of mathematics at that time, except for the neglect of the early Indian mathematics contained in the *Sulbasutras* ('The Rules of the Cord'), belonging to the period between 800 and 500 BC, which would make them at least as old as the earliest-known Greek mathematics. Thibaut's translations of these works, made around 1875, were known to historians of mathematics at the turn of the century. The mathematics contained in the *Sulbasutras* is discussed in Chapter 8.

The second statement, however, ignores a considerable body of research evidence pointing to the development of mathematics in Mesopotamia, Egypt, China, pre-Columbian America, India and the Arab* world that had come to light in the intervening period. Subsequent chapters will bear testimony to the volume and quality of the mathematics developed in these areas. But in both these quotations mathematics is perceived as an exclusive product of European civilization. And that is the central message of the Eurocentric trajectory described in Figure 1.1.

This comforting rationale for European dominance has become increasingly untenable for a number of reasons. First, there is the full acknowledgement given by the ancient Greeks themselves of the intellectual debt they owed the Egyptians. There are scattered references from Herodotus (*c.* 450 BC) to Proclus (*c.* AD 400) of the knowledge acquired from the Egyptians in fields such as astronomy, mathematics and surveying, while other commentators even considered the priests of Memphis to be the true founders of science.

To Aristotle (*c.* 350 BC), Egypt was the cradle of mathematics. His teacher, Eudoxus, one of the notable mathematicians of the time, had studied in Egypt before teaching in Greece. Even earlier, Thales (d. 546 BC), the legendary founder of Greek mathematics, and Pythagoras (*c.* 500 BC), one of the earliest and greatest of Greek mathematicians, were reported to have travelled widely

---

*The term 'Arab' is used in this book to refer to a civilization which contained a number of other ethnic, religious and linguistic groups. The Arab civilization included 'non-Arab' lands, such as present-day Iran, Turkey, Afghanistan and Pakistan, all of which have distinctive Islamic cultures. For further discussion of some of the difficulties of choosing 'labels' to describe different mathematical traditions, see the notes for this chapter in 'Reflections' (pp. 349–52).

in Egypt and Mesopotamia and learnt much of their mathematics from these areas. Some sources even credit Pythagoras with having travelled as far as India in search of knowledge, which may explain some of the close parallels between Indian and Pythagorean philosophy and religion.*

A second reason why the trajectory described in Figure 1.1 was found to be untenable arose from the combined efforts of archaeologists, translators and interpreters, who between them unearthed evidence of a high level of mathematics practised in Mesopotamia and in Egypt at the beginning of the second millennium BC, providing further confirmation of Greek reports. In particular, the Babylonians (a generic term that is often used to describe all inhabitants of ancient Mesopotamia) had invented a place-value number system, knew different methods of solving quadratic equations (which would not be improved upon until the sixteenth century AD) and understood (but had not proved) the relationship between the sides of a right-angled triangle which came to be known as the Pythagorean theorem. Indeed, as we shall see in later chapters, this theorem was stated and demonstrated in different forms all over the world.

A four-thousand-year-old Babylonian clay tablet, kept in a Berlin museum, gives the value of $n^3 + n^2$ for $n = 1, 2, \ldots, 10, 20, 30, 40, 50$, from which it has been surmised that the Babylonians may have used these values in solving cubic equations after reducing them to the form $x^3 + x^2 = c$. A remarkable solution in Egyptian geometry, which is found in the so-called Moscow Papyrus, dating from the Middle Kingdom (*c.* 2000–1800 BC), follows from the correct use of the formula for the volume of a truncated square pyramid. These examples and other milestones will be discussed in the relevant chapters of this book.

The neglect of the Arab contribution to the development of European intellectual life in general and mathematics in particular is another serious drawback of the 'classical' view. The course of European cultural history and the history of European thought are inseparably tied up with the activities of Arab scholars during the Middle Ages and their seminal contributions to

---

*These parallels include: (a) a belief in the transmigration of souls; (b) the theory of four elements constituting matter; (c) the reasons for not eating beans; (d) the structure of the religio-philosophical character of the Pythagorean fraternity, which resembled Buddhist monastic orders; and (e) the contents of the mystical speculations of the Pythagorean schools, which bear a striking resemblance to the Hindu *Upanishads*. According to Greek tradition, Pythagoras, Thales, Empedocles, Anaxagoras, Democritus and others undertook journeys to the East to study philosophy and science. While it is far-fetched to assume that all these individuals reached India, there is a strong historical possibility that some of them became aware of Indian thought and science through Persia.

mathematics, the natural sciences, medicine and philosophy. In particular, we owe to the Arabs in the field of mathematics the bringing together of the technique of measurement, evolved from its Egyptian roots to its final form in the hands of the Alexandrians, and the remarkable instrument of computation (our number system) which originated in India; and the supplementing of these strands with a systematic and consistent language of calculation which came to be known by its Arabic name, algebra. An acknowledgement of this debt in more recent books contrasts sharply with a failure to recognize other Arab contributions to science.*

Finally, in discussing the Greek contribution, there is a need to recognize the differences between the Classical period of Greek civilization (i.e. from about 600 to 300 BC) and the post-Alexandrian period (i.e. from about 300 BC to AD 400). In early European scholarship, the Greeks of the ancient world were perceived as an ethnically homogeneous group, originating from areas which were mainly within the geographical boundaries of present-day Greece. It was part of the Eurocentric mythology that from the mainland of Europe had emerged a group of people who had created, virtually out of nothing, the most impressive civilization of ancient times. And from that civilization had emerged not only the cherished institutions of present-day Western culture, but also the mainspring of modern science. The reality, however, is different and more complex.

The term 'Greek', when applied to times before the appearance of Alexander (356–323 BC), really refers to a number of independent city-states, often at war with one another, but exhibiting close ethnic or cultural affinities and, above all, sharing a common language. The conquests of Alexander changed the situation dramatically, for at his death his empire was divided among his generals, who established separate dynasties. The two notable dynasties from the point of view of mathematics were the Ptolemaic dynasty of Egypt, and the Seleucid dynasty which ruled over territories that included the earlier sites of the Mesopotamian civilization. The most famous centre of learning and trade

---

*They include: (a) an early description of pulmonary circulation of the blood, by ibn al-Nafis, usually attributed to Harvey, though there are records of an even earlier explanation in China; (b) the first known statement about the refraction of light, by ibn al-Hayatham, usually attributed to Newton; (c) the first known scientific discussion of gravity, by al-Khazin, again attributed to Newton; (d) the first clear statement of the idea of evolution, by ibn Miskawayh, usually attributed to Darwin; and (e) the first exposition of the rationale underlying the 'scientific method', found in the works of ibn Sina, ibn al-Haytham and al-Biruni, but usually credited to Francis Bacon. A general discussion of the Western debt to the Middle East is given by Savory (1976), while detailed references to specific contributions of Arab science are given by Gillespie (1969–).

became Alexandria in Egypt, established in 332 BC and named after the Conqueror. From its foundation, one of its most striking features was its cosmopolitanism – part Egyptian, part Greek, with a liberal sprinkling of Jews, Persians, Phoenicians and Babylonians, and even attracting scholars and traders from as far away as India. A lively contact was maintained with the Seleucid dynasty. Alexandria thus became the meeting-place for ideas and different traditions. The character of Greek mathematics began to change slowly, mainly as a result of the continuing cross-fertilization between different mathematical traditions, notably the algebraic and empirical basis of Babylonian and Egyptian mathematics interacting with the geometric and anti-empirical traditions of Greek mathematics. And from this mixture came some of the greatest mathematicians of Antiquity, notably Archimedes and Diophantus. It is therefore misleading to speak of Alexandrian mathematics as Greek, except in so far as the term indicates that Greek intellectual and cultural traditions served as the main inspiration and the Greek language as the medium of instruction and writing in Alexandria.

A modified Eurocentric trajectory

Figure 1.2 takes on board some of the objections raised about the 'classical' Eurocentric trajectory. The figure acknowledges that there is some awareness of the existence of mathematics before the Greeks, and of their debt to earlier mathematical traditions, notably those of Babylonia and Egypt. But this awareness is all too likely to be tempered by a dismissive rejection of their importance in relation to Greek mathematics: the 'scrawling of children just learning to write as opposed to great literature' (Kline, 1962, p. 14).

The differences in character of the Greek contribution before and after Alexander are also recognized to a limited extent in Figure 1.2 by the separation of Greece from the Hellenistic world (in which the Ptolemaic and Seleucid dynasties became the crucial instruments of mathematical creation). There is also some acknowledgement of the Arabs, but mainly as custodians of Greek learning during the Dark Ages in Europe. Their role as transmitters and creators of knowledge is ignored; so are the contributions of other civilizations – notably China and India – which have been perceived either as borrowers from Greek sources, or as having made only minor contributions that play an insignificant role in mainstream mathematical development (i.e. the development eventually culminating in modern mathematics).

Figure 1.2 is therefore still a flawed representation of how mathematics developed: it contains a series of biases and remains quite impervious to new

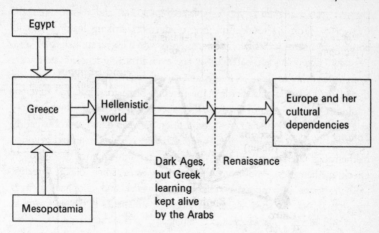

Figure 1.2 A modified Eurocentric trajectory.

evidence and arguments. With minor modifications, it is nevertheless the model to which many recent books on the history of mathematics conform. It is interesting that a similar Eurocentric bias exists in other disciplines as well: for example, diffusion theories in anthropology and social geography indicate that 'civilization' spreads from the Centre ('Greater' Europe) to the Periphery (the rest of the world). And the theories of modernization or evolution developed within some Marxist frameworks are characterized by a similar type of Eurocentrism. In all such conceptual schemes, the development of Europe is seen as a precedent for the way in which the rest of the world will follow — a trajectory whose spirit is not dissimilar to the one suggested in Figures 1.1 and 1.2.

An alternative trajectory for the Dark Ages

If we are to construct an unbiased alternative to Figures 1.1 and 1.2, our guiding principle should be to recognize that different cultures in different periods of history have contributed to the world's stock of mathematical knowledge. Figure 1.3 presents such a trajectory of mathematical development, but confines itself to the period between the fifth and fifteenth centuries AD — the period represented by the arrow labelled in Figures 1.1 and 1.2 as the 'Dark Ages' in Europe. The choice of this trajectory as an illustration is deliberate: it serves to highlight the variety of mathematical activity and exchange between a number of cultural areas that went on while Europe was in deep slumber. A

Figure 1.3 An alternative trajectory for the 'Dark Ages'.

trajectory for the fifteenth century onwards would show that mathematical cross-fertilization and creativity were more or less confined to countries within Europe until the emergence of the truly international character of modern mathematics during the twentieth century.

The role of the Arabs is brought out in Figure 1.3. Scientific knowledge which originated in India, China and the Hellenistic world was sought out by Arab scholars and then translated, refined, synthesized and augmented at different centres of learning, starting at Jund-i-Shapur in Persia around the sixth century (even before the coming of Islam), and then moving to Baghdad, Cairo, and finally to Toledo and Córdoba in Spain, from where this knowledge spread into Western Europe. Considerable resources were made available to the scholars through the benevolent patronage of the caliphs – the Abbasids (the rulers of the eastern Arab empire, with its capital at Baghdad) and the Umayyads (the rulers of the western Arab empire, with its capital first at Damascus and later at Córdoba).

The role of the Abbasid caliphate was particularly important for the future development of mathematics. The caliphs, notably al-Mansur (754–75), Harun al-Rashid (786–809) and al-Mamun (809–33), were in the forefront of promoting the study of astronomy and mathematics in Baghdad. Indian scientists were invited to Baghdad. When Plato's Academy was closed in 529, some of its scholars found refuge at Jund-i-Shapur, which a century later became part of the

Arab world. Greek manuscripts from the Byzantine empire, the translations of the Syriac schools of Antioch and Damascus, and the remains of the Alexandrian library in the hands of the Nestorian Christians at Edessa were all eagerly sought by Arab scholars, aided by the rulers who had control over or access to men and materials from the Byzantine empire, Persia, Egypt, Mesopotamia and places as far east as India and China.

Caliph al-Mansur built at Baghdad a *Bait al-Hikma* (House of Wisdom) which contained a large library for the manuscripts that had been collected from various sources; an observatory which became a meeting place of Indian, Babylonian, Hellenistic and, probably, Chinese astronomical traditions; and a university where scientific research continued apace. A notable member of the institution, Muhammad ibn Musa al-Khwarizmi (*c*. AD 825), wrote two books which were of crucial importance to the future development of mathematics. One of them, the Arabic text of which is extant, is entitled *Hisab al-jabr w'al-muqabala* (which may be loosely translated as 'Calculation by Restoration and Reduction'). The title refers to the two main operations in solving equations: 'reunion', the transfer of negative terms from one side of the equation to the other, and 'reduction', the merging of like terms on the same side into a single term. In the twelfth century the book was translated into Latin under the title *Liber algebrae et almucabola*, thus giving a name to a central area of mathematics. A traditional meaning of the Arabic word *jabr* is 'the setting of a broken bone' (and hence 'reunion' in the title of al-Khwarizmi's book). A few decades ago, it was not an uncommon sight on Spanish streets to come across the sign 'Algebrafista y Sangrador' ('bonesetting and bloodletting') at the entrance of barbers' shops.

Al-Khwarizmi wrote a second book, of which only a Latin translation is extant: *Algorithmi de numero indorum*, which explained the Indian number system. While al-Khwarizmi was at pains to point out the Indian origin of this number system, subsequent translations of the book attributed not only the book but the numerals to the author. Hence, in Europe any scheme using these numerals came to be known as an 'algorism' or, later, 'algorithm' (a corruption of the name al-Khwarizmi) and the numerals themselves as Arabic numerals.

Figure 1.3 shows the importance of two areas of Southern Europe in the transmission of mathematical knowledge to Western Europe. Spain and Sicily were the nearest points of contact with Arab science and had been under Arab hegemony, Córdoba succeeding Cairo as the centre of learning during the ninth and tenth centuries. Scholars from different parts of Western Europe congregated in Córdoba and Toledo in search of ancient and contemporary knowledge. It is reported that Gherardo of Cremona (*c.* 1114–87) went to Toledo, after its recapture by the Christians, in search of Ptolemy's *Almagest*, an astronomical work of great importance produced in Alexandria during the second century AD. He was so taken by the intellectual activity there that he stayed for twenty years, during which time he was reported to have copied or translated eighty manuscripts of Arab science or Greek classics, which were later disseminated across Western Europe. Gherardo was just one of a number of European scholars, including Plato of Tivoli, Adelard of Bath and Robert of Chester, who flocked to Spain in search of knowledge.

The main message of Figure 1.3 is that it is dangerous to characterize mathematical development solely in terms of European developments. The darkness that was supposed to have descended over Europe for a thousand years before the illumination that came with the Renaissance did not interrupt mathematical activity elsewhere. Indeed, as we shall see in later chapters, the period saw not only a mathematical renaissance in the Arab world, but also high points of Indian and Chinese mathematics.

MATHEMATICAL SIGNPOSTS AND TRANSMISSIONS ACROSS THE AGES

Alternative trajectories to the ones shown in Figures 1.1 and 1.2 should highlight the following three features of the plurality of mathematical development:

1 the global nature of mathematical pursuits of one kind or another,
2 the possibility of independent mathematical development within each cultural tradition, and
3 the crucial importance of diverse transmissions of mathematics across cultures, culminating in the creation of the unified discipline of modern mathematics.

However, to construct a feasible diagram we must limit the number of geographical areas of mathematical activity we wish to include. Selection inevitably introduces an element of arbitrariness, for some areas which may merit inclusion are excluded, while certain inclusions may be controversial. Two considerations have influenced the choice of the cultural areas represented in Figure 1.4. First, a judgement was made, on the basis of existing evidence, as to which places saw significant developments in mathematics. Second, an assessment of the nature and direction of the transmission of mathematical knowledge also helped to identify the areas of interest.

On the basis of these two criteria, ancient Egypt and Mesopotamia, Greece, the Hellenistic world, India, China, the Arab world and Europe were selected as being important in the historical development of mathematics. For one cultural area, the application of the two selection criteria produced conflicting results: from existing evidence, the Maya of Central America were isolated from other centres of mathematical activity, yet their achievements in numeration and calendar construction were quite remarkable by any standards. I therefore decided to include the Maya in Figure 1.4, and to examine their contributions briefly in Chapter 2.

The limited scope of this book and the application of the above criteria make it impossible to examine the mathematical experience of Africa, Korea and Japan in any detail. However, Chapter 2 contains a discussion of the Ishango Bone and the Yoruba numerals, and Chapter 3 a detailed examination of Egyptian mathematics, all of which were products of Africa. The 'Reflections' at the end of this book will provide some information on the mathematical traditions not covered in these chapters.

Figure 1.4, together with its detailed legend, emphasizes the following features of mathematical activity through the ages:

1 the continuity of mathematical traditions until the last few centuries in most of the selected cultural areas,

2 the extent of cross-transmissions between different cultural areas which were geographically or otherwise separated from one another, and

3 the relative ineffectiveness of cultural barriers (or 'filters') in inhibiting the transmission of mathematical knowledge. In a number of other areas of human knowledge, notably in philosophy and the arts, the barriers are often insurmountable unless filters can be devised to make foreign 'products' more palatable.

In both Egypt and Mesopotamia there existed well-developed written number systems as early as the third millennium BC. The peculiar character of the Egyptian hieroglyphic numerals led to the creation of special types of algorithm for basic arithmetic operations. Both these developments and subsequent work in the area of algebra and geometry, especially during the period between 1800 and 1600 BC, will form the subject-matter of Chapter 3. Figure 1.4 brings out another impressive aspect of Egyptian mathematics – the continuity of a tradition for over three thousand years, culminating in the great period of Alexandrian mathematics around the beginning of the Christian era. We shall not examine the content and personalities of this mature phase of Egyptian mathematics in any detail since its coverage in standard histories of mathematics is more than adequate. There is, however, a widespread tendency in many of these texts to view Alexandrian mathematics as a mere extension of Greek mathematics, in spite of the distinctive character of the

Figure 1.4 The spread of mathematical ideas down the ages.

EGYPT
1 *Pre-dynastic period*: Appearance of the earliest forms of writing and hieroglyphic numerals
2 *Middle Kingdom to New Kingdom*: Egyptian mathematics mainly contained in the Moscow and Ahmes Papyri
3 *Greek and Roman Period*: Flowering of mathematics at Alexandria

GREECE
1 *Classical Period*: Beginnings of deductive geometry and number theory
2 *Hellenistic Period*: Growing synthesis of Classical, Egyptian and Babylonian mathematics

MESOPOTAMIA
1 *Sumerian Period*: Beginnings of cuneiform numerals
2a *First Babylonian Dynasty*: Early algebra, commercial arithmetic and geometry from clay tablets
2b *New Babylonian and Persian Periods*: Mathematics and astronomy
3 *Seleucid Dynasty*: Hellenistic mathematics

INDIA
1 *Harappan Period*: Proto-mathematics from bricks, baths, etc.
2 *Vedic Period*: Ritual geometry
3 *Classical Period*: Indian numerals, computing algorithms, algebra and trigonometry
4 *Medieval Period*: Kerala mathematics

ARAB WORLD
1 Preservation and synthesis of mathematical traditions from different areas, laying the foundations of modern mathematics

CHINA
1 *River Valley Civilization*: Beginnings of practical mathematics
2 *Shang and Chou Dynasties*: Rod numerals, *Chou Pei* (the earliest extant mathematics textbook)
3 *Han to Tang Dynasties*: 'Arithmetic in Nine Sections' (the most important text in Chinese mathematics)
4 *Sung to Ming Dynasties*: Golden age of Chinese mathematics

MAYAN EMPIRE
1 Construction of a highly accurate calendar and the development of a place-value number system (base 20) with zero

EUROPE
1 Development of modern mathematics, building on mathematics from other sources

═══ (Hellenistic cultural areas (Egypt, Greece, Mesopotamia)

→ *Confirmed lines of transmission (two-way)*
(i) Harappan culture *and* First Babylonian dynasty
(ii) Han to Tang dynasties *and* Classical period (India)

→ *Confirmed lines of transmission (one-way)*
(i) Middle Kingdom to New Kingdom (Egypt) *to* Classical period (Greece)
(ii) New Babylonian and Persian periods *to* Classical period (Greece)
(iii) Hellenistic cultural areas *to* Classical period (India)
(iv) Hellenistic cultural areas *to* Arab world
(v) Classical period (India) *to* Arab world
(vi) Arab world *to* Europe and India

--→ *Unconfirmed or tentative lines of transmission (two-way)*
(i) Pre-dynastic period (Egypt) *and* Sumerian period (Mesopotamia)
(ii) Classical period (Greece) *and* Vedic period (India)
(iii) Vedic period (India) *and* Shang and Chou dynasties (China)

--→ *Unconfirmed or tentative lines of transmission (one-way)*
(i) Sumerian culture *to* Harappan culture
(ii) Sung to Ming dynasties *to* Europe
(iii) Kerala mathematics *to* Europe

mathematics of Archimedes, Heron, Diophantus and Pappus, to mention a few notable names of the Alexandrian period.

The other early contributor to mathematics was the civilization that grew around the twin rivers, the Tigris and the Euphrates, in Mesopotamia. There mathematical activity flourished, given impetus by the establishment of a place-value sexagesimal (i.e. base 60) system of numerals which must surely rank as one of the most significant developments in the history of mathematics. However, the golden period of mathematics in this area (or at least the period for which considerable written evidence exists) came during the First Babylonian period (c. 1800–1600 BC), which saw not only the introduction of further refinements to the existing numeral system, but the development of an algebra more advanced than that in use in Egypt. The period is so important that the mathematics that developed in Mesopotamia is often simply referred to as Babylonian mathematics. As with Egypt, the next period of significant advance followed Alexander's conquest and the establishment of the Seleucid dynasty. Babylonian mathematics (a term that will be used hereafter to describe the mathematics of this cultural area) is discussed in Chapter 4.

While there is no direct evidence of any mathematical exchange between Egypt and Mesopotamia before the Hellenistic period, it is a reasonable conjecture that, given their proximity and the records we have of their economic and political contacts, there could have been transmissions between these areas. Parker (1972) has examined the evidence for a spread of Mesopotamian algebra and geometry to Egypt. He points out certain parallel developments in both geometry and algebra to provide at least some support for links between the two cultural areas. It is because of the unconfirmed and tentative nature of present evidence that we represent the contacts between Egypt and Mesopotamia by a broken two-headed arrow in Figure 1.4.

There is, however, ample evidence of the great debt that Greece owed to Egypt and Mesopotamia for her earlier mathematics and astronomy. We have mentioned the acknowledgement of this debt by the Greeks themselves, who believed that mathematics originated in Egypt. The travels of the early Greek mathematicians such as Thales, Pythagoras and Eudoxus to Egypt and Mesopotamia in search of knowledge have been attested to both by their contemporaries and by later historians writing on the period. The period of greatest Egyptian influence on the Greeks may have been the first half of the first millennium BC. The Greek colonies scattered across the Mediterranean provided a wide channel of interchange. It is at the time of their heyday that we hear of Anaximander of Miletus (c. 600 BC) introducing the gnomon (a geometric shape of both mathematical and astronomical significance) from

Babylon. During the same period, contacts with the Greeks were maintained through the campaigns of the Assyrian king Sargon II (722–705 BC), and later through Ashurbanipal's occupation of Egypt and his meeting with Gyges of Lydia towards the middle of the seventh century BC. Even when Assyria ceased to exist, the Jewish captivity played a significant part in disseminating Babylonian learning. This was followed by the Persian invasion of Greece at the beginning of the fifth century and the final defeat of the Persians at the end of the fifth century. Thus continuous contacts were maintained throughout a period in which Greek mathematics was still in its infancy, as the foundations were being laid for the flowering of Greek creativity in a couple of centuries. In the next five hundred years, the pupil would learn and develop sufficiently to teach the teachers.

The transmissions to Greece from the two areas are shown in Figure 1.4 by the arrows from 2 in Egypt and 2a in Mesopotamia to 1 in Greece. All three areas then became part of the Hellenistic world, and during the period between the third century BC and the third century AD, partly due to the interaction between the three mathematical traditions, there emerged one of the most creative periods in mathematics. We associate this period with names such as Euclid, Archimedes, Apollonius and Diophantus. These links are represented by the double lines between 3 in Egypt, 2 in Greece and 3 in Mesopotamia.

The geographical location of India made her throughout history an important meeting-place of nations and cultures. This enabled her from the very beginning to play an important role in the transmission and diffusion of ideas. The traffic was often two-way, with Indian ideas and achievements travelling abroad as easily as those from outside entered her own consciousness. Archaeological evidence shows both cultural and commercial contacts between Mesopotamia and the Indus Valley. While there is no direct evidence of mathematical exchange between the two cultural areas, certain astronomical calculations of the longest and shortest day included in the *Vedanga Jyotisa*, the oldest extant Indian astronomical text, have close parallels with those used in Mesopotamia. And hence the tentative link, shown by broken lines in Figure 1.4, between 1 in Mesopotamia and 1 in India.

By the sixth century BC, with the appearance of the vast Persian empire which touched Greece at one extremity and India at the other, contacts between India and the West were close and productive. Tributes from Greece and from the frontier hills of India found their way to the same Imperial treasure-houses at Ecbatana or Susa. Soldiers from Mesopotamia, Greek cities of Asia Minor and India served in the same armies. The word 'indoi' for Indians made its first appearance in Greek; allusions to India begin to appear in Greek literature. Certain interesting parallels between Indian and Pythagorean philosophy have already been pointed out. Indeed, according to some Greek sources, Pythagoras had ventured as far afield as India in his search for knowledge.

By the time Ptolemaic Egypt and Rome's Eastern empire had established themselves just before the beginning of the Christian era, Indian civilization was already well developed, having founded three great religions – Hinduism, Buddhism and Jainism – and expressed in writing some subtle currents of religious thought and speculation as well as fundamental theories in science and medicine. There are scattered references to Indian science in literary sources from countries to the west of India after the time of Alexander. The Greeks had a high regard for Indian 'gymnosophists' (i.e. philosophers) and Indian medicine. Indeed, there are various expressions of nervousness about the Indian use of poison in warfare. In a letter Aristotle wrote to his pupil Alexander in India, he warns of the danger posed by intimacy with a 'poison-maiden', who had been fed on poison from her infancy so that she could kill merely by her embrace!

There is little doubt that the Babylonian influence on Indian astronomy continued into the Hellenistic period, when the astronomy and mathematics of the Ptolemaic and Seleucid dynasties became important forces in Indian science, readily detectable in the corpus of astronomical works known as *Siddhantas*, written around the beginning of the Christian era. Material evidence of such contacts has been found in places such as Jund-i-Shapur in Persia dating from between AD 300 and 600. Jund-i-Shapur was an important meeting-place of scholars from a number of different areas, including Indians and, later, Greeks who sought refuge there with the demise of Alexandria as a centre of learning and the closure of Plato's Academy. All such contacts are shown in Figure 1.4 by lines linking 2 in India to 1 in Greece and 3 in India to the Hellenistic cultural areas.

By the second half of the first millennium AD, the most important contacts for the future development of mathematics were those between India and the Arab world. This is shown by the arrow from 3 in India to 1 in the Arab world. As we saw in Figure 1.3, the other major influence on the Arab world was from the Hellenistic cultural areas, and the nature of these influences has been discussed in some detail. As far as Indian influence via the Arabs on the future course of development of mathematics is concerned, it is possible to identify three main areas:

1 the spread of Indian numerals and their associated algorithms, first to the Arabs and later to Europe,

2 the spread of Indian trigonometry, especially the use of the sine function, and

3 the solutions of equations in general, and of indeterminate equations* in particular.

These contributions will be discussed in Chapters 8 and 9, which deal with Indian mathematics.

We have already looked briefly at the contributions of the Arabs as producers, transmitters and custodians of mathematical learning. Their role as teachers of mathematics to Europe is not sufficiently acknowledged. The arrow from 1 in the Arab world to 1 in Europe represents the crucial role of the Arabs in the creation and spread of mathematics which culminated in the birth of modern mathematics. The contributions of the Arabs will be discussed in Chapter 10 of this book.

Figure 1.4 shows another important cross-cultural contact, between India and China. There is very fragmentary evidence (as shown by the broken line between 2 in India and 2 in China) of contacts between the two countries before the spread of Buddhism into China. After this, from around the first century AD, India became the centre for pilgrimage of Chinese Buddhists, opening up the way for a scientific and cultural exchange which lasted for several centuries. In a catalogue of publications during the Sui dynasty (*c.* 600), there appear Chinese translations of Indian works on astronomy, mathematics and medicine. Records of the Tang dynasty indicate that from 600 onwards Indian astronomers were hired by the Astronomical Board of Changan to teach the principles of Indian astronomy. The solution of indeterminate equations, using the method of *kuttaka* in India and of *qiuyishu* in China, was an abiding passion in both countries. The nature and direction of transmission of mathematical ideas between the two areas is a complex but interesting problem, one to which we shall return in later chapters. The two-headed arrow linking 3 in India with 3 in China is a recognition of the existence of such transmission. Also, there is some evidence of a direct transmission of mathematical ideas between China and the Arab world, around the beginning of the second millennium AD. Numerical methods of solving equations of higher order such as quadratics and cubics, which attracted the interest of later Arab mathematicians, notably al-Kashi (*c.* 1400), may have been influenced by Chinese work in this area. There is every likelihood that some of the important

---

*An example of an indeterminate equation in two unknowns (x and y) is $3x + 4y = 50$, which has a number of positive whole-number (or integer) solutions for (x, y). For example, $x = 14$, $y = 2$ satisfies the equation, as do the solution sets $(10, 5)$, $(6, 8)$ and $(2, 11)$.

trigonometric concepts introduced into Chinese mathematics around this period may have an Arab origin.

There are broken lines of transmission in Figure 1.4 which need some explanation. One of the conjectures posed in Chapter 9 is the possibility that mathematics from medieval India, particularly from the southern state of Kerala, may have had an impact on European mathematics of the sixteenth and seventeenth centuries. While this cannot be substantiated by existing evidence, the fact remains that around the beginning of the fifteenth century Madhava of Kerala derived infinite series for π and for certain trigonometric functions, thereby contributing to the beginnings of mathematical analysis about two hundred and fifty years before European mathematicians such as Leibniz, Newton and Gregory were to arrive at the same results from their work on infinitesimal calculus. The possibility of medieval Indian mathematics influencing Europe is indicated by the arrow linking 4 in India with 1 in Europe.

During the medieval period in India, especially after the establishment of Mughal rule in North India, the Arab and Persian mathematical sources became better known there. From about the fifteenth century onwards there were two independent mathematical developments taking place: one Sanskrit-derived and constituting the mainstream tradition of Indian mathematics, then best exemplified in the work of Kerala mathematicians in the South, and the other based in a number of Muslim schools (or *madrassas*) located mainly in the North. We recognize this transmission by constructing an arrow linking 1 in the Arab world to 4 in India.

The medieval period also saw a considerable transfer of technology and products from China to Europe which has been thoroughly investigated by Lach (1965) and Needham (1954). The fifteenth and sixteenth centuries witnessed the culmination of a westward flow of technology from China that had started as early as the first century AD. It included, from the list given by Needham (1954, pp. 240–41): the square-pallet chain pump, metallurgical blowing engines operated by water power, the wheelbarrow, the sailing-carriage, the wagon-mill, the crossbow, the technique of deep drilling, the so-called Cardan suspension for a compass, the segmental arch-bridge, canal lock-gates, numerous inventions in ship construction (including watertight compartments and aerodynamically efficient sails), gunpowder, the magnetic compass for navigation, paper and printing, and porcelain. The conjecture here is that with the transfer of technology went certain mathematical ideas, including different algorithms for extracting square and cube roots, the 'Chinese remainder theorem', solutions of cubic and higher-order equations by what is known as Horner's method, and indeterminate analysis. Such a trans-

Figure 1.5 Cultures whose mathematics form the subject of this book.

mission from China need not have been a direct one, but may have taken place through India and the Arabs. We shall return to the question of influences and transmission from China to the rest of the world in Chapter 7.

During the first half of the first millennium of the Christian era, the Central American Mayan civilization attained great heights in a number of different fields including art, sculpture, architecture, mathematics and astronomy. In the field of numeration, the Maya shared in two fundamental discoveries: the principle of place value and the use of zero. Present evidence indicates that the principle of place value was discovered independently four times in the history of mathematics. At the beginning of the second millennium BC, the Babylonians were working with a place-value notational system to base 60. Around the beginning of the Christian era, the Chinese were using positional principles in their rod numeral computations. Between the third and fifth centuries AD, Indian mathematicians and astronomers were using a place-value decimal system of numeration which would eventually be adopted by the whole world. And finally, the Maya – apparently cut off from the rest of the world – had developed a positional number system to base 20. As regards zero, there are only two original instances of its modern use in a number system: by the Maya, and by the Indians around the beginning of the Christian era.

But mathematics was not the only area in which the Maya surprise us. With the most rudimentary instruments at their disposal, they undertook astronomical observations and calendar construction with a precision that went beyond anything available in Europe at that time. They had accurate estimates of the duration of solar, lunar and planetary movements. They estimated the synodic period of Venus (i.e. the time between one appearance at a given point in the sky and its next appearance at that point) to be 584 days, which is an underestimate of 0.08 days. They achieved these discoveries with no knowledge of glass or, consequently, of any sort of optical device. Neither did they apparently have any device for measuring the passage of time, such as clocks or sand-glasses, without which it would now seem impossible to produce astronomical data.

Figure 1.5 shows the geographical areas whose mathematics form the subject-matter of this book. I am conscious of not having examined in sufficient detail the mathematical pursuits of other groups, notably the Africans south of the Sahara and the Amerindians of North America, although the topics treated in Chapter 2, and the notes in 'Reflections' (pp. 349), should go some way in making up for this neglect.

2   Mathematics from Bones, Strings and Standing Stones

◇◇

It is taking an unnecessarily restrictive view of the history of mathematics to confine our study to written evidence. Mathematics initially arose from a need to count and record numbers. As far as we know there has never been a society without some form of counting or tallying (i.e. matching a collection of objects with some easily handled set of markers, whether it be stones, knots or inscriptions such as notches on wood or bone). If we define mathematics as any activity that arises out of, or directly generates, concepts relating to numbers or spatial configurations together with some form of logic, we can then legitimately include in our study proto-mathematics, which existed when no written records were available.

BEGINNINGS: THE ISHANGO BONE

High in the mountains of Central Equatorial Africa, on the borders of Uganda and Zaïre, lies Lake Edward, one of the furthest sources of the River Nile. It is a small lake by African standards, about 80 kilometres long and 50 wide. Though the area is remote and sparsely populated today, about 20 000 years ago by the shores of the lake lived a small community that fished, gathered food or grew crops depending on the season of the year. The settlement had a relatively short life span of a few hundred years before being buried in a volcanic eruption. These neolithic people have come to be known as the Ishango, after the place where their remains were found.

Archaeological excavations at Ishango have unearthed human remains and tools for fishing, hunting and food production (including grinding and pounding stones for grain). Harpoon heads made from bone may have served as prototypes for tools discovered as far away as northern Sudan and West Africa. However, the most interesting find, from our point of view, is a bone tool handle (Figure 2.1) which is now at the Musée d'Histoire Naturelle in Brussels. The original bone may have petrified or undergone chemical change through the action of water and other elements. What remains is a dark brown object on which some markings are clearly visible. At one end is a sharp, firmly fixed piece of quartz which may have been used for engraving, tattooing or even writing of some kind.

The markings on the Ishango Bone, as it is called, consist of series of notches arranged in three distinct columns. The asymmetrical grouping of these notches, as shown in Figure 2.1, would make it unlikely that they were put there merely for decorative purposes. Row (a) contains four groups of notches with 9, 19, 21 and 11 markings. In Row (b) there are also four groups, of 19, 17, 13 and 11 markings. Row (c) has eight groups of notches in the following order: 7, 5, 5, 10, 8, 4, 6, 3. The last pair (6, 3) is spaced closer together, as are (8, 4) and (5, 5, 10), suggesting a deliberate arrangement in distinct sub-groups.

If these groups of notches were not decorative, why were they put there? An obvious explanation is that they were simply tally marks. Permanent records of counts maintained by scratches on stones, knots on strings, or notches on sticks or bones have been found all over the world, some going back to the very early history of human habitation. During an excavation of a cave in the Lebembo Mountains on the borders of Swaziland in southern Africa, a small section of the fibula of a baboon was discovered, with 29 clearly visible notches, dating to about 35 000 BC. This is the earliest evidence we have of a numerical recording device. An interesting feature of this bone is its resemblance to the 'calendar sticks' still used by some inhabitants of Namibia to record the passage of time. From about five thousand years later we have the shin-bone of a young wolf, found in Czechoslovakia, which contains 57 deeply cut notches grouped in 5's. It was probably a record kept by a hunter of the number of kills to his credit. Such artefacts represent a distinct advance, a first step towards constructing a numeration system, where the counting of objects in groups is supplemented by permanent records of these counts.

However, the Ishango Bone appears to have been more than a simple tally. Certain underlying numerical patterns may be observed within each of the rows marked (a) to (c) in Figure 2.1. The markings on rows (a) and (b) each add up to 60: $9 + 19 + 21 + 11 = 60$ and $19 + 17 + 13 + 11 = 60$, respectively. Row (b) contains the prime numbers between 10 and 20. Row (a) is quite consistent with a numeration system based on 10, since the notches are grouped as $20 + 1$, $20 - 1$, $10 + 1$ and $10 - 1$. Finally, row (c), where sub-groups (5, 5, 10), (8, 4) and (6, 3) are clearly demarcated, has been interpreted as showing some appreciation of the concept of duplication or multiplying by 2. De Heinzelin (1962), the archaeologist who helped to excavate the Ishango Bone, wrote that it 'may represent an arithmetical game of some sort, devised by a people who had a number system based on 10 as well as a knowledge of duplication and of prime numbers' (p. 111). Further, from the existing evidence of the transmission of Ishango tools, notably harpoon heads, northwards up to the frontiers of Egypt, de Heinzelin considered the possibility that the Ishango numeration system may have travelled as far as Egypt and

Figure 2.1 The Ishango Bone (Courtesy of Dr J. de Heinzelin).

influenced the development of its number system, the earliest decimal-based system in the world.

The African origins of Egyptian civilization are well attested to by archaeological and early written evidence. Herodotus wrote of the Egyptian people and culture having strong African roots, coming from the lands of the 'long-lived Ethiopians', which meant in those days the vast tract of inner Africa inhabited by black people. However, de Heinzelin's speculations about the state of mathematical knowledge of the Ishango, based as they are on the evidence of a single bone, seem far-fetched. A single bone with suggestive markings raises interesting possibilities of a highly developed sense of arithmetical awareness; it does not provide conclusive evidence.

There is, however, another answer, more firmly rooted in the cultural environment, to the puzzle of the Ishango Bone. Rather than attribute the development of a numeration system to a small group of neolithic settlers living in relative isolation on the shores of a lake, apparently cut off from other traceable settlements of any size and permanence, a more plausible hypothesis is that the bone markings constitute a system of sequential notation – a record of different phases of the moon. Whether this is a convincing explanation would depend in part on establishing the importance of lunar observations in the Ishango culture, and in part on how closely the series of notches on the bone matches the number of days contained in successive phases of the moon.

Archaeological evidence of seasonal changes in the habitat and activity of the Ishango highlights how important it was to maintain an accurate lunar calendar. At the beginning of the dry season, the Ishango moved down to the lake from the hills and valleys that formed their habitat during the rains. For those who were permanently settled along the shores of the lake, the onset of the dry season brought animals and birds to the lake in search of water. Now assume, for the sake of argument, that migration took place around the full moon or a few days before the full moon. About six months later the rainy season would begin and the water levels of the lake would rise. Between the beginning of the dry season and the onset of the rainy season, there might be festivities that coincided with particular phases of the moon. And such events might very well be what is recorded notationally on the bone. Activities such as gathering and processing of nuts and seeds, or hunting, both of which archaeological evidence suggests were important in the Ishango economy, could be incorporated sequentially into the lunar calendar represented by the Ishango Bone. Similarly, religious rituals associated with seasonal and other festivities could be recorded on the bone. Such a scenario is still conjectural, but quite consistent with what we know of present-day peoples who still follow the hunter-gatherer life style of the Ishango.

A cursory examination of the pattern of notches on the Ishango Bone shows no obvious regularity that one can associate with lunar phenomena. Two of the rows add up to 60, so that each of these rows may be said to represent two lunar months. The third row contains only 48 notches, which would only account for a month and a half. But a mere count of the notches would ignore the possible significance of the different sizes and shapes of the markings as well as the sequencing of the sub-groups demarcated on the bone.

Marshack (1972) carried out a detailed microscopic examination of the Ishango Bone and found markings of different indentations, shapes and sizes. He concluded that there was strong evidence of a close fit between different phases of the moon and the sequential notation contained on the bone, once the additional markings – visible only through the microscope – were taken into account. Also, the different engravings represented by markings of various shapes and sizes may have been a calendar of events of a ceremonial or ritual nature.

These conjectures about the Ishango Bone highlight three important aspects of proto-mathematics. First, the close link between mathematics and astronomy has a long history and is tied up with the need felt even by early man to record the passage of time, out of curiosity as well as practical necessity. Second, there is no reason to believe that early man's capacity to reason and conceptualize was any different from that of his modern counterpart. What has changed dramatically over the years is the nature of the facts and relationships with which man has to operate. Thus the creation of a complicated system of sequential notation based on a lunar calendar was well within the capacity of prehistoric man, whose desire to keep track of the passage of time and changes in seasons was translated into observations of the changing aspect of the moon. Finally, in the absence of records, conjectures about the mathematical pursuits of early man have to be examined in the light of their plausibility, the existence of convincing alternative explanations and the quality of evidence available. A single bone may well collapse under the heavy weight of conjectures piled upon it.*

---

*While we shall not be discussing in this book the inferences that have been made by Thom (1967), van der Waerden (1983) and others on the mathematical attainments of the constructors of megalithic monuments, such as Stonehenge in England, with these inferences too it is worth sounding the same note of caution: it is extremely unlikely that the neolithic life style of the builders of these monuments would have generated the demands or supplied the resources required for developing the 'advanced' mathematics attributed to them by such writers.

KNOTTED STRINGS FROM SOUTH AMERICA: THE INCA *QUIPU*

Knots as aids to memory

Human memory is remarkable for both its capacity and its complexity. It can store an incredible amount of information, but as a storage device it is often unreliable and not particularly well organized. Therefore from early times all types of mnemonic devices, including notches and knots, have been used as aids to memory. Compared with writing, the use of knots is a clumsy device, though for a preliterate culture it would have had its advantages: easy to use, convenient to carry around, and with familiar associations with everyday pursuits such as sewing or fishing. The knots served one primary purpose: to record and preserve information.

There are a host of anecdotes and legends about knots used for recording the passage of time. To take just one story: at the turn of the present century a German, Karl Weule, reported a conversation he had had with an old inhabitant of the Makonde Plateau in East Africa. At the beginning of a journey the old man would present his wife a piece of bark string with eleven knots. She would be asked to untie a knot each day. The first knot represented the day of his departure, the next three knots the period of his journey, the fifth knot the day he reached his destination, the sixth and seventh knots the days he spent conducting his business, and the next three knots the period of his return journey. So when she had untied the tenth knot, she would know that he was returning home the next day.

It is important not to confuse the purpose of these simple mnemonic knots with that of the *quipu*, to which we shall now turn. Such confusion may arise from failure to distinguish a straightforward numerical magnitude represented by tally marks or knots from the ordinal representation possible on a *quipu* or with a written number system. The fact that the *quipu* cannot be manipulated for calculations, while a written number system can, does not affect the argument. Tying knots in a cord to show a certain numerical quantity is no different from writing the same number on a piece of paper using some widely accepted symbols. This point will become clearer as we proceed.

The *quipu*: Appearance and history

Quipu is a Quechua (the language of the Incas) word meaning 'knot'. A *quipu* resembles a mop which has seen better days. It consists of a collection of cords, often dyed in one or more colours, and containing knots of different types, but not, apparently, arranged in any systematic fashion. *Quipus*, of which there are about four hundred authenticated examples, are to be found in the museums of Western Europe and the Americas. They were initially thought of as primitive artefacts with little aesthetic appeal. About fifty of these objects have now been carefully studied, the credit for unravelling part of their mystery going to Leyland Locke (1912, 1923). From a close study of statements made by Spanish chroniclers of the sixteenth century and a detailed examination of some of the *quipus*, Locke concluded that the *quipu* was basically a device for recording numbers in a decimal base system.

At its height during the last decades of the fifteenth century AD, the Inca empire occupied an area that today would include all of Peru and parts of Bolivia, Chile, Ecuador and Argentina. In this vast and difficult terrain lived a culturally diverse population of about six million. It was a well-organized society, cooperative in character, its material culture the creation of a number of different groups that the Inca state was able to organize and control during its short 150-year period of dominance. Yet, despite the level of their material culture, the Incas seem to have lacked the three widely accepted basics of early civilizations: the wheel, beasts of burden and a written language. Yet the high level of organization required the keeping of detailed accounts and records. In the absence of a system of writing, they used *quipus*.

There is, of course, no contemporary written evidence on the nature and uses of a *quipu* from the society that used it. There are, however, the chronicles of Spanish soldiers, priests and administrators. The most reliable and unbiased of these chroniclers was a soldier, Cieza de León, who began keeping a record in 1547 – fifteen years after the Spanish conquest – and stopped writing three years later. It provides a fascinating account, both of the flora and fauna of the vast territory and of the society there.

There was one aspect of the former Inca state that Cieza found impressive. Across the imperial highways, many of them more substantial than the Roman roads of his native land, were to be found small post-houses, the staging posts for runners who carried messages across the difficult mountainous terrain, impossible for any animal to negotiate. These trained runners, called *chasquis*, were stationed in pairs at intervals of about a mile along the highways. Running at top speed and handing their *quipus* on from one runner to the next, as in a relay race, they could transmit a message to the imperial capital Cuzco from three hundred miles away in twenty-four hours. Given the terrain, Cieza noted that this method of carrying messages was superior to using horses and mules, with which he was more familiar.

In two words, Cieza summed up the strengths of the former Inca empire and its ability later to withstand to some degree the havoc brought about by Spanish plunder: order and organization. And the essential prerequisite for maintaining good order and efficient organization was the existence of detailed and up-to-date information (or government statistics, as we would describe such information today) which the state could call upon whenever necessary. Records of all such information were kept on *quipus*.

A whole inventory of resources which included agricultural produce, live-stock and weaponry — as well as people — was maintained and updated regularly by a group of special officials known as *quipucamayus* ('quipu-keepers'). Each district under the rule of the Incas had its own specially trained *quipucamayu*, and larger villages had as many as thirty. For information on the role and status of the *quipucamayus* we have a set of remarkable drawings by one Guaman Poma de Ayala, a Peruvian, which form part of a 1179-page letter to the King of Spain sent in about 1600, some eighty years after the Spanish conquest. Apart from being one of the most searing indictments of Spanish rule, it contains a series of illustrations in which the Inca bureaucracy figures prominently. Seven of these drawings show people carrying *quipus*; two of them are reproduced in Figure 2.2.

The inscription in Figure 2.2(a) indicates that the figure holding the *quipu* is none other than the Secretary to the Inca and his Council. Figure 2.2(b) shows the Chief Treasurer to the Inca. There is little doubt that '*quipu* literacy' was widespread among government officials, of whom the *quipucamayus* were important members enjoying high social status.

Figure 2.2(b) contains another interesting feature, apart from the blank (i.e. unknotted) *quipu* held by the Inca's Treasurer. At the bottom left of the drawing is a rectangle divided into twenty cells, in each of which there is a systematic arrangement of small circles and dots probably representing seeds, stones or similar objects. The Inca abacus, as it has been nicknamed, may have been the device on which computations were worked out before the results were recorded on the *quipu*. We shall return to an explanation of how computations may have been carried out on this 'abacus' in a later section. But we begin by looking at how a *quipu* was constructed and used for storing numerical data.

The construction and interpretation of a *quipu*

A *quipu* is constructed by joining together different types of cord. Each cord is at least two-ply, with one end looped and the other tapered and tied with a small knot. Four different kinds of cord can be distinguished. The first type, which is thicker than the rest, is termed the *main cord*. From it are attached like a fringe a number of other cords, most of which hang down and are known as the *pendant cords*, and a few whose knotted ends are directed upwards, the *top cords*. In some *quipus* there may be an additional cord whose looped end is connected to the looped end of the main cord and tightened, which explains its name – the *dangle end cord*. To any of these cords suspended from the main cord, there may be attached *subsidiary cords*. And this process of attachment may be carried further, so that a subsidiary may be connected to a subsidiary of a subsidiary, and so on. Also, it is possible that some of the pendant cords may be drawn together by means of a single top cord to form a distinct group. What we have after the process is complete is a blank *quipu*, rather like the one in Figure 2.2(b), which apparently has no top cords. A blank *quipu* can have as few as three cords – or as many as two thousand.

There is a further dimension to the construction of a *quipu* – colour. The predominant colours of the cords in the *quipus* that have survived are dull white and varying shades of brown. It is not clear whether the small differences in the shades of brown are simply a reflection of the age of the *quipus* rather than real colour differences. However, early chroniclers of the Inca culture refer to the use of symbolic colour representation for different things: white for silver, yellow for gold, red for soldiers, and so on. The symbolic use of colours is common in many societies. The use of red and green in traffic lights, for example, conveys a meaning which cuts across cultural barriers. Even in those societies where red is not traditionally associated with danger, its appearance in a traffic signal is sufficient to produce a rapid response, much more readily than if the warning were in the form of printed words. And in any case, colour being more recognizable than print over a longer distance would clinch the argument for its adoption. But the use of colour codes to distinguish between mathematical quantities or operations is unusual in modern mathematics (though not in modern book-keeping). Yet, as we shall see later, the ancient Egyptians used red ink to represent 'auxiliaries' which they calculated as part of arithmetical operations with fractions; the Chinese distinguished between positive and negative numbers by using red and black rods, respectively; and the Indians called algebraic unknowns by the names of different colours. In a *quipu*, colour was used primarily to distinguish between different attributes. Each *quipu* had a colour coding system to relate some of the cords to one another and at the same time to distinguish them from other cords. The range and subtlety of colour-coding was extended by using different combinations of coloured yarns.

There have been suggestions that the colours had some numerical significance, but we cannot be certain. What we do know is that numerical representation

Figure 2.2 Two Inca officials holding *quipus*; an 'Inca abacus' can be seen at the bottom left in (b) (Poma de Ayala, 1936, pp. 358 and 360).

con tador y tegouro   con tador

(b)

on a *quipu* was achieved by means of knots. Contemporary records clearly indicate that the Incas used a decimal system of numeration. According to Garcilaso de la Vega (b. 1539), whose mother was the niece of the last king of the Incas (Inca Huayna Capac) and whose father was a Spaniard, the knots indicated a system of notation by position:

> According to their position, the knots signified units, tens, hundreds, thousands, ten thousands and, exceptionally, hundred thousands, and they are all as well aligned on their different cords as the figures that an accountant sets down, column by column, in his ledger.

On each cord except the main cord, clusters of knots were used to represent a certain number. A number shown on one of the pendant cords could be read by counting the number of knots in the cluster of knots closest to the main cord which represented the highest-valued digit, and proceeding along the cord to the next cluster of knots representing the next positional digit (i.e. the next lowest power of 10) as far as the 'units' cluster, at the other end of the pendant cord. To distinguish the units cluster of knots from the other clusters representing higher positional digits, a different knot was used. Generally a long knot with four turns indicated the units position unless a 1 occurred in the units position, in which case a figure-of-eight (or Flemish) knot was used instead. For all other positions single (or short) knot(s) were used. The absence of a knot indicated zero in any of the positions.

An illustration will be useful at this point. Figure 2.3 shows how the numbers 1351, 258 and 807 may be represented on a *quipu*, with L, S and F denoting long, single and Flemish knots respectively. The left-hand pendant cord contains four knot clusters, of one single knot (1S), three single knots (3S), five single knots (5S) and one Flemish knot (1F), reading downwards from the main cord. This may be read as

$$(1 \times 1000) + (3 \times 100) + (5 \times 10) + (1 \times 1) = 1351$$

In a similar manner, the other two pendant cords may be read as

$$(2 \times 100) + (5 \times 10) + (8 \times 1) = 258$$
$$(8 \times 100) + (0 \times 10) + (7 \times 1) = 807$$

The spacing of the knot clusters is crucial here. For example, the pendant cord on the right is read as representing 807 because of the considerable space without any knots that exists between the cluster of eight knots (or eight 100's, shown as 8S in Figure 2.3) and the cluster of seven knots (or seven 1's, shown

Figure 2.3 Recording numbers on a *quipu*.

as 7L). If there were the usual space between the two clusters of knots, this pendant cord would read as 87.

The knot clusters on a top cord usually represent the sum of the numbers of the pendant cords. So in Figure 2.3 the knots on the top cord may be interpreted as

$$(2 \times 1000) + (4 \times 100) + (1 \times 10) + (6 \times 1) = 2416 = 1351 + 258 + 807$$

The same principles apply to interpreting numbers on subsidiary cords, of which there is only one (with knots representing 302) in Figure 2.3. This example is only a simple illustration of one way of using a *quipu*. There are other ways of forming cord groups, by using different colours or by distinguishing between different sub-groups of pendant cords to extend the versatility of the *quipu*.

As an illustration of another use of the *quipu*, we have the report of the early chronicler Garcilaso de la Vega:

> The ordinary judges gave a monthly account of the sentences they imposed to their superiors, and they in turn reported to their immediate superiors, and so on finally to the Inca or those of his Supreme Council. The method of making these reports was by means of knots, made of various colours, where knots of such and such colours denote that such and such crimes had been punished. Smaller threads attached to thicker cords were of different colours to signify the precise nature of the punishment that had been inflicted. By such a device was information stored in the absence of writing.

The mathematics of the *quipu*

The *quipu* served as a device for storing ordered information, cross-referenced and summed within and between categories. One of the few real-life examples known to us is a *quipu* that was used to record data from a household census of an Andean population in 1567 (Ascher and Ascher, 1981; Murra, 1968). We shall look at this example in some detail, for it serves to bring out clearly the versatility of the *quipu* as a recording device.

Data for the Andean population of Lupaqa are given for seven provinces whose households were classified into two ethnic groups (Alasaa and Maasaa). Each of the two groups are further divided into two sub-groups (Uru and Aymara). However, for two of the seven provinces the only information available is the total number of Uru and Aymara households. How was this information fitted into a logical structure, involving cross-categorization and summation, so that it could be recorded on a *quipu*?

We can see that the data consist of 26 independent items of information:

1 The populations of the five provinces for which complete information is available, divided into Alasaa and Maasaa groups and further subdivided into Uru and Aymara (making a total of 20 items of information).
2 The populations of the two provinces for which the only information available is the number of Uru and Aymara households in each (4 items of information).

3 The total population of households in the two provinces for which information is incomplete (2 items of information).

From the same data it is possible to obtain 20 derived items of information:

1 The grand total of all households: 1 item of information.

2 The total number of Uru and Aymara households: 2 items of information.

3 The number of households in each province: 7 items of information.

4 The number of Alasaa and Maasaa households in each province: 10 items of information.

These are the 46 items of information – a mixture of given and derived values, and partial and total summations – to be represented on a *quipu*.

The simplest way would be to represent along the main cord each item of information on a pendant cord, equally spaced, of different colours to distinguish categories. But this is a most uneconomical method of formatting information for it takes no account of the relationships that exist among a number of these items. A more efficient construction, but not optimal in any sense, is to proceed as follows. The information is arranged in seven groups, each group having seven pendant cords. The first four groups relate to the number of Uru and Aymara households in Alasaa and Maasaa for the seven provinces. The first eight pendant cords are blank (i.e. unknotted) since they represent the two provinces for which information is not available separately for Alasaa and Maasaa. The other twenty pendant cords would have all the relevant information in the form of clusters of knots. Partial sums and total sums – the 20 items of derived information listed above – can be shown by proper positioning of top cords.

This is one of a number of possible arrangements of cords on a *quipu*. The better the *quipucamayu* considers the pattern of distribution, taking account of the relative sizes and positions of different cords, the better the logical structure of the final representation. Cord placement, colour coding and number representation are the basic constructional features, repeated and recombined to define a format and convey a logical structure. This search for a coherent numerical/logical structure is mathematical thinking.

The Inca abacus

The *quipu* could not have been used as a calculating device. While results of summations and other simple arithmetical operations were recorded on the *quipu*, the computations were worked out elsewhere. How did the Incas carry out these calculations? The clue may lie in a passage from a book written by

Figure 2.4 The Inca abacus: recording numbers.

Figure 2.5 The Inca abacus: possible multiplication procedure.

Father José de Acosta, a Spanish priest, who lived in Peru from 1571 to 1586:

> To see them use another kind of *quipu* [?], with maize kernels, is a perfect joy. In order to carry out a very difficult computation for which an able computer would require paper and pen, these Indians make use of their kernels. They place one here, three somewhere else and eight, I know not where. They move one kernel here and there and the fact is that they are able to complete their computation without making the smallest mistake. As a matter of fact, they are better at practical arithmetic than we are with pen and ink. Whether this is not ingenious and whether these people are wild animals, let those judge who will! What I consider as certain is that in what they undertake to do they are superior to us. (de Acosta, 1596)

Is the priest here describing a form of counting-board similar in appearance to Poma's drawing given on Figure 2.2(b) and reproduced in Figure 2.4(a)? There can of course be no conclusive answer. But Wassen (1931), who was among the first to describe and interpret Guaman Poma's drawings, had an interesting explanation. He interpreted the row values of Figure 2.4(a), from bottom to top, as successive powers of ten. More controversial is his explanation of the column values of the counting-board: that, from left to right, they represent the values 1, 5, 15 and 30. According to this interpretation the number represented by the dark circles on the counting-board, worked out in Figure 2.4(b), is

$$47 + 21(10) + 20(100) + 36(1000) + 37(10\,000) = 408\,257$$

There is no other evidence to substantiate this idiosyncratic interpretation of the columns of the counting-board. Indeed, it would appear a strange choice given the use of a decimal base. A more plausible explanation would be that all column values are equal to 1, as in Figure 2.4(c), so that the number represented is

$$6 + 3(10) + 6(100) + 3(1000) + 5(10\,000) = 53\,636$$

How might the Incas have used this counting-board for computations? Addition and subtraction present few problems. We can only conjecture as to how multiplications may have been carried out before the results were recorded on a *quipu*. Suppose an astronomer-priest were faced with the need to multiply 116 days (to the nearest whole number, the synodic period of the planet Mercury) by 52 (a number of considerable astronomical significance to both the Maya and the Incas). The multiplication could have proceeded as in Figure 2.5. Figure 2.5(a) shows the number 1160 (i.e. 116×10). A successive process of repeatedly adding 1160 to itself five times will give the result 5800 (i.e.

1160 × 5), as shown in Figure 2.5(b). To complete the multiplication we need only add twice 116 to 5800 to obtain 6032, shown in Figure 2.5(c).

It is not part of my argument that the Incas used this precise method of multiplication, or even that this representation on the board is the correct one. What is clear, however, is that before a *quipu* could be used for storing information, some calculations had to be made, and these were probably done on a device like the Inca abacus.

THE EMERGENCE OF WRITTEN NUMBER SYSTEMS: A DIGRESSION

It is possible to view the appearance of a written number (or numeral) system as a culmination of earlier developments. First was the recognition of the distinction between *more* and *less* (a capacity we share with certain other animals). From this developed first simple counting, then the different methods of recording the counts as tally marks, of which the Ishango Bone is one example. This progression continued with the emergence of more and more complex means of recording information, culminating in the construction of devices such as the *quipu*. Before the appearance of such devices, there must have emerged an efficient system of spoken numbers founded on the idea of a base to enable numbers to be arranged into convenient groups. It was then only a matter of time before a system of symbols was invented to represent different numbers.

There is ample historical and anthropological evidence indicating that a variety of bases have been used over the ages and around the world. The numbers 2, 3 and 4 may have served as the earliest and simplest bases. Anthropologists have drawn our attention to certain groups in Central Africa who operated until recently with a rudimentary binary base, so that their spoken numbers would proceed thus: one, two, two and one, two twos, many. Similar systems using base 3 or 4 have been reported to be used by remote communities in South America.

The sheer variety of counting systems that have existed at some time is brought out in Figure 2.6. The main systems included:

1 *Counting by 2's* A typical example was found among an indigenous Australian group, the Gumulgal. Their counting proceeded as follows:

$$1 = urapon$$
$$2 = ukasar$$
$$3 = ukasar\text{--}urapon$$

Figure 2.6 Some counting systems (after Open University, 1975).

2 - count

4 - count

20 - count, and mixed 5 and 20

10 - count, and mixed 5 and 10

4 = *ukasar–ukasar*

5 = *ukasar–ukasar–urapon*

6 = *ukasar–ukasar–ukasar*

7 = *ukasar–ukasar–ukasar–urapon*

Clearly, this system becomes increasingly inefficient as the number-words become longer. A version of the 2-count system modified to take account of longer word-numbers uses special words for 3 and 4, so 6 and 8 become 'twice three' and 'twice four'. Both versions of the 2-count system were found to coexist in adjacent areas in Africa, in southern Australia and in South America, probably indicating a form of evolution from the less to the more efficient version.

2 *Counting by 5's* This suggests the use of finger-counting, and proved a more efficient method since it avoided the repetitions of the 2-count systems. It was also a more productive procedure since it could be more easily extended to include the use of 'two-hands', 'three-hands', and so on for increasing multiples of five. It was only a matter of time before this counting system gave way to counting by 10's.

3 *Counting by 10's* It is clear from Figure 2.6 that this counting system has been the most widespread, probably because it is associated with the use of fingers on both hands. It was, according to the accounts of the Spaniards, the base used by the Inca in their numeration. The etymology of number-words in the 10-count system may bring out its close association with finger-counting. As an illustration, the meanings of the words for the numbers from one to ten in the Zulu language are given in Table 2.1. Note the use of the subtraction principle in forming the words for 8 and 9: the word for 9, for example, means 'leave out one finger' (from ten fingers).

Two other bases have been either popular or mathematically important. The vigesimal scale (i.e. base 20) had its most celebrated development as a number system during the first millennium of the present era among the Maya of Central America. There have also been other base 20 number systems, of which the Yoruba system from West Africa is one of the better-known examples. Both systems will be discussed in later sections of this chapter. Base 20 systems may have originated in finger- and toe-counting among early societies.

The origins of the sexagesimal scale (i.e. base 60), first developed in Mesopotamia about five thousand years ago, cannot be traced to human physiology. This scale has certain computational advantages arising from the number of integral fractional parts, and survives in certain time and angular measurements even today. There was also another scale (duodecimal, or base

Table 2.1 The Zulu words for one to ten.

Number	Zulu word	Meaning
1	*Nyi*	State of being alone
2	*Bili*	Raise a separate finger
3	*Tatu*	To pick up
4	*Ne*	?
5	*Hlanu*	(All fingers) together
6	*Tatisitupa*	Take the right thumb
7	*Ikombile*	Point with forefinger of right hand
8	*Shiya'ngalombile*	Leave out two fingers
9	*Shiya'ngalolunye*	Leave out one finger
10	*Shumi*	Make all (fingers) stand

12) which must have enjoyed some popularity in the past because it survives in astronomical quantities such as (twice) the number of hours in a day or the approximate number of lunar months in a year. Other remnants of its usage are found in British units of measurement (12 inches = 1 foot), old money (12 pence = 1 shilling), and also in terms such as 'dozen' and 'gross' (=12 dozen). It had some powerful advocates in the past. So convinced was Charles XII of Sweden (1682–1718) of the superiority of this scale over the decimal, that he tried – though unsuccessfully – to ban the latter.

The Yoruba counting system

The origins of the Yoruba people of south-western Nigeria are lost in the mists of time. Oral traditions indicate that they came from the east, and a number of similarities between the customs and practices of ancient Egyptians and those of the Yoruba would support this. Their more recent history began with the foundation of the Oyo state around the early centuries of the second millennium AD. Commercial and other contacts with the north provided an important stimulus to scientific and cultural activity in the region. Later centuries saw the establishment of the vast Benin empire, independent of the Oyo kingdom, both of which were finally dissolved by the British at the end of the nineteenth century.

The Yoruba system of numeration is essentially a base 20 counting system, its most unusual feature being a heavy reliance on subtraction. The subtraction principle operates in the following way. As in our system, there are different names for the numbers one (*okan*) to ten (*eewa*). The numbers eleven (*ookanla*) to fourteen (*eerinla*) are expressed as compound words which may be translated as 'one more than ten' to 'four more than ten'. But once fifteen (*aarundinlogun*) is reached the convention changes, so that fifteen to nineteen (*ookandinlogun*) are expressed as 'twenty less five' to 'twenty less one', respectively, where twenty is known as *oogun*. Similarly, the numbers twenty-one to twenty-four are expressed as additions to twenty, and twenty-five to twenty-nine as deductions from thirty (*ogbon*). At thirty-five (*aarundinlogoji*), however, there is a change in the way the first multiple of twenty is referred to: forty is expressed as 'two twenties' (*ogoji*) while higher multiples are named *ogota* ('three twenties'), *ogerin* ('four twenties'), and so on to 'ten twenties', for which a new word *igba* is used. It is in the naming of some of the intermediate numbers that the subtraction principle comes into its own. To take a few examples, the following numbers are given names which indicate the decomposition shown on the right:

$$45 = (20 \times 3) - 10 - 5$$
$$50 = (20 \times 3) - 10$$
$$108 = (20 \times 6) - 10 - 2$$
$$300 = 20 \times (20 - 5)$$
$$318 = 400 - (20 \times 4) - 2$$
$$525 = (200 \times 3) - (20 \times 4) + 5$$

All the numbers from 200 to 2000 (except those that can be directly related to 400, or *irinwo*) are reckoned as multiples of 200. From the name *egbewa* for 2000, compound names are constructed for numbers in excess of this figure, using subtraction and addition wherever appropriate, similar to the above examples.

The origin of this unusual counting system is uncertain. One conjecture is that it grew out of the widespread practice of using cowrie shells for counting and computation. A description of the cowrie shell counting procedure given by Mann in 1887 is interesting. From a bag containing a large number of shells, the counter draws four lots of five to make twenty. Five twenties are then combined to form a single pile of one hundred. The merging of two piles of one hundred shells gives the next important unit of Yoruba numeration, two hundred. As a direct result of counting in fives, the subtraction principle comes into operation: taking 525 as an illustration, we begin with three piles of *igba* (200), remove four smaller piles of *oogun* (20) and then add five (*aarun*) cowrie shells to make up the necessary number.

This amazingly complicated system of numeration, in which the expression of certain numbers involves considerable feats of arithmetical manipulation, runs counter to the widespread view that indigenous African mathematics is primitive and unsophisticated. But does it have any intrinsic merit for computation? As an example of a calculation which exploits Yoruba numeration to the full, consider the multiplication 19×17. The cowrie calculator begins with twenty piles of twenty shells each. From each pile, one shell is removed (-20). Then three of the piles, now containing nineteen shells each, are also removed. The three piles are adjusted by taking two shells from one of them, and adding one each to the other two piles to bring them back to twenty ($-20 \times 2 - (20 - 3)$). At the end of these operations, we have

$$400 - 20 - (20 \times 2) - (20 - 3) = 323$$

While the Yoruba system shows what is possible in arithmetic without a written number system, it is clearly impractical for more difficult multiplications. It is a cumbersome method requiring a good deal of recall and mental arithmetic. Its peculiar characteristics, the base twenty and the subtraction principle of reckoning, seem to have had only a limited impact on other counting systems, even within West Africa.

Constructing a written number system

Once a base (say b) has been chosen, one of the simpler ways of constructing a number system is to introduce separate symbols to represent $b^0, b^1, b^2, \ldots$, where these symbols, repeated if necessary, may be used additively to represent any number. Possibly the earliest and certainly the best-known example of such a system with $b = 10$ is the Egyptian hieroglyphic number system, dating back to 3500 BC. (We shall be discussing the principles underlying the construction and arithmetical operations with these numerals in the next chapter.) The Aztecs of Central America later developed a system of numerals similar in principle to the Egyptian number system.

The Aztecs were a people who migrated to Mexico from the north in the early thirteenth century AD and founded a large tribute-based empire, ruled from their capital city Tenochtitlán, which reached the height of its power in the fifteenth and early sixteenth centuries. The empire's prosperity was founded on a highly centralized agricultural system in which existing land was intensively cultivated, irrigation systems were built, and swampland was reclaimed. The staple crop, maize, figures prominently in the Aztec number system.

This number system was vigesimal (base 20) and used four different symbols. The unit symbol was a 'blob' representing a maize seed-pod; the symbol for 20 was a flag, commonly used to mark land boundaries; 400 was represented by a schematic maize plant; and the symbol for 8000 is thought to be a 'maize dolly', similar to the decorative figures traditionally woven from straw in some European countries. The four symbols were:

The Aztecs would represent the number 9287 as

They also developed an intricate system of counting in which the bases depended on the type of objects being counted. Cloths or tortillas would be counted in twenties, while round objects such as eggs or oranges would be counted in tens.

Often an additive grouping system may evolve into a 'ciphered' number system. Here new symbols are introduced not only for the powers of the base b, but also for

$$1, 2, 3, \ldots, b - 1; 2b, 3b, \ldots, b(b - 1); b^2, 2b^2, 3b^2, \ldots, b^2(b - 1); \ldots$$

While such a system calls for a greater effort to memorize many more symbols, the representation of numbers is obviously more compact and computation more efficient than with a simple additive system. Examples of such systems include later Egyptian numerals (i.e. hieratic and demotic), Ionic (or Greek) alphabetic numerals, early Arabic numerals and the Indian Brahmi numerals. Some of these number systems will be discussed in later chapters.

There are a few instances of a simple additive grouping system developing into a multiplicative system. In such a system, after a base b has been selected separate sets of symbols are used for

$$1, 2, \ldots, (b - 1); b, b^2, b^3, \ldots$$

To represent any number greater than b, the symbols of both sets are used together. One of the best known of the multiplicative number systems is the standard Chinese system, discussed in Chapter 6.

It is likely that a positional or place-value number system such as ours, which evolved from the Indian number system, may have its origin in an earlier multiplicative or even a ciphered number system. In a positional system, after the base b has been selected any number can be represented by b different symbols. For example, our decimal numeral system requires symbols (or digits) to represent the numbers from zero to nine: $0, 1, 2, \ldots, 9$. With these symbols we can represent any integer N uniquely as

$$N = c_n b^n + c_{n-1} b^{n-1} + \cdots + c_2 b^2 + c_1 b + c_0$$

where $c_0, c_1, \ldots, c_{n-1}, c_n$ represent the basic symbols (in our number system, $0, 1, 2, \ldots, 9$). (Note that $0 \leqslant c_i < b$ for $i = 0, 1, 2, \ldots, n$ and $n \leqslant 9$.) For example,

$$1385 = (1 \times 10^3) + (3 \times 10^2) + (8 \times 10) + (5 \times 1)$$

This is a number system which is both economical in representation, given the requirement of only ten symbols to write any number, and less taxing on memory than a ciphered system. But its chief advantage is its immense computational efficiency when it comes to working with paper and pencil. It is not that arithmetic was impossible in non-positional number systems, as we shall soon see when we discuss arithmetical operations with Egyptian numerals; but rather that computations were often cumbersome or relied on a mechanical device such as an abacus or a counting-board, or even cowrie shells. The crucial advantage of our number system is that it gave birth to an arithmetic which could be done by people of average ability, not just an elite.

There are historical records of only three other number systems which were based on the positional principle. Predating all other systems was the Babylonian, which must have evolved during the third millennium BC. A sexagesimal scale was employed, with a simple collection of the correct number of symbols employed to write numbers less than 60. Numbers in excess of 60 were written according to the positional principle, though the absence of a symbol for zero until the early Hellenistic period limited the usefulness of the system for computational and representational purposes. The Chinese rod numeral system is essentially a base 10 system. The numbers $1, 2, \ldots, 9$ are represented by rods whose orientation and location determine the place value of the number represented and whose colour shows whether the quantity is positive or negative. We shall be discussing this system in detail in Chapter 6. The third positional system was the Mayan, essentially a vigesimal ($b = 20$) system incorporating a symbol for zero. It is to the details of this system that we now turn.

MAYAN NUMERATION

There is an area of Central America whose history and culture were shaped by the Mayan civilization. On the eve of the Spanish conquest of 1519–20, it occupied over 300 000 square kilometres (approximately the area of the British Isles) and covered present-day Belize, central and southern Mexico, Guatemala, El Salvador and parts of Honduras. Some cultural similarities helped to unite this vast territory: hieroglyphic writing, lip ornaments, positional numeration, and a calendar built around a year consisting of eighteen months of twenty days and a final month of five extra days. The main agency for the spread of this distinctive culture and retention of cultural links between different regions of Central America was the remarkable civilization of the Maya, which reached its classical phase between the third and tenth centuries AD. Unfortunately, this great civilization has left only the following scientific evidence:

1 Hieroglyphic inscriptions cover the stelae (upright pillars or slabs) found scattered around the region. They were constructed by the Maya every twenty years and span at least fifteen centuries. These stone monuments generally recorded the exact day of their erection, the principal events of the previous twenty years and the names of prominent nobles and priests.

2 The walls of some Mayan ruins and caves contain paintings and hieroglyphs which provide valuable evidence of not only their everyday life but their scientific activities.

3 A few manuscripts escaped the destruction of the Spanish conquerors. The most notable are the Dresden Codex, the Codex

Figure 2.7 Mayan 'snake numbers' (after Spinden, 1924, p. 27).

Peresianus and the Troano-Cortesian Codex. These books were made of durable paper from the fibre of the plant *Agave americana* and then covered with size (a gelatinous solution used to glaze paper) before the hieroglyphs were recorded in various colours.

Many hieroglyphs remain undeciphered even today, though some notable investigative work towards the end of the nineteenth century shed some light on part of the mysteries, especially those inscriptions relating to astronomical or calendar data. It was in the course of this work that the remarkable achievement of the Maya in the field of numeration was discovered.

Mayan numerals

The Mayan system of notation was one of the most economical systems ever devised. In the form that was used mainly by the priests for calendar computation from as early as 400 BC, it required only three symbols. A dot was used for one, and a bar for five; these symbols are thought to represent a pebble and a stick. Larger numbers were represented by a combination of these symbols up to nineteen. To write twenty, they introduced a symbol for zero which resembled a snail's shell. A few examples are:

$$(1 \times 7200) + (18 \times 360) + (5 \times 20) + 0 = 13\,780$$

The Dresden Codex, an important source of written evidence on Mayan numerals, contains a representation of the 'snake numbers' which were of such significance in Mayan cosmology; it is reproduced here in Figure 2.7. There are two sets of numerals represented on the coils: one in black, and one in red (shown as white in the figure). Reading from bottom to top, the Mayan numerals can be transliterated as in Table 2.2. It is immediately apparent that a departure from a strictly vigesimal system occurs at the second number group, where instead of $b^2 = 20^2 = 400$ we have $18b = 360$. Subsequent groups are of the form $18b^i$, where $i = 2, 3, \ldots$. This anomaly reduces efficiency in arithmetical calculation. For example, one of the most useful facilities with our number system is the ability to multiply a given number by 10 by adding a zero to the end of it. An addition of a Mayan zero to the end of a number would not in general multiply the number by twenty because of the mixed base system employed.

Table 2.2 The numbers represented in Figure 2.7.

Black number			Red (white) number		
$4 \times 18(20)^4$	$=$	11520000	$4 \times 18(20)^4$	$=$	11520000
$6 \times 18(20)^3$	$=$	864000	$6 \times 18(20)^3$	$=$	864000
$9 \times 18(20)^2$	$=$	64800	$1 \times 18(20)^2$	$=$	7200
$15 \times 18(20)$	$=$	5400	$9 \times 18(20)$	$=$	3240
12×20	$=$	240	15×20	$=$	300
19×1	$=$	19	0×1	$=$	0
		12454459			12394740

What is the reason for the presence of this curious irregularity in Mayan numeration? To understand the anomaly, we need to appreciate the social context in which the number system was used. As far as we know, this form of writing numbers was used only by a tiny elite – the priests who were responsible for carrying out astronomical calculations and constructing calendars, and it is the exigencies of these tasks that lie at the root of the explanation.

Before we examine the Mayan calendars, it is worth noting that the Maya had an alternative notation, shown in Figure 2.8, which often occurred in inscriptions alongside the 'dot and dash' numerals. This was the 'head variant' system, relying on a series of distinct anthropomorphic deity head glyphs to represent zero and one to nineteen. The heads are those of the thirteen deities of the Superior World, with six variants, whose significance is revealed in the next section.

Mayan calendars

The Maya had three kinds of calendar. The first, known as the *tzolkin* or 'sacred year', was specially devised for carrying out certain religious rituals. It contained 260 days, in twenty cycles of thirteen days each. Superimposed on each of the cycles was an unchanging series of twenty days, each of which was considered a god to whom prayers and other supplications were to be made. For example, the first of these was known as Imix, which represented the god associated with a crocodile or a water lily. One of the most auspicious days was the last of the twenty days (or gods), known as Ahau and associated with the Sun. To complicate matters still further, each of the twenty days was in turn assigned a number from one to thirteen. So once the fourteenth day of a series of twenty was reached, it was allocated the number 1 in a new cycle; thus the

Figure 2.8 Mayan 'head variant' numerals (Closs, 1986, p. 335).

twentieth day became 7 Ahau. This procedure for assigning the basic twenty days to the thirteen numbers continued indefinitely. Now, after the date 1 Imix, 260 days would have to pass before it recurred (as there are $13 \times 20 = 260$ possible combinations of the twenty basic days and the first thirteen numbers). Thus a particular day in the religious year of 260 days could be indicated uniquely by adding to the hieroglyph associated with one of the twenty basic days a number corresponding to it from the series of thirteen numbers. Each of the thirteen numbers could represent one of the thirteen gods of the Superior World or one of the thirteen gods of the Inferior World. Even today, among certain descendants of the Maya in Guatemala, a child takes the name and persona of the god associated with its date of birth.

This religious calendar was of limited utility to people like farmers who needed to keep track of the passing seasons for their livelihood. For them there was a second calendar, a true solar calendar known variously as the civil, secular or vague calendar. This calendar had 360 days, grouped into 18 monthly periods of twenty days, and an extra 'month' consisting of five days. The regular months were known as *uinals*, and the additional period was called *uayeb* (which means a period without a name; it was shown by a hieroglyph that represented chaos, corruption and disorder). Anyone born during this most unlucky period of the civil year was supposed to have been cursed for life.

There was yet a third calendar, known as the *tun*, mainly used for 'long' counts and with a unit of 360 days. This calendar was also based on a vigesimal system but the third order, as we saw earlier, was irregular since it consisted of $18 \times 20 = 360$ *kins* (days). The calendar took the following form:

$$20 \text{ } kins = 1 \text{ } uinal \text{ or } 20 \text{ days}$$
$$18 \text{ } uinals = 1 \text{ } tun \text{ or } 20 \times 18 = 360 \text{ days}$$
$$20 \text{ } tuns = 1 \text{ } katun \text{ or } 20^2 \times 18 = 7200 \text{ days}$$
$$20 \text{ } katuns = 1 \text{ } baktun \text{ or } 20^3 \times 18 = 144\,000 \text{ days}$$
$$20 \text{ } baktuns = 1 \text{ } piktun \text{ or } 20^4 \times 18 = 2\,880\,000 \text{ days}$$
$$20 \text{ } piktuns = 1 \text{ } calabtun \text{ or } 20^5 \times 18 = 57\,600\,000 \text{ days}$$
$$20 \text{ } calabtuns = 1 \text{ } kinchiltun \text{ or } 20^6 \times 18 = 1\,152\,000\,000 \text{ days}$$
$$20 \text{ } kinchiltuns = 1 \text{ } alautin \text{ or } 20^7 \times 18 = 23\,040\,000\,000 \text{ days}$$

For each of these units there was a special head-variant hieroglyph, the head taking one of various forms – man, animal, bird, deity or some mythological creature. These hieroglyphs were accompanied by the bars and dots standing for the numerals we discussed earlier. On a stela at Quirigua in Guatemala, shown in Figure 2.9, is inscribed the date on which it was built, using a calendar system of 'long counts'. The number represented in Figure 2.9 reads:

9 *baktuns*	17 *katuns*
0 *tuns*	0 *uinals*
0 *kins*	

This corresponds to

$$[9 \times 18(20)^3] + [17 \times 18(20)^2] = 1\,418\,400 \text{ days}$$

One of the oldest stelas found in Ires Zapotes bears the number $8 + [16 \times 20] + [6 \times 18(20)^2] + [16 \times 18(20)^3] + [7 \times 18(20)^4]$, which corresponds to 31 BC (Closs, 1986, p. 327).

The presence of the irregularity in the operation of the place-value system arose from the need to make all three calendars compatible with one another. Given the central importance in Mayan culture of the measurement of time, the curious anomaly in the third place-value position becomes more comprehensible. But this inconsistency inhibited the development of further arithmetical operations, particularly those involving fractions. Yet one of the more amazing aspects of Mayan astronomy was the high degree of accuracy that was obtained without ever working with fractions, rational or decimal. The Mayan estimate of the duration of a solar year, expressed in modern terms, is 365·242 days; the currently accepted value is 365·242 198 days. A similar degree of accuracy was obtained for the average duration of a lunar month. According to the Mayan astronomers 149 lunar months lasted 4400 days. This is equivalent to an average lunar month of 29·5302 days – with all our present-day knowledge and technology the figure we get is 29·530 59 days!

In this chapter we have travelled two continents in search of proto-mathematics. The intellectual bases of the examples given are mathematical in that they all consist of manipulating numerical systems in order to create some form of number record. Objections to considering these examples as mathematical are mainly on the following grounds:

1 The nature and scope of the mathematical ideas contained in these examples are fairly trivial, or unimportant in the long-term development of mainstream mathematics. The first part of this argument, dismissing something as trivial, is to ignore the socio-economic environment in which it developed – a reflection of an ahistorical bias which cannot be taken too seriously. The second part, the charge of insignificance, is valid only if one perceives mathematical development as an essentially linear or autonomous process. It is not, as we have sought to show in Chapter 10.

2 A more serious point centres on devices such as *quipus*. While *quipus* were important to the Inca state, they were little more than mnemonics and therefore, it is argued, hardly fruitful in generating mathematical concepts or algorithms. How valid is this argument? From our earlier discussion, it is clear that a *quipu* is more than a set of knots to jog the memory: it is a unique device in which numerical, logical and spatial relationships are brought together for the purpose of recording information and showing correlations between data. As such it throws an interesting light on the nature of different types of categorizations and summations as well as providing elementary exercises in

Long-count
introduction glyph
(glyph of the deity who
is patron of the
month Cumku on the
sacred calendar)

9 *baktuns*
$9 \times 18(20)^3 =$
1 296 000 days

17 *katuns*
$17 \times 18(20)^2 =$
122 400 days

0 *tuns*

0 *uinals*

0 *kins*

13 Ahau
(day on the sacred
calendar reached by
counting forward the
total number of days
on the long-count calendar
from the starting point of
the Mayan calendar)

Figure 2.9 The upper section of the stela at Quirigua in Guatemala, showing dating according to the 'long count' calendar (Midonick, 1968, p. 143).

55

combinatorials. A *quipu* to whose top cord are attached pendant cords from which are suspended subsidiary cords resembles an inverted tree. Questions of mathematical interest then arise: both specific ones, such as how many paths are possible along a particular tree, and general ones, such as how many trees can be constructed with a given number of cords. In modern mathematics the concept of a 'tree' is first found in the work of Gustav Kirchhoff, who in 1847 applied it to the study of electrical networks. Arthur Cayley used the same concept in his study of chemical isomers in 1857. From these attempts to answer questions that were essentially similar to those that confronted the makers of the *quipu* emerged a field of modern mathematics known as graph theory. Gerdes (1988a) has another illustration: traditional Angolan sand-drawings, where the skill lies in constructing figures without retracing a line or removing one's finger from the sand. It was a similar problem, about the bridges of Königsberg, that led the Swiss mathematician Leonhard Euler to achieve the breakthrough in graph theory around the middle of the eighteenth century.

3 There is also the argument that the term 'mathematics' should be used only for the study of numerical and spatial concepts for their own sake, rather than for their applications. This is a highly restrictive view of mathematics, often attributed to the Greeks – that mathematics devoid of a utilitarian bent is in some sense a nobler or better mathematics.* Where this attitude has percolated into the mathematics curriculum in schools and colleges, it engenders a sense of remoteness and irrelevance associated with the subject in many who study it, and an ingrained elitism in many who teach it.

4 Finally, the most substantive criticism of proto-mathematics points to a danger that inferences about mathematical activities of the past, especially from artefacts, may be unsound because of a natural tendency to attribute to the ancients our modern modes of thought and knowledge. This is a legitimate concern, and in the course of this chapter I have pointed out the need for caution and for a search for corroborative evidence before any definitive conclusions are drawn.

---

*This is brought out well in a passage from Plutarch's description of the Roman siege of Syracuse in 214 BC at which Archimedes lost his life. After stating that Eudoxus and Archytas had shown considerable ingenuity in devising mechanical demonstrations of some of the more difficult geometric theorems, Plutarch adds: 'Plato was indignant at these developments, and attacked both men for having corrupted and destroyed the ideal purity of geometry. He complained that they had caused [geometry] to forsake the realm of disembodied and abstract thought for that of material objects and to employ instruments which required much base and manual labour' (quoted by Fauvel and Gray, 1987, p. 173).

3 The Beginnings of Written Mathematics: Egypt

◇◇

THE URBAN REVOLUTION AND ITS AFRICAN ORIGINS

In the last chapter we began our examination of early evidence of mathematical activity with an artefact found in the middle of Africa. For the next stage of our journey we remain on the same continent, but move north to Egypt. Egypt is generally recognized as the homeland of one of the four early civilizations that grew up along the great river-valleys of Africa and Asia over five thousand years ago, the other three being in Mesopotamia, India and China. Egyptian civilization did not emerge out of the blue as a full-blown civilization without any African roots. Carbon dating of the remains of barley and einkorn wheat found at Kubbaniya, near Aswan in Upper Egypt, shows the beginnings of agriculture to have been around 16 000 BC, and this evidence is supported by the large concentrations of agricultural implements from around 13 000 BC found during the UNESCO-led operations to salvage the ancient monuments of Nubia.

Although there are no tangible traces of the origins of these neolithic communities, recent archaeological discoveries indicate that they may have belonged to groups from the once fertile Sahara region who were forced to migrate, initially to the areas south and east, as the desert spread. So, just as Egypt was a 'gift of the Nile' (in the words of Herodotus), the culture and people of Egypt were at least initially a 'gift' of the heartlands of Africa, the inhabitants of which were referred to then as 'Ethiopians'. This is borne out by the historian Diodorus, who wrote around 50 BC that the Egyptians 'are colonists sent out by the Ethiopians . . . And the large part of the customs of the Egyptians . . . are Ethiopian, the colonists still preserving their ancient manners' (Davidson, 1987, p. 7).

It is important that the African roots of the Egyptian civilization are emphasized so as to counter the still deeply entrenched view that the ancient Egyptians were racially, linguistically and even geographically separated from

The Crest of the Peacock

Africa.* Recent work, well summarized by Bernal (1987) and Davidson (1987), lays bare the flimsy scholarship and ideological bias of those who persist in regarding ancient Egypt as a separate entity, plucked out of Africa and replanted in the middle of the Mediterranean Sea.

What were the origins of the urban revolution that transformed Egypt into one of the great ancient civilizations? It is not possible to give a definitive answer. All we can do is surmise that the gradual development of effective methods of flood control, irrigation and marsh drainage contributed to a significant increase in agricultural yield. But each of these innovations required organization. An irrigation system calls for digging canals and constructing reservoirs and dams. Marsh drainage and flood control require substantial cooperation among what may have been quite scattered settlements. Would it be too fanciful to conjecture that, before the emergence of the highly centralized government of Pharaonic Egypt, a form of *ujamaa*† may have come into existence as an institutional back-up for these agricultural innovations? This may eventually have led to the establishment of administrative centres that grew into cities.

Between 3500 and 3000 BC the separate agricultural communities along the banks of the Nile were gradually united, first to form two kingdoms – Upper and Lower Egypt – which were then brought together, in about 3100 BC, as a single unit by a legendary figure called Menes, who came from Nubia (part of present-day Sudan). Menes founded a long line of Pharaohs, 32 dynasties in all, who ruled over a stable but relatively isolated society for the next three thousand years.

It is worth remembering that up to 1350 BC the territory of Egypt covered not only the Nile valley but also parts of modern Israel and Syria. Control over such a wide expanse of land required an efficient and extensive administrative system. Censuses had to be taken, taxes collected, and large armies maintained. Agricultural requirements included not only drainage, irrigation and flood control, but also the parcelling out of scarce arable land among the peasantry and the construction of silos for storing grain and other

---

*Davidson (1987, pp. 1–2), writing about public reactions to a television series that he presented on the history of the Africans, points out that what a number of viewers in Europe and North America found particularly difficult to accept were the 'black' origins of the ancient Egyptians: 'To affirm this, of course, is to offend nearly all established historiographical orthodoxy.'

†*Ujamaa* is a Swahili word meaning 'brotherhood' that was used to describe a Tanzanian government initiative in the late 1960s and early 1970s to encourage scattered rural homesteads to form villages, which would then serve both as pools of labour for communal activities and as units for meeting social needs in health, education, communication and water supply.

produce. Herodotus, the Greek historian who lived in the fifth century BC, wrote that

> Sesostris [Pharaoh Ramses II, c. 1300 BC] divided the land into lots and gave a square piece of equal size, from the produce of which he exacted an annual tax. [If] any man's holding was damaged by the encroachment of the river . . . The King . . . would send inspectors to measure the extent of the loss, in order that he might pay in future a fair proportion of the tax at which his property had been assessed. Perhaps this was the way in which geometry was invented, and passed afterwards into Greece. (Herodotus, *The Histories*, p. 169)

He also tells of the obliteration of the boundaries of these divisions by the overflowing Nile, regularly requiring the services of surveyors known as *harpedonaptai* (literally 'rope-stretchers'). Their skills must have impressed the Greeks, for Democritus (c. 410 BC) wrote that 'no one surpasses me in the construction of lines with proofs [?], not even the so-called rope-stretchers among the Egyptians'. One can only suppose that 'lines with proofs' in this context refers to constructing lines with the help of a ruler and a compass.

There were other pursuits requiring practical arithmetic and mensuration. As the Egyptian civilization matured, there evolved financial and commercial practices demanding numerical facility. The construction of calendars and the creation of a standard system of weights and measures were also products of an evolving numerate culture serviced by a growing class of scribes and clerks. And the high point of this practical mathematical culture is well exemplified in the construction of ancient Egypt's longest-lasting and best-known legacy – the pyramids.

SOURCES OF EGYPTIAN MATHEMATICS

Time has been less kind to Egyptian mathematical sources recorded on papyri than to the hard clay tablets from Babylonia. The exceptional nature of the climate and the topography along the Nile made the Egyptian civilization one of the more agreeable and peaceful of the ancient world. In this it contrasted sharply with its Mesopotamian neighbours, who not only had a harsher natural environment to contend with, but were often at the mercy of invaders from surrounding lands. Yet the very dryness of most of Babylonia, as well as the unavailability of any natural writing material, resulted in the creation of a writing medium which has stood the test of time far better than the Egyptian papyrus. However, it must be remembered that papyrus is quite a bit more durable than the palm leaves, bark or bamboo used as writing materials by the ancient Chinese and Indians.

There are two major sources and a number of minor ones on Egyptian mathematics. Most of the minor sources preserve the mathematics of a later period, the Hellenistic (332 BC to 30 BC) or Roman (30 BC to AD 395) periods of Egyptian history. The most important major source is the Ahmes (or Ahmose) Papyrus, named after the scribe who composed it, in about 1650 BC, from a work three centuries older. It is also known as the Rhind Mathematical Papyrus, after the British collector who acquired it in 1858 and subsequently donated it to the British Museum. (Since we know in this instance who penned the document, it would be more proper to name it after the composer than the collector.) The second major source is the Moscow Papyrus, written in about 1850 BC; it was brought to Russia in the middle of the last century, finding its way to the Museum of Fine Arts in Moscow. Between them, the Ahmes and Moscow Papyri contain a collection of 112 problems with solutions. At the time of the receipt of the Ahmes Papyrus by the British Museum in 1864, it was highly brittle with sections missing. A fortunate discovery of the missing fragments in the possession of the New York Historical Association in 1922 helped to restore it to its original form.

Of the minor sources, there is an Egyptian mace head from the beginning of the third millennium BC which contains a record of the spoils of war of the Pharaoh Narmer: 120 000 prisoners, 400 000 oxen and 1 422 000 goats. To record these numbers required a highly developed system of numerals which allowed counting to continue indefinitely by the introduction of a new symbol wherever necessary. Other minor sources include the Egyptian Mathematical Leather Roll, from the same period as the Ahmes Papyrus, which is a table consisting of twenty-six decompositions into unit fractions; the Berlin Papyrus, which contains two problems in simultaneous equations, one of second degree; the Reisner Papyri, from around 1800 BC, consisting of four worm-eaten rolls which record volume calculations relating to temples; and the Kahun Papyrus, also from around 1800 BC, containing six scattered mathematical fragments, not all of which have been deciphered. There will be further references to these sources later in this chapter.

Ahmes tells us that his material is derived from an earlier document belonging to the Middle Kingdom (2000–1800 BC), and hints at the possibility that the knowledge may ultimately have been derived from Imhotep (*c.* 2650 BC), the legendary architect and physician to Pharaoh Zoser of the Third Dynasty. Its opening sentence claims it to be 'a thorough study of all things, insight into all that exists, knowledge of all obscure secrets'. While an examination of the Ahmes Papyrus does not bear this out, it remains, with eighty-seven problems and their solutions, the most comprehensive source of early Egyptian mathematics. The Moscow Papyrus was composed by a less competent scribe, who remains unknown. It shows little order in the arrangement of topics covered, which are not very different to those in the Ahmes Papyrus. It contains twenty-five problems, among them two notable results of Egyptian mathematics: the formula for the volume of a truncated square pyramid (or frustum), and a remarkable solution to the problem of finding what some interpreters consider to be the curved surface area of a hemisphere. Before looking in detail at the mathematics in these two main sources, we begin with a discussion of the Egyptian system of numeration.

NUMBER RECORDING AMONG THE EGYPTIANS

There is an impression, fostered (no doubt inadvertently) by many textbooks on the history of mathematics, that only one scheme of numeration was used in ancient Egypt: the hieroglyphic. This impression is quite consistent with a view of Egyptian civilization as stable and unchanging, with mathematics primitive yet sufficient to serve the economic and technological needs of the time. The truth is very different from this view. It is possible to distinguish three different notational systems – hieroglyphic (pictorial), hieratic (symbolic) and demotic (popular) – the first two of which made their appearance quite early in Egyptian history. The hieratic notation was employed in both the Ahmes and Moscow Papyri. The demotic variant was a popular adaptation of the hieratic notation and became important during the Greek and Roman periods of Egyptian history.

The hieroglyphic system of writing was a pictorial script where each character represented an object, some easily recognizable. Special symbols were used to represent each power of 10 from 1 to 10^7. Thus a unit was commonly written as a single vertical stroke, though when rendered in detail it was shown as a short piece of rope. The symbol for ten was a longer piece of rope, in the shape of a horseshoe. A hundred was a coil of rope. In all these cases, the overriding motif seems to have been a rope whose length and shape determined the magnitude of the number represented, quite in keeping with the important role of the 'rope-stretchers' in ancient Egypt. The pictograph for a thousand resembled a lotus flower, though the plant sign formed the initial of *khaa*, 'measuring cord'. Ten thousand was represented by a crooked finger, which probably had some obscure phonetic or allegorical connotation. The stylized tadpole for a hundred thousand may have been a general symbol for large numbers. A man with arms upraised, as if in astonishment, possibly signifying vastness or eternity, represented one million; and a rising sun for ten million could have been associated with one of the more powerful of the Egyptian deities – Ra, the Sun-god.

Thus the earliest Egyptian number system was based on the following symbols:

1	10	10^2	10^3	10^4	10^5	10^6	10^7

Any number can be written using the above symbols additively, for example:

$$12013 = 3 + 1(10) + 2(10^3) + 1(10^4) = $$

No difficulties arose from not having a zero or place-holder in this number system. It is of little consequence in what order the hieroglyphs appeared, though the practice was generally to arrange them from right to left in descending order of magnitude, as in the example above. Addition and subtraction posed few problems. In adding two numbers, one made a collection of each set of symbols that appeared in both numbers, replacing them with the next higher symbol as necessary. Subtraction was merely the reversal of the process for addition, with decomposition achieved by replacing a larger hieroglyph with ten of the next lower symbol.

This form of notation served as a useful means of making clear imprints on stone or metal. Indeed, there is evidence to suggest that even when these numerals were replaced for all practical purposes by the hieratic notation, they continued to be used occasionally (rather as Roman numerals sometimes appear today, on clock-faces for example).

The hieratic representation was similar in that it was additive and based on powers of ten. But it was far more economical, as a number of identical hieroglyphs were replaced with fewer symbols, or just one symbol. For example, the number 57 was written in hieroglyphic notation as

But the same number would be written in hieratic notation as ⇀ 乙 , where ⇀ and 乙 represent 50 and 7 respectively. It is clear that the idea of a ciphered number system, which we discussed in the last chapter, is already present here.

While this notation was no doubt more taxing on memory, its economy, speed, conciseness and greater suitability for writing with pen and ink must have been the main reasons for its fairly early adoption in ancient Egypt. For example, to represent the number nine hundred and ninety-nine would take altogether twenty-seven symbols in hieroglyphics, compared with six marks in hieratic representation or three in present-day numerals. And from the point of view of the history of mathematics, the hieratic notation may have inspired, at least in its formative stages, the development of the alphabetic Greek number system around the middle of the first millennium BC.

The gradual replacement of the hieroglyphic by the hieratic notation may have been given impetus by the growing popularity of papyrus as a writing medium. From as early as the First Dynasty, thin sheets of whitish 'paper' were produced from the interior of the stem of a reed-like plant which grew in the swamps along the banks of Nile. Fresh stems were cut, the hard outer parts removed, and the soft inner pith was laid out and beaten until it formed into sheets, the natural juice of the plant acting as the adhesive. Once dried in the sun, the writing surface was scraped smooth and gummed into rolls, of which the longest known measures over 40 metres. On these rolls the Egyptians wrote with a brush-like pen, using for ink either a black substance made from soot or a red substance made from ochre.

EGYPTIAN ARITHMETIC

The method of duplation and mediation

One of the great merits of the Egyptian method of multiplication or division is that it requires prior knowledge of only addition and the two times table. A few simple examples will illustrate how the Egyptians would have done their multiplication and division. Only in the first example will the operation be explained in terms of both the hieroglyphic and present-day notation.

EXAMPLE 3.1 *Multiply 17 by 13.*

Solution The scribe had first to decide which of the two numbers was the multiplicand – the one he would multiply by the other. Suppose he chose 17. He would proceed by successively multiplying 17 by 2 (i.e. continuing to double each result), and stopping before he got to a number on the left-hand side of the 'translated' version below which exceeded the multiplier, 13:

IIII∩ III	I	→ 1	17
II∩∩ II	II	2	34
IIII∩∩∩ IIII∩∩∩	II	→ 4	68
IIII∩∩9 II ∩	IIII IIII	→ 8	136
I∩∩99 ⌣	IIII∩	1 + 4 + 8 = 13	17 + 68 + 136 = 221

The hieroglyph ⌣, resembling a papyrus roll, meant 'the result is the following'. The numbers to be added to obtain the multiplier 13 are arrowed.

If this method is to be used for the multiplication of any two integers, the following rule must apply:

Every integer can be expressed as the sum of integral powers of 2.

Thus

$$15 = 2^0 + 2^1 + 2^2 + 2^3$$
$$23 = 2^0 + 2^1 + 2^2 + 2^4$$

It is not known whether the Egyptians were aware of this general rule, though the confidence with which they approached all forms of multiplication by this process suggests that they were.

This ancient method of multiplication provides the foundation for Egyptian calculation. It was widely used, with some modifications, by the Greeks and continued well into the Middle Ages in Europe. In a modern variation of this method, still popular among rural communities in Russia, Ethiopia and the Near East, there are no multiplication tables and the ability to double and halve numbers (and to distinguish odd from even) is all that is required.

EXAMPLE 3.2 *Multiply 225 by 17.*
Solution

→ 225	17	Inspect the left-hand column for odd numbers
112	34	or 'potent' terms (ancient lore in many
56	68	societies imputed 'potency' to odd numbers).
28	136	Add the corresponding terms in the right-
14	272	hand column to get the answer.
→ 7	544	
→ 3	1088	
→ 1	2176	

$$17 + 544 + 1088 + 2176 = 3825$$

This method, known in the West as the 'Russian peasant method', works by expressing the multiplicand, 225, as the sum of integral powers of 2:

$$225 = 1(2^0) + 0(2^1) + 0(2^2) + 0(2^3) + 0(2^4) + 1(2^5) + 1(2^6) + 1(2^7)$$

Adding the results of multiplying each of these components by 17 gives the answer.

In Egyptian arithmetic, the process of division was closely related to the method of multiplication. In the Ahmes Papyrus a division x/y is introduced by the words 'reckon with y so as to obtain x'. So an Egyptian scribe, rather than thinking of 'dividing 696 by 29', would say to himself, 'Starting with 29, how many times should I add it to itself to get 696?' The procedure he would set up to solve this problem would be similar to a multiplication exercise:

EXAMPLE 3.3 *Divide 696 by 29.*
Solution

		The scribe would stop at 16, for
1	29	the next doubling would take
2	58	him past the divisor, 29. Some
4	116	quick mental arithmetic on the
8	→ 232	numbers in the right-hand
16	→ 464	column shows that the sum of
16 + 8 = 24	232 + 464 = 696	232 and 464 would give the

The scribe would stop at 16, for the next doubling would take him past the divisor, 29. Some quick mental arithmetic on the numbers in the right-hand column shows that the sum of 232 and 464 would give the exact value of the dividend, 696. Taking the sum of the corresponding numbers in the left-hand column gives the answer.

Where a scribe was faced with the problem of not being able to get any combination of the numbers in the right-hand column to add up to the value of the dividend, fractions had to be introduced. And here the Egyptians faced constraints arising directly from their system of numeration: their method of writing numerals did not allow any unambiguous way of expressing fractions. But the way they tackled the problem was quite ingenious.

Egyptian representation of fractions

In 1927, the unrolling of a leather manuscript as old as the Ahmes Papyrus aroused a great deal of excitement among Egyptologists. Many expected important disclosures. Imagine their disappointment when all that was revealed was a collection of twenty-six simple identities, such as $1/10 + 1/40 = 1/8$, recorded by an unknown and inexperienced scribe. Glanville (1927), the first translator of the manuscript, suggested wryly that it might be of some value for the fresh insights it could offer into the technology of leather-making of the period. However, such a pessimistic judgement has not been borne out by further examination, for the contents of the Leather Roll have helped to clarify a number of calculations contained in the Ahmes Papyrus.

Operating with unit fractions is a singular feature of Egyptian mathematics, and is absent from almost every other mathematical tradition. A substantial proportion of surviving ancient Egyptian calculations makes use of such operations – of the eighty-seven problems in the Ahmes Papyrus, only six do not. Two reasons may be suggested for this great emphasis on fractions. In a society that did not use money, where transactions were carried out in kind, there was a need for accurate calculations with fractions, particularly in practical problems such as division of food, parcelling out land and mixing different ingredients for beer or bread. We shall see later that a number of problems in the Ahmes Papyrus deal with such practical concerns.

A second reason arose from the peculiar character of Egyptian arithmetic. The process of halving in division often led to fractions. Consider how the Egyptians solved the following problem (No. 25) from the Ahmes Papyrus:

EXAMPLE 3.4   *Divide 16 by 3.*
Solution

1	$\rightarrow$	3
2		6
4	$\rightarrow$	12
2/3		2
1/3	$\rightarrow$	1
$1 + 4 + \frac{1}{3} = 5 + \frac{1}{3}$		16

As $12 + 3 = 15$ falls one short of 16, the Egyptian scribe would proceed by working out 2/3 of 1 and then halving the result. These steps are shown on the left. Now, $3 + 12 + 1 = 16$. The sum of the corresponding figures in the left-hand column gives the answer, $5\frac{1}{3}$.

Two important features of Egyptian calculations with fractions are highlighted here:

1 Perverse as it may seem to us today, to calculate a third of a number a scribe would first find two-thirds of that number and then halve the result. This was standard practice in all Egyptian computations.

2 Apart from two-thirds (represented by its own hieroglyph, either $\stackrel{\frown}{\cdot\cdot}$ or $\bigoplus$), Egyptian mathematics had no compound fractions: all fractions were decomposed into a sum of unit fractions (fractions such as 1/4 and 1/5).

To represent a unit fraction the Egyptians used the symbol $\bigcirc$, meaning 'part', with the denominator underneath. Thus $\frac{1}{5}$ and $\frac{1}{40}$ would appear as $\stackrel{\frown}{\cdots}$ and $\stackrel{\frown}{\cap\cap}$ respectively.

The $2/n$ table: its construction

The dependence on unit fractions in arithmetical operations, together with the peculiar system of multiplication, led to a third aspect of Egyptian computation. Every multiplication and division involving unit fractions would invariably lead to the problem of how to double unit fractions. Now, doubling a unit fraction with an even denominator is a simple matter of halving the denominator. Thus doubling 1/2, 1/4, 1/6 and 1/8 yields 1, 1/2, 1/3 and 1/4. Doubling 1/3 raised no difficulty, for 2/3 had its own hieroglyphic or hieratic symbol. But it was in doubling unit fractions with other odd denominators that difficulties arose. For some reason unknown to us, it was not permissible in Egyptian computation to write two times $1/n$ as $1/n + 1/n$. Thus the need arose for some form of ready reckoner which would provide the appropriate unit fractions that summed to $2/n$, where $n = 5, 7, 9, \ldots$.

At the beginning of the Ahmes Papyrus there is a table of decompositions of $2/n$ into unit fractions for all odd values of n from 3 to 101. A few of its

Table 3.1 Some entries from the Ahmes Papyrus $2/n$ table.

$2/n$	Unit fractions
2/5	1/3 + 1/15
2/7	1/4 + 1/28
2/9	1/6 + 1/18
2/15	1/10 + 1/30
2/17	1/12 + 1/51 + 1/68
2/47	1/30 + 1/141 + 1/470
2/49	1/28 + 1/196
2/51	1/34 + 1/102
2/55	1/30 + 1/330
2/57	1/38 + 1/114
2/59	1/36 + 1/236 + 1/531
2/97	1/56 + 1/679 + 1/776
2/99	1/66 + 1/198
2/101	1/101 + 1/202 + 1/303 + 1/606

entries are given in Table 3.1. The usefulness of this table for computations cannot be overemphasized: it may quite legitimately be compared in importance to the logarithmic tables that were used before the advent of electronic calculators. The table is interesting for a number of reasons. It does not contain a single arithmetical error, in spite of the long and highly involved calculations that its construction must have entailed; it may be a final corrected version of a number of earlier attempts that have not survived.

There is an even more remarkable aspect to this table. With the help of a computer, it has been worked out that there are about 28 000 different combinations of unit fraction sums that can be generated for $2/n$, $n = 3, 5, \ldots, 101$. The constructor of this table arrived at a particular subset of 50 unit fraction expressions, one for each value of n. Gillings (1972) argued that the unit fractions in this subset are optimal or near-optimal as judged by the following criteria of computational efficiency:*

1 a preference for small denominators, and none greater than 900,
2 a preference for combinations with only a few unit fractions (no expression contains more than four), and
3 a preference for even numbers, especially as the denominator of the first unit fraction in each expression, even though they are larger or might increase the number of terms in the expression.

To take an example, according to Gillings's calculations the fraction 2/17 can be decomposed into unit fraction summations in just one way if there are two unit fraction terms, eleven ways with three unit fraction terms and 467 ways with four unit fraction terms. Table 3.1 shows that the constructor opted for one of the three unit fraction groups, $2/17 = 1/12 + 1/51 + 1/68$, rather than the solitary two unit fraction group, $2/17 = 1/9 + 1/153$. It would seem that criteria (1) and (3) prevailed in this instance.

Multiplication and division with unit fractions

The main purpose of constructing the table was to use it for multiplication and division. Let us consider one example of each to illustrate its use. First, multiplication:

---

*Bruckheimer and Salamon (1977) have argued that in a number of cases the selection criteria put forward by Gillings are inapplicable.

EXAMPLE 3.5 *Multiply $1\frac{8}{15}$ by $30\frac{1}{3}$ (or $1 + 1/3 + 1/5$ by $30 + 1/3$).*
Solution

1	$1+1/3+1/5$
→ 2	$2+2/3+2/5=2+2/3+1/3+1/15=3+1/15$
→ 4	$6+2/15=6+1/10+1/30$
→ 8	$12+1/5+1/15$
→ 16	$24+2/5+2/15=24+1/3+1/15+1/10+1/30$
2/3	$2/3+2/9+2/15=2/3+1/6+1/18+1/10+1/30$
→ 1/3	$1/3+1/12+1/36+1/20+1/60$
$2+4+8+16+1/3=30+1/3$	$46+1/5+1/10+1/12+1/15+1/30+1/36$

The product of the two numbers using modern multiplication would be $46\frac{23}{45}$, which is exactly equivalent to the Egyptian result given in the last row. In the course of multiplication we have taken the unit fraction terms of 2/5, 2/15 and 2/9 from Table 3.1. And because Egyptian multiplication was based on doubling, only Table 3.1 was required. The sheer labour and tedium of this form of multiplication should not make us forget how modest is the 'toolkit' required. The ability to double, halve and work with the fraction 'two-thirds', together with the $2/n$ table, is sufficient.

To illustrate division with fractions, we take one of the more difficult problems of its kind from the Ahmes Papyrus, Problem 33, which may be restated as follows:

EXAMPLE 3.6 *The sum of a certain quantity together with its two-third, its half and its one-seventh becomes 37. What is the quantity?*
Solution   In the language of modern algebra, this problem is solved by setting up an equation of the first degree in one unknown. Let the quantity be x. The problem is then to solve

$$(1 + 2/3 + 1/2 + 1/7)x = 37$$

to give

$$x = 37 \div (1 + 2/3 + 1/2 + 1/7) = 16\frac{2}{97}$$

The problem restated *Divide 37 by 1 + 2/3 + 1/2 + 1/7.*

1	1 + 2/3 + 1/2 + 1/7
2	4 + 1/3 + 1/4 + 1/28 (2/7 = 1/4 + 1/28 from the 2/*n* table)
4	8 + 2/3 + 1/2 + 1/14
8	18 + 1/3 + 1/7
→16	36 + 2/3 + 1/4 + 1/28

At this point in the procedure two questions arise:

1 It is easily seen that the right-hand side of the last row is close to 37, which is the dividend. What must be added to 2/3 + 1/4 + 1/28 to make up 1? With our present method, we easily find the answer, 1/21.

2 The next question is: By what must the divisor 1 + 2/3 + 1/2 + 1/7 be multiplied to get 1/21? The answer is again easily obtained, as 2/97, or in unit fractions 1/56 + 1/679 + 1/776, from Table 3.1.

So the solution is

$$37 \div (1 + 2/3 + 1/2 + 1/7) = 16 + 1/56 + 1/679 + 1/776$$

Egyptian division: The use of 'red auxiliaries'

The real question remains: How would the Egyptians, working within the constraints of their arithmetic, have dealt with the problems raised by (1) and (2) in Example 3.6? A study of some of the problems in the Ahmes Papyrus provides us with the answer. Problems 21 to 23 are commonly known as 'problems in completion', since they are expressed as:

Complete   2/3   1/15   to   1   (Problem 21)

Complete   1/4   1/8   1/10   1/35   1/45   to   3   (Problem 23)

These problems are identical to the one in question (1) above, which may also be expressed in this way:

Complete   2/3   1/4   1/28   to   1

The Egyptians adopted a method of solution which is analogous (but not equivalent) to the present-day method of least common denominator. First they took the denominator of the smallest unit fraction as a reference number, and then multiplied each of the fractions by this number to obtain 'red auxiliaries' (so named because the scribe wrote these numbers in red ink). They proceeded to calculate by how much the sum of these auxiliaries fell short of the reference number. This shortfall quantity was then expressed as a fraction of the reference number to obtain the desired complement. If the shortfall quantity turned out to be an awkward fraction, a further search was made for a reference number which would result in more manageable auxiliaries. So, how was question (1) tackled the Egyptian way?

EXAMPLE 3.7   *Complete 2/3 1/4 1/28 to 1.*
Solution

$$2/3 + \quad 1/4 \quad + \quad 1/28 \quad + \text{(some fraction)} = \quad 1$$

$$28 + (10 + 1/2) + (1 + 1/2) + \qquad 2 \qquad = 42$$

The denominator of the smallest fraction, 28, is not a suitable reference number given the auxiliaries that result. Instead, 42 is chosen for ease of calculation because it is important that the sum of the auxiliaries belonging to the divisor in Example 3.6 is an integer. Thus 42 is the lowest common multiple of the numbers 1, 3, 2 and 7. However, the reference number chosen in Egyptian computation was not necessarily the lowest common multiple. So what fraction(s) of 42 will give 2? The answer is 1/21.

The next step is to find by what fraction the divisor $1 + 2/3 + 1/4 + 1/7$ must be multiplied to get 1/21. In other words, we have to divide 1/21 by $1 + 2/3 + 1/2 + 1/7$:

→ 1	21
→ 2/3	14
→ 1/2	10 + 1/2
→ 1/7	3
$1 + 2/3 + 1/2 + 1/7$	$48 + 1/2$

Now,

$$1 \div (48 + 1/2) = 2/97 = 1/56 + 1/679 + 1/776$$

(obtained from the $2/n$ table)

Hence

$$37 \div (1 + 2/3 + 1/2 + 1/7) = 16 + 1/56 + 1/679 + 1/776$$

We have not followed the scribe all the way in his solution to the problem, for the reason that at one stage his approach requires an addition of 16 unit fractions, the last six of which are 1/1164, 1/1358, 1/1552, 1/4074, 1/4753 and 1/5432! We can only assume that the scribe was either an incredible calculator, or that he had a battery of tables which he could consult when called upon to add different combinations of unit fractions. The latter explanation seems more likely: the Egyptians were inveterate table-makers, and the summation table of unit fractions contained in the Leather Roll and the decomposition table of $2/n$ in the Ahmes Papyrus are prime examples.

It is unlikely that the original problem (Example 3.6) had any practical import. In an attempt probably to illustrate, for the benefit of trainee scribes, the solution of simple equations of this type, an unfortunate choice of numbers led to difficult sets of unit fractions with the attendant cumbersome operations, which the scribe accomplished without faltering. One is again struck by the mental agility of the scribes who could perform such feats with a minimum of mathematical tools to call upon. The use of red auxiliaries is further evidence of the high level of Egyptian achievement in computation, since they enabled any division, however complicated, to be performed.

Applications of unit fractions: Distribution of loaves

As has been suggested, the exclusive use of unit fractions in Egyptian mathematics also had a practical rationale. This is brought out quite clearly in the first six problems of the Ahmes Papyrus, which are concerned with sharing out n loaves among 10 men, where $n = 1, 2, 6, 7, 8, 9$. As an illustration let us consider Problem 6, which relates to the division of 9 loaves among 10 men. A present-day approach would be to work out the share of each man, i.e. 9/10 of a loaf, and then divide the loaves so that the first nine men would each get 9/10 cut from one of the nine loaves. The last man, left with the nine pieces of 1/10 remaining from each loaf, might well regard this method of distribution as less than satisfactory. The Egyptian method of division avoids such a difficulty. It consists of first looking up the decomposition table for $n/10$ and discovering that $9/10 = 2/3 + 1/5 + 1/30$. The division would then proceed as shown in Figure 3.1: seven men would each receive 3 pieces of bread, consisting of 2/3, 1/5 and 1/30 of a loaf. The other three men would each receive 4 pieces consisting of two 1/3 pieces, a single 1/5 piece and a single 1/30 of a loaf. Justice is not only done, but *seen* to be done!

Applications of unit fractions: Remuneration of temple personnel

In a non-monetary economy, payment for both goods and labour is made in kind. Often the choice of the goods which act as measures or standards of value provides interesting insights into the character of the society. In Egypt, bread and beer were the most common standards of value for exchange. A number

Figure 3.1 Problem 6 from the Ahmes Papyrus: sharing nine loaves among ten men (after Gillings, 1962, p. 67).

of problems in the Ahmes Papyrus concern these goods, dealing with their distribution among a given number of workers, and also with measuring the strength (*pesu*) of different types of these two commodities. We shall be examining one of the *pesu* problems later in this chapter. But first we look at an example, brought to our attention by Gillings (1972), which sheds some interesting light.

Table 3.2, adapted from Gillings's book, is a record of payments to various temple personnel at Illahun around 2000 BC. The payments were made in loaves of bread and two different types of beer (referred to here as Beer A and Beer B). The temple employed 21 persons, and had 70 loaves, 35 jugs of Beer A and $115\frac{1}{2}$ jugs of Beer B available for distribution every day. The unit of distribution was taken to be 1/42 of a portion of each of these items, which was worked out as $1 + 2/3$ of bread, $2/3 + 1/6$ of Beer A and $2 + 2/3 + 1/10$ of Beer B.

The table is interesting for a number of reasons. It contains a rare example of an arithmetical error on the part of a scribe: the unit of distribution of Beer B was wrongly worked out as $2 + 2/3 + 1/10$. The scribe proceeded to use the incorrect figure in working out the shares of different personnel, but did not apparently check his calculations by adding all the shares. He wrote down the total as $115\frac{1}{2}$ jugs, which is what it should have been, whereas the portions in the table add up to $114\frac{1}{2}$ jugs. Also, the table bears ample testimony to the facility with which the Egyptians could handle fractions. Given all the limitations of their number system, they proved to be extremely adept at computations. Further, the minute fractional division of both beer and bread suggests a highly developed system of weights and measures: it is intriguing how 2 + 1/6 + 1/18 jugs of Beer A were shared equally among eight 'other workers'.

From the table we have some indication of the relative status of the personnel at the temple. At the top was the high priest, or temple director, often a member of the royal family. Among his duties was to pour out the drink-offering to the gods and to examine the purity of the sacrificial animals. It was only after he had 'smelt' the blood and declared it pure that pieces of flesh could be laid on the table of offerings. Hence Ue'b, meaning 'pure', was the name by which he was known. Perhaps more important than the Ue'b from a ritual point of view was the head reader, or reciter-priest, whose duty it was to recite from the holy books. Since magical powers were attributed to these texts it was generally believed that the reciter-priest was a magician, making him in status and remuneration second only to the high priest. After him came other classes of the priesthood, the largest of which was known as the 'servants of God'. Some of them were prominent in civil life; others were appointed to

Table 3.2 Remuneration of the personnel of Illahun Temple (units of distribution per person).

Status of personnel	Number of portions received	Commodity		
		Bread (1 + 2/3 loaves)	Beer A (2/3 + 1/6 jugs)	Beer B* (2 + 1/2 + 1/4 jugs)
Temple director	10	16 + 2/3	8 + 1/3	27 + 1/2
Head reader	6	10	5	16 + 1/2
Usual reader	4	6 + 2/3	3 + 1/3	11
Head lay priest	3	5	2 + 1/2	8 + 1/4
Priests, various (7)	14	23 + 1/3	11 + 1/3	37 + 1/2
Temple scribe	1 + 1/3	2 + 1/6 + 1/18	1 + 1/9	3 + 2/3
Clerk	1	1 + 2/3	2/3 + 1/6	2 + 1/2 + 1/4
Other workers (8)	2 + 2/3	4 + 1/3 + 1/9	2 + 1/6 + 1/18	7 + 1/3
Totals	42	70	35	$115\frac{1}{2}$

Adapted from Table 11.2 in Gillings (1972).

*The scribe made an error in working out the amount in one portion of Beer B, 1/42 of $115\frac{1}{2}$, which he estimated as 2 + 2/3 + 1/10 instead of 2 + 1/2 + 1/4. This mistake has been rectified along the lines indicated by Gillings.

serve particular gods. Their job included washing and dressing statues of their assigned deities and making offerings of food and drink to them at certain times of the day. The scribes came quite low down on the list, though this was not the case in other walks of life — most scribes, particularly those associated with the royal court, enjoyed considerable status and power.

EGYPTIAN ALGEBRA: THE BEGINNINGS OF RHETORICAL ALGEBRA

It is sometimes claimed that Egyptian mathematics consisted of little more than applied arithmetic, and that one cannot therefore talk of Egyptian algebra or geometry. We shall come to the question of Egyptian geometry, but first we consider the existence or otherwise of an entity called Egyptian algebra.

The rules devised by mathematicians for solving problems about numbers of one kind or another may be classified into three types. In the early stages of mathematical development these rules were expressed verbally, and consisted of detailed instructions about what was to be done to obtain the solution to a problem, for which reason this approach is referred to as 'rhetorical algebra'. In time, the prose form of rhetorical algebra gave way to the use of abbreviations for recurring quantities and operations, heralding the appearance of 'syncopated algebra'. Traces of such algebra are to be found in the works of the Alexandrian mathematician Diophantus (c. AD 250), but it achieved its fullest development — as we shall see in later chapters — in the work of Indian and Arab mathematicians during the first millennium AD. During the past five hundred years there has developed 'symbolic algebra' where, with the aid of letters and signs of operation and relation $(+, -, \times, \div, =)$, problems are stated in such a form that the rules of solution may be applied consistently and systematically. The transformation from rhetorical to symbolic algebra marks one of the most important advances in mathematics. It had to await

1 the development of a positional number system which allowed numbers to be expressed concisely and with which operations could be carried out efficiently, and

2 the emergence of administrative and commercial practices which helped to speed the adoption, not only of such a number system, but also of symbols representing operators.

It is taking too narrow a view to equate the term 'algebra' just with symbolic algebra.

Let us look at Problem 72 of the Ahmes Papyrus.

EXAMPLE 3.8   *100 loaves of pesu 10 are to be exchanged for a certain number of loaves of pesu 45. What is this certain number?*

(The word *pesu* may be loosely defined as a measure of the 'weakness' of a commodity. Here it can be taken to be the ratio of the number of loaves produced to the quantity of grain used in the production of a given amount of bread; thus the higher the *pesu*, the weaker the bread.)

Solution   We would tackle the above problem today as one of simple proportions, obtaining the number of loaves as $45/10 \times 100 = 450$. The solution prescribed in the Egyptian text is quite involved. It is interesting from our point of view because it contains the germs of algebraic reasoning. Below are the Egyptian solution and a restatement of the same steps in modern symbolic terms.

EGYPTIAN EXPLANATION	MODERN EXPLANATION
	Let x and y be the loaves of p and q *pesu*, respectively. Find y if x, p, q are known.
1 Find excess of 45 over 10: Result 35. Divide this 35 by 10: Result $3 + 1/2$.	$(q - p)/p$
2 Multiply this $(3 + 1/2)$ by 100: Result 350. Add 100 to 350: Result 450.	$[(q - p)/p]x + x$
3 Then the exchange is: 100 loaves of 10 *pesu* for 450 loaves of 45 *pesu*.	$y = [(q - p)/p]x + x$ $\quad = (q/p)x$

What is important here is not whether the scribe arrived at this method of solution by any thought process akin to ours, but that what we have here from four thousand years ago is a form of algebra, dependent on knowing that $y/x = q/p$ and $(y - x)/x = (q - p)/p$. In other words, the beginnings of rhetorical algebra.

Solutions of simple and simultaneous equations

To find topics which are represented in modern elementary algebra, we have to turn to Problems 24–34 of the Ahmes Papyrus. One of these problems, Problem 26, will serve as an illustration.

EXAMPLE 3.9   *A quantity and its quarter added become 15. What is the quantity?*

Solution   In terms of modern algebra, the solution is straightforward and involves finding the value of x, the unknown quantity, from an equation:

$$x + \tfrac{1}{4}x = 15, \quad \text{so} \quad x = 12$$

The scribe, however, reasoned as follows: If the answer were 4, then $1 + 1/4$ of 4 would be 5. The number that 5 must be multiplied by to get 15 is 3. If 3 is now multiplied by the assumed answer (which is clearly false), the correct answer will result:

$$4 \times 3 = 12$$

The scribe was using the oldest and probably the most popular way of solving linear equations before the emergence of symbolic algebra – the method of false assumption (or false position). It was still in common use in Europe until about a hundred years ago.

Other problems involving simple equations using variants of the method just described are found in the Moscow and Kahun Papyri. The Berlin Papyrus contains two problems which appear to involve non-linear simultaneous equations (i.e. equations with terms like x^2 and xy). It is badly mutilated in places, so the solution offered below is a reconstructed one.

EXAMPLE 3.10   *It is said to thee [that] the area of a square of 100 [square cubits] is equal to that of two smaller squares. The side of one is $1/2 + 1/4$ of the other. Let me know the sides of the two unknown squares.*

Solution 1: *The symbolic algebraic approach*   Let x and y be the sides of the two smaller squares. From the information given above, we can derive the following set of equations:

$$x^2 + y^2 = 100$$
$$4x - 3y = 0$$

The solution set, obtained by the method of substitution, is $x = 6$ and $y = 8$. Solution 2: *The Egyptian rhetoric algebraic approach* (restatement of the original solution) Take a square of side 1 cubit (i.e. a false value of y equal to 1 cubit). Then the other square will have side $1/2 + 1/4$ cubits (i.e. $x = 1/2 + 1/4$). The areas of the squares are 1 and $1/2 + 1/16$ square cubits respectively. Adding the areas of the two squares will give $1 + 1/2 + 1/16$ square cubits. Take the square root of this sum: $1 + 1/4$. Take the square root of 100 square cubits: 10. Divide this 10 by this $1 + 1/4$. This gives 8 cubits, the side of one square. (So from the false assumption $y = 1$, we have deduced that $y = 8$.) At this point, the papyrus is so badly damaged that the rest of the solution has to be reconstructed. One can only assume that the side of the smaller square was calculated as $1/2 + 1/4$ of the side of the larger square, which was 8 cubits. So the side of the smaller square is 6 cubits.

Geometric and arithmetic series

A series is the sum of a sequence of terms. The most common types are the arithmetic and geometric series. The terms of the former are an arithmetic progression (AP), a sequence in which each term after the first (usually denoted by a) is obtained by adding a fixed number, called the common difference (usually denoted by d), to the preceding term. For example, $1, 3, 5, 7, 9, \ldots$ is an AP with $a = 1$ and $d = 2$. In a geometric progression (GP) each term after the first (a) is formed from the preceding term by multiplying by a fixed number called the common ratio (usually denoted by r). For example, $1, 2, 4, 8, 16, \ldots$ is a GP with $a = 1$ and $r = 2$. From early times, mathematicians have sought ways of obtaining rules for the sum of the first n terms of arithmetic and geometric series.

The Egyptian method of multiplication leads naturally to an interest in such series, since it is based on operations with the basic GP $1, 2, 4, 8, \ldots$ and an understanding that any multiplier may be expressed as the sum of elements of this sequence. It would follow that Egyptian interest would focus on finding rules which made it easier to add up certain elements of such sequences. Here is another problem from the Ahmes Papyrus.

EXAMPLE 3.11   The actual statement of the problem in the Ahmes Papyrus is uncharacteristically ambiguous. It presents the following information, and nothing else:

Houses	7
Cats	49
Mice	343
Heads of barley	2401
Hekats of barley	16807
Total	19607

1	2801
2	5602
4	11204
Total	19607

This curious set of data, with no accompanying explanation, has led to some interesting suggestions. It was first believed that the problem was merely a statement of the first five powers of 7, along with their sum; and that the words 'houses', 'cats' and so on were really a symbolic terminology for the first, second and third powers, and so on. Since no such terminology occurs elsewhere, this explanation is unconvincing. Moreover, it does not account for the other set of data on the right.

A more plausible interpretation is that we have here an example of a geometric series, where the first term (a) and common ratio r are both 7, which shows that the sum of the first five terms of the series is obtained as $7[1 + (7 + 49 + 343 + 2401)] = 7 \times 2801$. We now see that the second set of data in the problem is merely the multiplication of 7 by 2801 in the Egyptian way.

A detailed solution to another problem in the Ahmes Papyrus gives some support to the view that the Egyptians had an intuitive rule for summing n terms of an arithmetic progression. Problem 64 may be restated as follows:

EXAMPLE 3.12 *Divide 10 hekats of barley among 10 men so that the common difference is 1/8 of a hekat of barley.*
Solution   The solution of the problem as it appears in the Papyrus is given on the left-hand side. On the right-hand side the algorithm is stated symbolically.

EGYPTIAN METHOD	SYMBOLIC EXPRESSION
	Let a be the first term, f the last term, d the common difference, n the number of terms and S the sum of n terms.
1 Average value: $10/10 = 1$.	**1** Average value of n terms $= S/n$.
2 Total number of common differences: $10 - 1 = 9$.	**2** Number of common differences $= n - 1$.
3 Find half the common difference: $1/2 \times 1/8 = 1/16$.	**3** Half the common difference $= d/2$.
4 Multiply 9 by $1/16$: $1/2 + 1/16$.	**4** Multiply $n - 1$ by $d/2$: $(n - 1)d/2$.
5 Add this to the average value to get the largest share: $1 + 1/2 + 1/16$.	**5** $f = S/n + (n - 1)d/2$.
6 Subtract the common difference $(1/8)$ nine times to get the lowest share: $1/4 + 1/8 + 1/16$.	**6** $a = f - (n - 1)d$.
7 Other shares are obtained by adding the common difference to each successive share, starting with the lowest. The total is 10 *hekats* of barley.	**7** Now form: $a, a + d, a + 2d, \ldots, a + (n - 1)d$. So $S = an + \frac{1}{2}n(n - 1)d$. Or $S/n = a + \frac{1}{2}(n - 1)d$.

The correspondence between the rhetorical algebra of the Egyptians and our symbolic algebra is quite close, though a word of caution is necessary here. It would not be reasonable to infer, on the basis of this correspondence, that the ancient Egyptians used anything like the algebraic reasoning on the right-hand side. It is more likely that they took a common-sense approach, listing the following sequence on the basis that the terms added to 10:

$$a, a + 1/8, a + 2/8, \ldots, a + 9/8$$

Each successive term gives the rising share of barley received by the ten men.

EGYPTIAN GEOMETRY

The practical character of Egyptian geometry has led a number of commentators to question whether it can be properly described as geometry, but that is to take too restrictive a view. The word itself comes from two Greek words meaning 'earth' and 'measure', indicating that the subject had its origin in land-surveying and other practical applications, and it was from the need to compute land areas and the volumes of granaries and pyramids that Egyptian geometry emerged with its peculiarly practical character. If there was any theoretical motivation, it was well hidden behind rules for computation.

Common problems of measurement based on the volumes and areas of the more familiar plane and solid figures were for the most part worked out correctly. Areas of rectangles, triangles and isosceles trapeziums were obtained correctly, probably by a process of 'decomposition and reassembly', similar to those found in Indian and Chinese geometry (to be discussed in later chapters).

When it is asked what the three major achievements of Egyptian geometry were, there is general agreement on two – the approximation to the area of the circle, and the derivation of the rule for calculating the volume of a truncated pyramid, but some disagreement over the third: did they indeed find the correct formula for the surface area of a hemisphere?

The area of a circle: The implicit value for π

Problem 50 from the Ahmes Papyrus reads:

EXAMPLE 3.13   *A circular field has diameter 9 khet. What is its area?*

(One *khet* is equal to 100 royal cubits, or approximately 50 metres. It is interesting that this problem is hardly a practical one since the area works out to be about 16 hectares (0·16 square kilometres) and the circumference of the field is nearly a kilometre and a half!)

The Ahmes solution   Subtract 1/9 of the diameter, namely 1 *khet*. The remainder is 8 *khet*. Multiply 8 by 8; it makes 64. Therefore, it contains 64 *setat* (square *khet*) of land.

In symbolic algebra this amounts to:

$$A^E = [d - (d/9)]^2 = (8d/9)^2$$

where d is the diameter. Worked out the modern way, the result is

$$A = \pi r^2 = (3·142)(4·5)^2 = 63·63$$

which is close to the value estimated by the scribe.

The implicit estimate of π contained in the Egyptian method of calculating the area of a circle can be worked out quite easily by equating A with A^E:

$$\pi d^2/4 = (8d/9)^2$$

from which we get

$$\pi = 4(8/9)^2 = 256/81 = (16/9)^2 \simeq 3{\cdot}1605$$

How was this rule derived by the Egyptians? Problem 48 may provide us with a clue. This is a unique problem in that it is the only one of the 87 problems in the Ahmes Papyrus which is expressed in the form of a labelled diagram, reproduced here as Figure 3.2(a). It shows a square with four isosceles triangles in the corners. In the middle of the square is the normal hieratic symbol for 9, written as $\digamma\backslash$. The removal of the triangles, each of area 9/2 square *khet* (or *setat*), would leave a regular octagon with side 3 *khet*, as in Figure 3.2(b). It is easily seen that the area of the octagon equals the area of the square minus the total area of the triangles cut off from the corners of the square:

$$A = 9^2 - 4(9/2) = 81 - 18 = 63$$

This is nearly the value that is obtained by taking $d = 9$ in the expression $A^E = (8d/9)^2$. So the octagon is a reasonable approximation to the circle inscribed in the square, as illustrated by Figure 3.2(c).

There is something rather contrived and unconvincing about this explanation, for it assumes an algebraic mode of reasoning which is not immediately apparent in Egyptian mathematics. For a better explanation, we turn to the geometric designs that were popular in ancient Egypt. A common motif found in burial chambers is the 'snake curve', which looks like a snake coiled around itself several times. (It is found today in areas of Africa as far apart as Mozambique and Nigeria.) This spiral motif also appeared in the design of objects. We are told that at the time of Ramses III (*c.* 1200 BC) the royal bread was baked in the shape of a spiral. Sisal mats were shaped in the form of a snake

Figure 3.2 Problem 48 from the Ahmes Papyrus: measuring a circle.

Figure 3.3 An alternative Egyptian way of measuring a circle, suggested by Gerdes (1985, p. 267).

curve, and are not uncommon objects in Africa even today. If one of these spiral mats (whose presence in ancient Egypt is well attested to in the drawings of the period), of diameter 9 units, were uncoiled and a square formed of side 8 units, the close correspondence in the areas of the two shapes would be easy to establish experimentally.

Yet another explanation (Gerdes, 1985), also based on material evidence, is related to a board game which is still widely played in many parts of Africa. In ancient Egypt it was played on a board with three rows of fourteen hollows with counters that were, as they are today, round objects such as seeds, beans or pebbles. Perhaps, then, the exigencies of the game prompted experimentation to find a square and a circle such that the same number of spherical counters could be packed into each as tightly as possible. Figure 3.3 shows that for both the circle of diameter 9 units and the square of side 8 units it is possible to pack in 64 small circles (or spherical counters). Thus the area of a circle with diameter 9 units is approximately equal to the area of a square of side 8 units, where the area is expressed as 64 small circles. An implicit value of π, given earlier, can easily be calculated from this approximate equivalence of the areas of the square and the circle.

The Egyptian rule for obtaining the area of a circle is applied in a few examples from the Ahmes and Moscow Papyri. In Problem 41 of the Ahmes Papyrus, the volume of a cylindrical granary of diameter 9 cubits and height 10 cubits is calculated. The solution offered may be expressed symbolically as

$$\text{Volume} = A^E h = (8d/9)^2 h = 640 \text{ cubic cubits}$$

Volume of a truncated square pyramid

It is generally agreed that the Egyptian knowledge of the correct formula for the volume of a truncated square pyramid is the zenith of Egyptian geometry. Bell (1940) referred to this achievement as the 'greatest Egyptian pyramid'. Problem 14 of the Moscow Papyrus is as follows:

EXAMPLE 3.14   *Example of calculating a truncated pyramid. You are told: a truncated pyramid of 6 cubits for vertical height by 4 cubits on the base by 2 cubits on the top. (Calculate the volume of this pyramid.)*
Solution   The solution, as it appears in the Papyrus, is given on the left-hand side below; on the right-hand side the algorithm is stated in symbolic terms (see Figure 3.4(a)).

EGYPTIAN METHOD	SYMBOLIC EXPRESSION
	Let h be the vertical height, a and b the sides of the two squares which bound the solid above and below, and V the volume of the solid.
1 Square this 4: Result 16.	**1** Find the area of the square base: a^2.
2 Square this 2: Result 4.	**2** Find the area of the square top: b^2.
3 Take 4 twice: Result 8.	**3** Find the product of a and b: ab.
4 Add together this 16, this 8 and this 4: Result 28.	**4** Find $a^2 + ab + b^2$.
5 Take 1/3 of 6: Result 2.	**5** Find 1/3 of the height: $h/3$.
6 Take 28 twice: Result 56.	**6** $(a^2 + ab + b^2)h/3$.
7 Behold it is 56!	**7** $V = (a^2 + ab + b^2)h/3$.

At the end of the solution in the Moscow Papyrus is a drawing of the trapezoid (see Figure 3.4(b)).

The Egyptian approach is clearly equivalent to the symbolic representation, and is correctly based on the formula

$$V = (a^2 + ab + b^2)h/3$$

(a)

(b)

Figure 3.4 Problem 14 from the Moscow Papyrus: the truncated pyramid, (a) in its modern version, and (b) the solution as given in the Papyrus (Eves, 1983, p. 42).

A number of attempts have been made to explain how the Egyptians may have arrived at the correct formula for the volume of a truncated pyramid. All explanations start with the assumption that they were aware of the formula for the volume of the complete pyramid, for otherwise it is difficult to explain the appearance of the factor 1/3 in the expression for the volume of a truncated pyramid. (The formula for the volume of the whole pyramid is in fact a special case of the more general formula for the truncated pyramid, since a substitution of $b = 0$ into the latter gives $V = a^2h/3$. Putting $a = b$ gives $V = a^2h$, the volume of a square prism.)

There are three main explanations. The first suggests that the truncated pyramid was cut up into smaller and simpler solids whose volumes were then estimated before putting them back together again. It is with the last part of this explanation that difficulties arise, since the reduction of the sum of the volumes of all the component solids to the final formula would require a degree of algebraical knowledge and sophistication which few would concede to the Egyptians.

The second explanation is that the Egyptians had discovered empirically that the volume of a truncated pyramid can be obtained as the product of the height of the frustum, h, and the Heronian mean* of the areas of the bases, a^2 and b^2. The only evidence to support this viewpoint is provided by Heron (or Hero), an Alexandrian mathematician of the first century AD, whose work contains a useful synthesis of Egyptian, Greek and Babylonian traditions. Book II of his *Metrica* has a detailed treatment of the volume mensuration of prisms, pyramids, cones, parallelepipeds and other solids. The inference is that his method for estimating the volume of a truncated pyramid, using the 'mean' named after him, derived directly from the Egyptian mathematical tradition.

Finally, there is the view that the volume was calculated as the difference between an original complete pyramid and a smaller one removed from its top. Gillings (1964) gives a detailed discussion of this explanation, which sounds the most plausible of the three, given the 'concrete' approach to geometry that the Egyptians favoured. Irrespective of how the Egyptians came to the discovery, the formula remains a lasting testimony to their mathematics.

The area of a curved surface: A semi-cylinder or a hemisphere?

Problem 10 of the Moscow Papyrus reads as follows:

---

*The Heronian mean of two positive numbers x and y is given by $(x + y + \sqrt{xy})/3$.

EXAMPLE 3.15 *Example of working out [i.e. calculating the area] of a basket. You are given a basket with a mouth [i.e. an opening] of 4 + 1/2 in preservation [presumably, diameter]. Let me know [the area] of its surface.*

Suggested solution

1 Take 1/9 of 9 since the basket is half an egg [i.e. a hemisphere]: Result 1.

2 Take the remainder which is 8. Take 1/9 of 8: Result 2/3 + 1/6 + 1/18.

3 Find the remainder of this 8 after [the subtraction of] 2/3 + 1/6 + 1/18: Result 7 + 1/9.

4 Multiply 7 + 1/9 by 4 + 1/2: Result 32.

Behold this is its surface [area]! You have found it correctly.

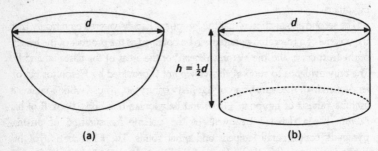

(a)   (b)

Figure 3.5 Problem 10 from the Moscow Papyrus. It may be asking for (a) the area of a hemisphere, or (b) the area of a semi-cylinder.

The problem restated: *Find the surface area of a hemisphere of diameter $4\frac{1}{2}$.* The suggested solution can be expressed symbolically as

$$A = 2d(8/9)(8/9)d = 2d^2(8/9)^2 = 2\pi r^2$$

where the Egyptian value of π is 256/81. This is identical to the modern formula for the curved surface of a hemisphere ($A = \frac{1}{2}\pi d^2$) with a different value for π.

If this interpretation of the Egyptian method is valid, then here is an even more remarkable achievement than the application of the correct formula for the volume of a truncated pyramid, for the very idea of a curved surface (not simply one that can be obtained by rolling up a plane surface) is quite an advanced mathematical concept. It would predate the innovative work of Archimedes (*c.* 250 BC) by about one thousand five hundred years. However, doubts have been raised over the interpretation of the term 'egg'. Peet (1931) has argued that this term could be taken to mean a semi-cylinder, in which case the suggested solution given above, expressed symbolically, would become (see Figure 3.5(b))

$$A = 2\pi rh, \quad \text{where } \pi = 256/81 \text{ and } h = \tfrac{1}{2}d \text{ is the height}$$

This would be the Egyptian counterpart of the modern formula for the area of the curved surface of a semi-cylinder.

Peet's translation and interpretation of the text has had its own critics. Gillings (1972) argued that if the original interpretation of the basket as a hemisphere is accepted, the rule could have arisen from the empirical observation that, in weaving a hemispherical basket whose radius is approximately equal to its height, the quantity of material required to make a circular lid is approximately half that required for the basket itself. If so, then the rule was a matter of simple deduction, and the Egyptians' 'greatest pyramid' remains the correct application of the formula for the volume of a truncated pyramid.

Egyptian mathematics has been discussed here in more detail than in many general textbooks on the early history of mathematics. (A joint assessment of Egyptian and Babylonian mathematics will be found in Chapter 5.) The treatment of Egyptian mathematics in most standard histories of the subject tends to be rather lopsided: Egyptian numeration is overemphasized, and consequently the rest of the mathematics receives less attention than it should. Where comparisons are made with contemporary or later mathematical traditions, the quality of Babylonian mathematics is stressed, but both the Egyptian and Babylonian contributions are judged to be meagre, or – more charitably – seen merely as a prelude to the 'Greek miracle'.

Is the overly critical attitude to Egyptian mathematics found in many textbooks an attempt to counteract the Greeks' (and others') generous acknowledgements of the great debt they owed these earlier civilizations? If the Greek dependence on Egypt and Babylonia for their early mathematical ideas is now recognized, the myth of the 'Greek miracle' will no longer be sustainable. This would undermine one of the central planks of the Eurocentric view of history and progress.

We have said little about the later phase of Egyptian mathematics, when Alexandria became the centre of mathematical activity. It was the creative synthesis of Classical Greek mathematics, with its strong geometric and deductive tradition, and the algebraic and empirical traditions of Egypt and Babylonia that produced some of the greatest mathematics and astronomy of Antiquity, best exemplified in the works of Archimedes, Ptolemy, Diophantus, Pappus and Heron. We shall not take up the story of Hellenistic mathematics, which has been extensively explored in general histories, such as those by Boyer (1968), Eves (1983) and Kline (1972), as well as in specialized works on Greek mathematics by Heath (1921), van der Waerden (1961) and others.

4 The Beginnings of Written Mathematics: Babylonia

◇◇◇

Studying ancient Mesopotamian history is rather like going on a long and unfamiliar journey: we are not sure whether we are on the right road until we reach our destination. The abridged chronology given in Table 4.1 will be of some help in plotting our course across this difficult terrain; the places mentioned in the table are shown in the accompanying map (Figure 4.1). The earliest written records begin around 3500 BC, and come to an end with the Persian conquest in 539 BC when Mesopotamia ceased to exist as an independent entity. The subsequent history of this region cannot be separated from the histories of other countries such as Persia, Greece, Arabia and, more recently, Turkey.

Table 4.1 Chronology of Ancient Mesopotamia from 3500 BC to 64 BC.

Dates	Historical features	Notable rulers	Mathematical developments
3500–2400	Small city-states of Sumeria (centres of power: Ur, Eridu, Lagash, Nippur)		Development of sexagesimal (base 60) numerals
2400–2200	Earliest Mesopotamian empire of Sumer and Akkad (centres of power: Ur, Agade)	Sargon I (c. 2350)	Construction of tables for multiplication, division and other operations
2200–1900	Conflicts and wars; rule by city-states (centres of power: Ur, Girsu)		

table continues

Table 4.1 *continued*

Dates	Historical features	Notable rulers	Mathematical developments
1900–1650	First or Old Babylonian empire (centre of power: Babylon)	Hammurabi (1792–1750)	Beginnings of algebra and geometry, and subsequent rapid developments
1600–885	Period of unrest and invasions by Hittites, Hurrians and Mitanni (centre of power: Babylon)		Astronomy
885–612	Assyrian empire (centre of power: Nineveh)	Sennacherib (705–681) Ashurbanipal (668–627)	Astronomy
612–539	Second or New Babylonian empire (Chaldeans) (centre of power: Babylon)	Nebuchadnezzar (604–561)	Revival of algebra
539	Persian invasion; end of ancient Mesopotamia (centres of power: Babylon, Susa)	Cyrus the Great (c. 525) Darius (521–485)	Considerable work on astronomy
311–64	Seleucid dynasty (centre of power: Antioch)		Work on astronomy and algebra continues; construction of extensive mathematical and astronomical tables

At the beginning of the 3000-year period represented in Table 4.1, we find small villages which gradually develop into larger towns and cities. Their prosperity attracts marauders, often led by a strong personality who makes it his mission to unite the city-states into an empire. The empire prospers under a centralized regime which undertakes massive public works, including the construction of ziggurats. At the death of the founder, dissensions arise and the empire is weakened, and a group of invaders waiting in the wings seize their opportunity to take control. The process of establishing a new empire begins afresh.

Figure 4.1 Map of Mesopotamia.

Around 3500 BC, a group of people living in the region of Sumer in southern Mesopotamia established a number of small city-states, of which Ur, Nippur and Lagash are among the best known. Biblical references to this period abound. The details of the great flood during Noah's time resemble legends of other great floods contained in the oldest extant work of literature, the *Epic of Gilgamesh* (c. 3000 BC). Abraham, a founding father of three world religions, was supposed to have originated from Ur. The Sumerians reached a high level of cultural and intellectual development in their independent city-states. But around 2400 BC they were attacked by a group of Akkadians from the surrounding desert, led by Sargon I, who established a large empire that extended from the Persian Gulf in the south to the Black Sea in the north, and from the Persian steppes in the east to the Mediterranean in the west. The Akkadians learnt much from the Sumerians, including their cuneiform (meaning wedge-shaped) script and their number system. But the Akkadian rule was short-lived, and by the beginning of the second millennium BC the city-states were independent once more.

Their regained independence did not last long, for around 1900 BC there arose the First (or Old) Babylonian empire, which successfully merged the Sumerian and Akkadian elements. The well-known King Hammurabi lived from 1792 to 1750 BC. The period 1900–1600 BC was one of prosperity and cultural renaissance in the region, and most of the mathematical records we have of the Mesopotamian civilization come from this period. This civilization has come to be associated with the capital city of Hammurabi's dynasty, and thus to be known as the Babylonian civilization.

The next invaders were the Hittites, who originated from a region in or near present-day Turkey. They established control over a disintegrating Babylonian empire in 1600 BC, no doubt helped by the novel use of iron rather than bronze weapons, and retained Babylon as their capital. But their rule was marked by considerable unrest, culminating in the successful invasion by the Hurrians and Mitanni around 1000 BC. However, these groups were soon displaced by the most feared of all invaders, the Assyrians, who established their empire in 885 BC. The Assyrian empire lasted only a relatively short time, but under its rulers – notably Sennacherib (705–681 BC) and Ashurbanipal (668–627 BC) – there was a revival of culture, architecture and science (notably astronomy) centred on their capital city, Nineveh.

In 612 BC the Assyrians were in turn conquered by a southern Mesopotamian people, called the Chaldeans. Under their famous ruler Nebuchadnezzar the capital city of Babylon reached its greatest period of glory, and so this empire is often called the New or Second Babylonian empire.

The Persian invasion by Cyrus the Great in 539 BC put an end to this empire, though its work in mathematics, astronomy and medicine continued. With the establishment in 312 BC of the Seleucid dynasty, named after one of Alexander the Great's generals, the Greek and Babylonian traditions came together to provide us with our second major source of evidence on the mathematics of this region, over a thousand years after the date of our first major source, from the Old Babylonian period. The central role of Babylon in both cases justifies the description of the mathematics produced by the Mesopotamian civilization as Babylonian; this is not to discount contributions by other peoples such as the Sumerians, Akkadians, Chaldeans and Assyrians, who all played an important part in shaping this civilization.

SOURCES OF BABYLONIAN MATHEMATICS

Of the half a million inscribed clay tablets that have been excavated, less than five hundred are of direct mathematical interest. Collections of these mathematical tablets are scattered around the museums of Europe and the universities of Yale, Columbia and Pennsylvania in the United States. Some of the more recent finds, notably from Tell Harmal and Tell Dhibayi in Iraq, are kept in the Iraqi Museum in Baghdad. The tablets vary in size from a postage stamp to a pillow. Some are inscribed only on one side, others on both sides, and a few even on their edges.

To make a tablet, clay that may have come from the banks of the Tigris or Euphrates was collected and kneaded into shape. It was then ready for recording. The scribe used a piece of reed about the size of a pencil, shaped at one end so that it made wedge-like impressions used in the soft, damp clay. He had to work fast for the clay dried out and hardened quickly, making corrections or additions difficult. Having completed one side, he might turn the tablet over and continue, using even the edges if required. When he had finished, the tablet was dried in the sun or baked in a kiln, leaving a permanent record for posterity.

The wedge-shaped cuneiform script of the Sumerians was deciphered as early as the middle of the nineteenth century through the pioneering efforts of George Frederick Grotefend (1775–1853) and Henry Creswicke Rawlinson (1810–95), but only since the 1930s have the mathematical texts been studied seriously. This delay may be partly explained by the different ways in which a mathematician and a philologist approach early literature. The average mathematician, unless presented with a text which falls within the limits of what 'mathematics' is perceived to be, has little time for the past; rarely is historical curiosity aroused by mathematical teaching. The philologist seeks to revive the past in order to explore the growth and decline of ancient civilizations; but, probably because of a lack of mathematical training, the philologist rarely takes an interest in ancient mathematics.* So the Babylonian mathematical texts lay undeciphered and uninterpreted until the pioneering work by Otto Neugebauer, who published his *Mathematische Keilschrift-Texte* in three volumes in the period 1935–7, and by François Thureau-Dangin, whose complete works, entitled *Textes mathématiques Babyloniens*, were brought out in 1938. Since then new evidence and interpretations have continued to appear, even in recent years.

There are three main sources for Babylonian mathematics. Some of the oldest texts, written in Sumerian cuneiform, date back to the third millennium BC. They contain information of a commercial and legal nature, as would now be found on invoices, receipts and mortgage statements, and details about weights and measures. The mathematical interest in these records is that they demonstrate a developing sexagesimal (base 60) positional system of numeration. There are few direct mathematical records until we come to the Old Babylonian period, during the first half of the second millennium BC. It has been estimated that between two-thirds and three-quarters of all the Babylonian mathematical texts that have been found belong to this period; in our subsequent discussion of Babylonian mathematics we shall concentrate on the evidence from this period. The remainder of the texts belong to a period beginning with the establishment of the New Babylonian empire of the Chaldeans, around 600 BC, after the destruction of Nineveh, and continuing well into the Seleucid era. This was also a period of considerable accomplishments in astronomy.

THE BABYLONIAN NUMBER SYSTEM

One of the most outstanding achievements of Babylonian mathematics, and one which helped to shape subsequent developments, was the invention of a place-value number system. Early clay tablets (c. 3000 BC) show that the Sumerians did not have a systematic positional system for all powers of 60 and their multiples. They used the following symbols:

1	10	60	600	3600
D	O	D	▢	◯

---

*To give him his due, Sir Henry Rawlinson *was* interested in mathematics, although he was no mathematician. He left many papers on the subject, but – ironically – they are rather difficult to decipher as the lack of sufficient and suitable writing-material in nineteenth-century India led him to fill every available space, even writing sideways in margins.

The symbols for the first three numbers were written with the lower end of a cylindrical stylus, held obliquely for 1 and 60 and vertically for 10. The symbol for 600 was a combination of those for 10 and 60; the large circle for 3600 was scratched with a sharp stylus. From around 2000 BC there evolved a sexagesimal place-value system using only two symbols: T for one and $\mathsf{\langle}$ for ten. In this system, the representation of numbers smaller than 60 was as straightforward and cumbersome as it was in the Egyptian notation. Thus

4: TTTT , 28: $\mathsf{\langle TTTT \atop \langle TTTT}$, 59: $\mathsf{\langle \atop \langle TTTT \atop \langle TTTT}\,\mathsf{T}$

If the Babylonians had merely continued to use these symbols on an additive basis, their numeration and computations would probably have developed along Egyptian lines. But, from as early as 2500 BC, we find indications that they realized they could double, triple, quadruple (and so on) the two symbols for one and ten by giving them values that depended on their relative positions. Thus the two symbols could be used to form numbers greater than 59:

$$60 = 60(1): \quad \mathsf{T}$$

$$95 = 60(1) + 35: \quad \mathsf{T}\ \mathsf{\langle \atop \langle}\mathsf{\langle}\ \mathsf{T_T T_T T}$$

$$120 = 60(2): \quad \mathsf{TT}$$

$$4002 = 60^2(1) + 60(6) + 42: \quad \mathsf{T}\ \mathsf{T_T T_T T}\ \mathsf{\langle\langle \atop \langle\langle}\ \mathsf{TT}$$

It was a relatively simple matter, though one of momentous significance, to extend this principle of positional notation to allow fractions to be represented:

$$1/2 = 60^{-1}(30) = 30/60: \quad \mathsf{\langle\langle\langle}$$

$$1/4 = 60^{-1}(15) = 15/60: \quad \mathsf{\langle}\ \mathsf{T_T T_T T}$$

$$1/8 = 60^{-1}(7) + 60^{-2}(30):$$

$$532\tfrac{3}{4} =$$

$$60(8) + 52 + 60^{-1}(45):$$

Two important features of Babylonian positional notation are highlighted by these examples: unlike our present-day system, there is no symbol for zero, and neither is there a symbol corresponding to our decimal point to distinguish between the integer and fractional parts of a number. There is also the more fundamental question of why the Babylonians should have constructed a number system on base 60 rather than the more 'natural' base 10 (i.e. the decimal system).

The absence of a symbol for a place-holder could lead to confusion over what number was being recorded. For example,

could be $60(2) + 40 = 160$, or $60^2(2) + 60(0) + 40 = 7240$, or it could be $2 + 60^{-1}(40) = 2\tfrac{2}{3}$, or even $60^{-1}(2) + 60^{-2}(40) = \tfrac{2}{45}$, since there is no 'sexagesimal point' place-holder to indicate that the number is a fraction. In the absence of a special symbol for zero, the number might be identifiable from the context in which it appeared, or a space might be left to indicate a missing sexagesimal place. There again, it could have been that the lack of a zero symbol in ancient Babylonia was of little practical consequence, for the existence of a large base, 60, would ensure that most numbers of everyday concern could be represented unambiguously. For example, it is unlikely that the prices of commodities in ordinary use would have exceeded 59 'units' (discounting inflation, of course!). Moreover, the relative positions of the two $\top$'s and the four $\langle$'s in the example above would indicate that, if the number were an integer, it would not be less than 160. This was because the Babylonians, unlike the Egyptians, wrote their numerals the same way as we do, from left to right.

It was not until the Seleucid period that a separate place-holder symbol was introduced to indicate an empty space between two digits inside a number. Thus the number 7240 would be written as

$$ \text{T T} \, {}_{\blacktriangle}^{\blacktriangle} \, {\large \langle\langle \atop \langle\langle} $$

where ${}_{\blacktriangle}^{\blacktriangle}$ serves as the place-holder symbol. The problem still remained of how to represent the absence of any units at the end of a number. Nowadays we use the symbol for zero in the terminal position. Without something like that, it is difficult to know whether the number

$$ \text{T T T} \, {\large \langle \atop \langle}{}^{\langle} $$

is $60(3) + 30 = 210$, or $60^2(3) + 60(30) = 12\,600$, or even $3 + 60^{-1}(30) = 3\frac{1}{2}$. It is therefore clear that, while the Babylonians were consistent in their use of place-value notation, they never operated with an absolute positional system. When, in the second century AD, Ptolemy of Alexandria began to use the Greek letter o (omicron) to represent zero, even in the terminal position of a number, there was still no awareness that zero was as much a number as any other and so, just like any other, could enter into any computation. Recognition of this fact – 'giving to airy nothing, not merely a local inhabitation and a name, a picture, a symbol, but also a helpful power' (Halstead, 1912) – was not to occur for another thousand years, in India and Central America.

If we are to make any further headway, we need a way of transliterating the Babylonian numerical representation into a notation more convenient for us. We shall adopt Neugebauer's convention of using a semicolon (;) to separate the integral part of a number from its fractional part, just as we use the decimal point today – the semicolon is in effect the 'sexagesimal point'. All other sexagesimal places are separated by a comma (,). Some examples, of numbers whose cuneiform representations have been given above, will make this convention clear:

$$
\begin{aligned}
60 &= 60(1): \quad 1,0 \\
95 &= 60(1) + 35: \quad 1,35 \\
120 &= 60(2): \quad 2,0 \\
4002 &= 60^2(1) + 60(6) + 42: \quad 1,6,42 \\
1/2 &= 60^{-1}(30) = 30/60: \quad 0;30 \\
1/8 &= 60^{-1}(7) + 60^{-2}(30): \quad 0;7,30 \\
532\tfrac{3}{4} &= 60(8) + 52 + 60^{-1}(45): \quad 8,52;45
\end{aligned}
$$

With this scheme, the ambiguity in the representation of 7240 in the Babylonian notation disappears: this number is now written as 2,0,40.

Different explanations have been offered for the origins of the sexagesimal system, which, unlike base 10, or even base 20, has no obviously anatomical basis. Theon of Alexandria, in the fourth century AD, pointed to the computational convenience of using the base 60. Since 60 is exactly divisible by 2, 3, 4, 5, 6, 10, 15, 20 and 30, it becomes possible to represent a number of common fractions by integers, thus simplifying calculations: the integers that correspond to the unit fractions 1/2, 1/3, 1/4, 1/5, 1/6, 1/10, 1/15, 1/20 and 1/30 are 30, 20, 15, 12, 10, 6, 4, 3 and 2, respectively. Of the unit fractions with denominators from 2 to 9, only 1/7 is not 'regular' (i.e. 60/7 gives a non-terminating number). It is therefore quite a simple matter to work with fractions in base 60. In a decimal base, though, only three of the nine fractions above produce integers, and none of 1/3, 1/6, 1/7 and 1/9 is regular. Indeed, while base 10 may be more 'natural', since we have ten fingers, it is computationally more inefficient than base 60, or even base 12.

However, this explanation for the use of base 60 is unconvincing because of its 'hindsight' character. It is highly unlikely that such considerations were taken into account when the base was chosen. A second explanation emphasizes the relationship that exists between base 60 and numbers that occur in important astronomical quantities. The length of a lunar month is 30 days. The Babylonian estimate of the number of days in a year was 360, based on the zodiacal circle of 360°, divided into twelve signs of the zodiac of 30° each. The argument goes that either 30 or 360 was first chosen as the base, later to be modified to 60 when the advantages of such a change were recognized. Here again, there is a suggestion of deliberate, rational calculation in the choice of the base which is not totally convincing. A more plausible explanation is that two bases (e.g. 5 and 12), favoured perhaps by two different groups, gradually merged, and that the advantages of base 60 for astronomical and computational work then came to be recognized.

The sexagesimal system continued to be used well into the fifteenth century AD. Sexagesimal fractions appeared in Ptolemy's *Almagest* in 150 AD. The Alfonsine Tables, astronomical tables prepared from Arab sources on the instruction of Alfonso X of Castile and written in Latin at the end of the thirteenth century AD, used a consistent sexagesimal place-value system. The Arab astronomer al-Kashi (d. AD 1429) determined 2π sexagesimally as 6;16,59,28,1,34,51,46,15,50, the decimal equivalent of which is accurate to sixteen places. And Copernicus's influential work in mathematical astronomy during the sixteenth century contained sexagesimal fractions. The current use of the sexagesimal scale in measuring time and angles in minutes and seconds is part of the Babylonian legacy.

Before going on to look at operations with Babylonian numerals, let us pause to compare the Babylonian way of representing numbers with other systems. In assessing a notational system, the following questions are pertinent:

1 Is the system easy to learn and write?

2 Is the system unambiguous?

3 Does the system lend itself readily to computation?

The Babylonian system scores well on questions 1 and 3. It is easily learnt, being one of the most economical systems in terms of the symbols used. The only other number system that operated with just two symbols (a dot and a dash) was the Mayan, though unlike the Babylonian system there was also a special sign for zero. If we compare the Babylonian with the Greek number system, which used twenty-seven symbols, the simplicity of the Babylonian notation is obvious. But one must contrast this simplicity with the awkwardness of representing a number such as fifty-nine, which in the unabridged Babylonian notation would require fourteen symbols, as against just two in the Greek notation.* The Babylonian system was also remarkable for its computational ease, which arose from its place-value principle and its base of 60. This base proved a distinct advantage in calculating with fractions. Until the emergence of decimal fraction representation, the Babylonian treatment of fractions remained the most powerful computational method available.

But the great disadvantage of Babylonian notation was its ambiguity, the consequence of having neither a symbol for zero nor a suitable device for separating the integral part of a number from its fractional part. It was not that the system of notation precluded the incorporation of these additional features, but that the Babylonians simply did not use them. (In the time of the New Babylonian empire, though, the place-holder symbol (▲) appeared.) All in all, compared with the Egyptian system, the Babylonian notation was computationally more 'productive' and symbolically more economical (since the place-value principle made it unnecessary to invent new symbols for large numbers), but it had the disadvantage of being ambiguous. The Egyptian

---

*There is some fragmentary evidence that the Babylonians made use of a subtraction sign (𝌆 ⊢) to relieve the tedium of their unabridged notation. In this scheme 39 would be represented as 40 − 1:

$$\langle\langle\,\langle\langle \quad T\!\vdash\ T$$

system had another advantage over the Babylonian: the order in which the symbols representing a number are written is of no consequence in Egyptian notation.

Operations with Babylonian numerals

With a positional system of numeration available, ordinary arithmetical operations with Babylonian numerals would follow along the same lines as modern arithmetic. To relieve the tedium of long calculations, the Babylonians made extensive use of mathematical tables. These included tables for finding reciprocals, squares, cubes, and square and cube roots, as well as exponential tables and even tables of values of $n^3 + n^2$, for which there is no modern equivalent. These tables account for a substantial portion of the sources of Babylonian mathematics available to us.

Multiplication and division were carried out largely as we would today. Division was treated as multiplication of the dividend by the reciprocal of the divisor (obtained from a table of reciprocals). To take a simple example:

EXAMPLE 4.1   *Divide 1029 by 64.*
Solution   In Neugebauer's notation, $1029 = 60^1(17) + 60^0(9)$ is written as 17,9. Also, 1/64 becomes 0;0,56,15 (since $1/64 = 60^{-2}(56) + 60^{-3}(15)$, found from a table of reciprocals). Therefore

$$17,9 \text{ multiplied by } 0;0,56,15 \text{ equals } 16;4,41,15$$

The long multiplication is carried out in the same way as we would today, apart from the sexagesimal base:

$$
\begin{array}{r}
0; \ 0,56,15 \\
17, \ 9 \\
\hline
8,26,15 \\
15;56,15 \\
\hline
16; \ 4,41,15
\end{array}
$$

The answer 16;4,41,15 can be converted to the decimal base:

$$16 + 60^{-1}(4) + 60^{-2}(41) + 60^{-3}(15) \simeq 16 \cdot 0781$$

A complete set of sexagesimal multiplication tables was available for each number from 2 to 20, and for 30, 40 and 50. This would be sufficient to carry out all possible sexagesimal multiplications, just as present-day multiplication

The Beginnings of Written Mathematics: Babylonia

tables for numbers from 2 to 10 are sufficient for all decimal products. Often, the tables of reciprocals were available only for those 'regular' integers up to 81 that are multiples of 2, 3 or 5. The reciprocals of 'irregular' numbers, or those containing prime numbers which are not factors of 60 (i.e. all prime numbers except 2, 3 and 5), would, in effect, have been non-terminating sexagesimal fractions. For example, the reciprocals of the 'regular' numbers 15, 40 and 81 are

$$1/15 = 0;4, \qquad 1/40 = 0;1,30, \qquad 1/81 = 0;0,44,26,40$$

The reciprocals of the 'irregular' numbers 7 and 11 are

$$1/7 = 0;8,34,17,8,34,17, \ldots, \qquad 1/11 = 0;5,27,16,21,49, \ldots$$

The tables of reciprocals found on the older tablets are all for 'regular' numbers, with the exception of 7. There is one tablet, from the period just before the Old Babylonian empire, which contains the following problem:

EXAMPLE 4.2   *Divide 5,20,0,0 by 7.*
Suggested solution   Multiply 5,20,0,0 by the reciprocal of 7 (i.e. 0;8,34,17,8) to get the answer: 45,42,51;22,40.

A later tablet from the Seleucid period gives the upper and lower limits on the magnitude of 1/7 as

$$0;8,34,16,59 < 1/7 < 0;8,34,18$$

Statements such as 'approximation given since 7 does not divide' from the earlier periods, and the later estimates of bounds, give us a tantalizing glimpse of the Babylonians taking the first step (though it is not clear whether they were fully aware of the implications) in coming to grips with the incommensurability of certain numbers.

A Babylonian masterpiece

Evidence that the Babylonians were ready to work with irrational numbers is found on a small tablet belonging to the Old Babylonian period which forms part of the Yale collection. It contains the diagram shown in Figure 4.2(a), and 'translated' in Figure 4.2(b). The number 30 indicates the length of the side of the square. Of the other two numbers, the upper one (if we assume that the 'sexagesimal point' (;) occurs between 1 and 24) is 1;24,51,10, which in decimal

Figure 4.2 A Babylonian masterpiece: (a) the original tablet (Neugebauer and Sachs, 1945, p. 42); (b) the modern 'translation'.

notation is

$$1 + 60^{-1}(24) + 60^{-2}(51) + 60^{-3}(10)$$
$$\simeq 1 + 0.4 + 0.014\,166\,67 + 0.000\,046\,3$$
$$= 1.414\,212\,97$$

To the same number of decimal places, the square root of 2 is $1.414\,213\,56$, so the Babylonian estimate is correct to five places of decimals. The lower number is easily seen to be the product of 30 (the side of the square) and the estimate of the square root of 2.

The interpretation is now clear. Let d be the diagonal of the square; applying the Pythagorean theorem then gives

$$d^2 = 30^2 + 30^2$$
$$d = \sqrt{2}(30) \simeq (1;24,51,10)(30) = 42;25,35$$

The number below the diagonal is therefore the length of the diagonal of a square whose side is 30.

The solution to this problem highlights two important features of Babylonian mathematics. First, over a thousand years before Pythagoras, the Babylonians knew and used the result now known under his name. (In the next section we discuss further applications of this result, as well as evidence that the Babylonians knew rules for generating Pythagorean triples a, b, c, where $a^2 + b^2 = c^2$.) Second, there is the intriguing question of how the Babylonians arrived at their remarkable estimate of the square root of 2, an estimate that would still be in use two thousand years later when Ptolemy constructed his table of chords.

One conjecture is that the method they used to extract square roots resembles the iterative procedure used by digital computers today. This procedure is as follows. Let x be the number whose square root you want to find, and the positive number a your rough guess of the answer. Then $x = a^2 + e$, where the difference (or 'error') e can be positive or negative. We now try to find a better approximation for the square root of x, which we denote by $a + c$. It is obvious that the smaller the error e, the smaller is c relative to e. Thus we impose the following condition on c:

$$x = (a + c)^2 = a^2 + e \qquad (4.1)$$

from which

$$2ac + c^2 = e$$

Now, if you made a sensible guess for a in the first place, c^2 will be much smaller than $2ac$ and can therefore be ignored. So (4.1) may be written as

$$c \simeq e/2a \qquad (4.2)$$

Hence, from (4.1) and (4.2), an approximation for the square root of x is

$$a + c \simeq a + (e/2a) = a_1$$

Now, taking $a_1 = a + (e/2a)$ as the new 'guesstimate', the process can be repeated over and over again to get $a_2, a_3, \ldots$ as better and better approximations.

Let us illustrate this approximation procedure with the example we started with – how did the Babylonians obtain their estimate for $\sqrt{2}$ as 1;24,51,10?

EXAMPLE 4.3 *Find the square root of 2 up to the fourth approximation (i.e. a_3), starting with $a_0 = 1$.*
Solution   The steps of the solution are summarized below:

i	a_i	e_i	c_i	a_{i+1} Modern	a_{i+1} Babylonian
0	1	1	1	1·5	1;30
1	1·5	−0·25	−0·0833	1·41667	1;28,0,7,12
2	1·41667	−0·00695	−0·00246	1·41421	1;24,51,10

The value obtained after two steps is exactly the value in Figure 4.2(b)!

This procedure for calculating square roots is widely known as Heron's method, after the Alexandrian mathematician who lived in the first century AD. A similar procedure is found in the earliest extant mathematical writings of the Indians, the *Sulbasutras*, which have been variously dated from 800 BC to 500 BC. The Indian approximation procedure is discussed in Chapter 8.

Other tables and their uses

Like the Egyptians, the Babylonians were inveterate table-makers. Apart from the multiplication and reciprocal tables already mentioned, there are two sets of tables which are worth examining. One contains the values of $n^3 + n^2$ for integers n from 1 to 20 and 30, 40 and 50. It was used for solving mixed cubic equations of the form

$$ax^3 + bx^2 = c \tag{4.3}$$

for if you multiply equation (4.3) through by a^2/b^3 you get

$$(ax/b)^3 + (ax/b)^2 = ca^2/b^3$$

from which

$$y^3 + y^2 = (a^2/b^3)c \qquad (4.4)$$

where $y = ax/b$. Equation (4.4) can be solved for y using the $n^3 + n^2$ table, and x obtained from y by using $x = by/a$. To illustrate:

EXAMPLE 4.4   *Solve $144x^3 + 12x^2 = 21$.*
Solution   Multiply both sides of the equation by 12 and substitute $y = 12x$:

$$(12x)(12x)(12x) + (12x)(12x) = (12)(21)$$
$$y^3 + y^2 = 252$$

from the table $y = 6$, so $x = 0.5$ (or 0;30).

The solution of cubic equations is a remarkable achievement in view of the high level of technical skill necessary to handle the algebraic concepts in the absence of a symbolic notation. With the benefit of modern symbolic algebra, it is easy to see that $(ax)^3 + (ax)^2 = c$ is equivalent to the equation $y^3 + y^2 = c$. Try to imagine the difficulties in recognizing this equivalence without the algebraic notation available to us, and you will appreciate the measure of the Babylonian achievement.

A number of tables in the collections at Berlin, Istanbul, the Louvre and Yale contain values for exponents a^n, where n is an integer taking values $1, 2, \ldots, 10$, and $a = 9, 16, 100$ and 225 – all perfect squares. What were these tables used for? The following problem, taken from a Louvre tablet of the Old Babylonian period, may provide the answer.

EXAMPLE 4.5   *Calculate how long would it take for a certain amount of money to double if it has been loaned at a compound annual rate of 0;12 [20%].*
Solution   Using modern symbolic notation, let P be the amount of the loan (the principal) and r the interest rate. The question may then be restated as: Find n, given $2P = P(1 + r)^n$. The problem is solved by first identifying from an exponential table that n must lie between 3 and 4 in order to satisfy the

equation $(1;12)^n = 2$ (or in modern terms $(1·2)^n = 2$). From the table,

$(1;12)^3 = 1;43,40,48$ (or 1·7280) and $(1;12)^4 = 2;4,25$ (or 2·0736)

Applying linear interpolation and working in modern notation would give

$$3 + \frac{2 - 1·7280}{2·0736 - 1·7280} = 3·7870$$

or, in Babylonian notation, 3;47,13,20 years — exactly the answer given on the Louvre tablet!

BABYLONIAN ALGEBRA

Unlike the Egyptians, who had been constrained by the absence of an efficient number system, basic computations posed few difficulties for the Babylonians with their place-value number system and their imaginative use of various tables. As a result, even as early as the Old Babylonian period they had developed sophisticated numerical methods of solving equations and systems of equations, within the framework of a rhetorical algebra. However, we should not be fooled by the general rhetorical nature of their algebra into overlooking traces of syncopated algebra present even then, and best exemplified by the use of certain geometric terms to denote unknown quantities. Analogous to the modern symbol x was the term 'length' (of a square or rectangle); the square of this unknown quantity was referred to by the term 'square'. Where there was need to refer to two unknowns, they were called 'length' and 'breadth', their product being described as 'area'. Three unknowns became 'length', 'breadth' and 'height', and their product 'volume'. Table 4.2 lists these terms and their modern equivalents.

Table 4.2 Symbolic notation in Babylonian algebra.

Modern symbol	Geometric term	Babylonian quantity
x	length	*ush*
y	breadth	*sag*
x^2	square	*lagab*
z	height	*sukud*
xy	area	*asha*
xyz	volume	*sahar*

It is this peculiar form of reference to unknown quantities in Babylonian algebra that led some earlier commentators to be dismissive of statements such as: 'I have subtracted the side of the square from the area, and the result is 14,30,' which we would now interpret as

$$x^2 - x = (14 \times 60) + (30 \times 1) = 870$$

Linear and non-linear equations in one unknown

The Babylonians' solution to equations of the form $ax = c$ was no different from ours, which is $x = (1/a)c$. They would have taken $1/a$ from a table of reciprocals, and obtained the product by referring to a multiplication table. If $1/a$ was not a regular sexagesimal fraction, they would have used a suitable approximation. An example from a mathematical text, found during excavations at Tell Harmal in 1949 and belonging to the Old Babylonian period, illustrates the approach. The statement of the problem and its solution are based on Taha Baqir's (1951) translation:

EXAMPLE 4.6   *If somebody asks you thus: If I add to the two-thirds of my two-thirds a hundred qa of barley, the original quantity is summed up. How much is the original quantity?*
Suggested solution
1 First multiply two-thirds by two-thirds: Result 0;26,40 (i.e. 4/9).
2 Subtract 0;26,40 from 1: Result 0;33,20 (i.e. 5/9).
3 Take the reciprocal of 0;33,20: Result 1;48 (i.e. 1 + 4/5).
4 Multiply 1;48 by 1,40 (i.e. 100): Result 3,0 (i.e. 180).
3,0 (*qa*) is the original quantity.
This procedure is identical to the one we would now use to solve a simple equation:

$$(2/3 \text{ of } 2/3)x + 100 = x$$
$$(4/9)x + 100 = x$$
$$(5/9)x = 100$$
$$x = 180$$

The Babylonians were able to solve different types of quadratic equation. The two types that occurred most frequently have the forms

$$x^2 + bx = c, \quad b > 0, c > 0 \qquad (4.5)$$

$$x^2 - bx = c, \quad b > 0, c > 0 \qquad (4.6)$$

In solving (4.5), the Babylonian approach was equivalent to the application of the formula

$$x = \sqrt{(b/2)^2 + c} - (b/2)$$

to get a positive solution. The corresponding formula for (4.6) gives the positive solution as

$$x = \sqrt{(b/2)^2 + c} + (b/2) \qquad (4.7)$$

As an illustration of how the Babylonians solved these quadratics, let us consider a problem from a tablet of the Old Babylonian period, now in the Yale collection.

EXAMPLE 4.7 *The length of a rectangle exceeds its width by 7. Its area is 1,0. Find its length and width.*
Solution   The solution, shown below, establishes a close correspondence between the Babylonian approach and its modern symbolic variant.

SOLUTION GIVEN ON THE TABLET	SOLUTION EXPRESSED IN MODERN NOTATION
1 Halve 7, by which length exceeds width: Result 3;30.	Let x be the width and y the length. Then $y = x + 7$, and $xy = 60$. Or, $x(x + 7) = 60$, from which $x^2 + 7x = 60$. Using equation (4.7) gives
2 Multiply together 3;30 by 3;30: Result 12;15.	
3 To 12;15 add 60, the product: Result 1, 12;15	$x = \sqrt{(3 \cdot 5)^2 + 60} - (3 \cdot 5) = 5$
4 Find the square root of 1, 12;15: Result 8;30	The length is then obtained by adding 3·5 to the square root, rather than adding 7 to the width.
5 Lay down 8;30 and 8;30. Subtract 3;30 from one (8;30) and add it to the other (8;30).	
6 12 is the length, 5 the width.	Hence 12 is the length, and 5 is the width.

The Yale tablet also gives examples of solutions to quadratics of a more general type, such as

$$ax^2 + bx = c$$

The technique here was to multiply throughout by a to get

$$(ax)^2 + b(ax) = ac$$

and then to substitute $y = ax$ and $e = ac$ to obtain the standard Babylonian form of quadratic:

$$y^2 + by = e, \quad e > 0$$

After solving for y, the value of y is divided by a to get the solution for x.

We have already discussed how the Babylonians handled cubic equations of the form $x^3 = c$ with the help of cube root tables, and equations of the form $x^2(x + 1) = c$ with the help of $n^3 + n^2$ tables. There is also a correct solution on the Yale tablet for a cubic equation of the form $x(10 - x)(x + 1) = c$, where $c = 2,48$. The correct solution is $x = 6$. It is a tribute to the level of abstraction and manipulative skills of Babylonian mathematicians that they were solving higher-order equations such as $ax^4 + bx^2 = c$ and $ax^8 + bx^4 = c$ by treating them as if they were 'hidden' quadratics in x^2 and x^4, respectively.

Linear and non-linear equations in two or three unknowns

The Babylonians approached these problems in two different ways. For a system of equations with two unknowns, they sometimes used the method of substitution, familiar to us, in which one of the equations is solved for one of the unknowns, and the value found is substituted into the other equation. There was, however, another approach which remained uniquely Babylonian until it was adopted by Hellenistic and, probably, Indian mathematicians around the beginning of the Christian era. The method has been called Diophantine, after the Greek mathematician Diophantus, who lived in Alexandria during the third century AD. It has been remarked that his algebraic methods have much in common with the Babylonian procedures.

The Diophantine method is particularly suitable for solving a system of two equations where one is of the form $x + y = s$, and the other may be any type of equation (linear or non-linear) in the two unknowns x and y. The procedure is as follows: If x and y were equal, $x + y = s$ would imply that $x = y = \frac{1}{2}s$. Now, if we assume that x is greater than $\frac{1}{2}s$ by a quantity w, then

$$x = \frac{1}{2}s + w \quad \text{and} \quad y = \frac{1}{2}s - w$$

If we substitute these expressions for x and y into another equation, which can be either linear or quadratic, we obtain an equation for w and can proceed to solve it. Next, we substitute the value for w into the above equations for x and y to obtain the solutions for the whole system of equations. An illustration is provided by a problem from a tablet found at Senkereh which dates back to the Hammurabi dynasty:

EXAMPLE 4.8   *Length (ush), width (sag). I have multiplied ush and sag, thus obtaining the area (asha). Then I added to asha, the excess of the ush over the sag: Result 3,3. I have added ush and sag: Result 27. Required [to know] ush, sag and asha.*

Solution

EXPLANATION IN THE TEXT	EXPLANATION IN MODERN NOTATION
1 One follows this method	Let x = length (*ush*),
$27 + 3,3 = 3,30$	y = width (*sag*)
$2 + 27 = 29$	Then the problem can be restated as:
2 Take one half of 29 (14;30) and square it:	
	$$xy + x - y = 183 \quad (4.8)$$
$14;30 \times 14;30 = 3,30;15$	$$x + y = 27 \quad (4.9)$$
3 Subtract 3,30 from the result:	Define $y' = y + 2$
$3,30;15 - 3,30 = 0;15$	(so $y = y' - 2$). Then
4 Take the square root of 0;15:	$$xy' = 210 \quad (4.10)$$
the square root of 0;15 is 0;30.	$$x + y' = 29 \quad (4.11)$$
5 Then	A general solution to the above set of equations may be expressed as
Length (*ush*) =	
$14;30 + 0;30 = 15$	$$xy' = p$$
$14;30 - 0;30 = 14$	$$x + y' = s$$
6 Subtract 2 (which has been added to 27) from 14:	or
	$$x = \tfrac{1}{2}s + w$$
width (*sag*) = $14 - 2 = 12$	$$y' = \tfrac{1}{2}s - w$$
7 I have multiplied 15 (*ush*) by 12 (*sag*) to get *asha*:	where w is the square root of $(\tfrac{1}{2}s)^2 - p$.

112

Area (*asha*) = 15 × 12 = 3,0

So if $s = 29$ and $p = 210$, then
$w = 0.5$.
Hence $x = 15$,
$y = y' - 2 = 14 - 2 = 12$,
and the area is 180.

Note that the transformations from equations (4.8) and (4.9) to equations (4.10) and (4.11) respectively are indicated in step 1 on the left-hand side. There is a close correspondence between the procedure as explained on the Babylonian tablet and the modern version given on the right.

There are a few examples in Babylonian algebra of the solution of a set of equations in three unknowns. In a problem text kept at the British Museum, we find (in modern notation):

$$x^2 + y^2 + z^2 = 1400, \qquad x - y = 10, \qquad y - z = 10$$

Its solution is unlikely to have caused many difficulties, for x and y can easily be expressed in terms of z as $z + 20$ and $z + 10$, respectively, so as to obtain a quadratic equation in z. The reader may wish to check the correctness of the solution set for (x, y, z), which is given as (30, 20, 10); this solution set holds for positive z.

BABYLONIAN GEOMETRY

Only a few years ago most historians of mathematics shared the view that, although the Babylonians excelled in algebra, they were inferior to the Egyptians in geometry. One cannot deny the notable achievements of the Egyptians in the field of mensuration of spherical objects and pyramids, which we discussed in the last chapter, or that the achievements of the Babylonians were modest in comparison. To take a specific example, it is true that most of the evidence points to the Babylonians calculating the area of a circle by taking three times the square of the radius, which, as we saw in Chapter 4, is a cruder approximation than the square of 8/9 of the diameter found in the Ahmes Papyrus. However, in an Old Babylonian tablet excavated in 1950 appears the direction that 3 must be multiplied by the reciprocal of 0;57,36 to get a more accurate estimate of the area. This gives a value for π of 3.125.

We shall not be taking a detailed look at the Babylonian knowledge of simple rules of mensuration. Suffice it to say that there is evidence of their familiarity with general rules for the areas of rectangles, right-angled triangles,

Figure 4.3 The Plimpton Tablet (Neugebauer and Sachs, 1945, Plate 25).

isosceles triangles, and trapeziums with one side perpendicular to the parallel sides. The most notable achievements of Babylonian geometry were in two areas where their algebraic skills could be given full rein: their work on the Pythagorean theorem and on similar triangles foreshadowed Greek work in these areas by over a thousand years.

Pythagorean triples or proto-trigonometry?

In 1945, Neugebauer and Sachs published their decipherment of a clay tablet (No. 322 in the Plimpton collection at the University of Columbia) made at some time between 1800 and 1650 BC. The tablet, as it appears today, is shown in Figure 4.3. Some think that it is the surviving section of a larger tablet which had several additional columns of numbers to the left, probably broken and lost when the tablet was excavated. As Figure 4.3 shows, the tablet is further marred by a deep chip in the middle of the right-hand edge and a flaked area at the top left-hand corner.

The tablet as we have it contains four columns of numbers arranged in fifteen rows, the last of these (Column 4) giving the number of the row. Table 4.3 presents the four columns that can definitely be deciphered. (Column 5 is included for illustrative purposes.)

There are errors in the original; asterisks indicate where correct values have been substituted. The break in Column 1 throws some doubt on the accuracy of the first few terms in each sequence of numbers: it is not certain whether the entries to the left of the dashed line were present on the original tablet. But what is beyond doubt is that a definite relationship exists between Columns 2 and 3, which becomes clearer if we examine Column 5. In terms of the right-angled triangle shown in Figure 4.4, $b^2 + h^2 = d^2$. Hence, from Row 1, $b = 1,59$ (119) and $d = 2,49$ (169). So

$$d^2 - b^2 = (169)^2 - (119)^2 = (120)^2$$

and therefore

$$h = 120 \quad (2,0)$$

One of the conjectures is that Column 1 represents the square of the ratio of the diagonal to the height (i.e. the square of d/h). Thus, in Row 1,

$$(d/h)^2 = (169/120)^2 \quad (1;59,0,15)$$

What was the purpose of Column 1? There have been a number of interesting suggestions. Some of the first to examine the Plimpton Tablet concluded, from an examination of the magnitude and pattern of decrease of numbers in the column, that it might have represented some form of 'proto-trigonometric' function. The first number in the first column is quite close to the square of the secant (i.e. the reciprocal of the sine) of 45°, and the last number in the column is approximately the square of the secant of 31°, with the intervening numbers decreasing one degree at a time. This interpretation is a trifle fanciful. There is no reason to believe that the Babylonians were familiar with the concept of a secant, or, for that matter, any other trigonometric function; indeed, neither they nor the Egyptians had any concept of an angle in the modern sense, which first occurs in the work of Indian and Hellenistic mathematicians around the beginning of the first millennium AD.

It would seem, then, that Column 1 served another function. It seems likely that this column, and indeed the tablet itself, have to do with the derivation of Pythagorean triples (e.g. b, h, d in Figure 4.4) for use in the construction of right-angled triangles with rational sides. It is improbable that the values inscribed on the tablet were derived by using trial-and-error methods, for these would have given simpler triples. But if the Babylonians had a more systematic method of deriving b, h and d to satisfy the equation $b^2 + h^2 = d^2$,

Table 4.3 The Plimpton Tablet deciphered.

Column 1 (?)	Column 2 (width, b)	Column 3 (diagonal, d)	Column 4 (Row no.)	Column 5 (height, h)
1;59, 0,15	1,59	2,49	1	2, 0
1;56,56, 58,14,50,6,15	56, 7	1,20,25*	2	57,36
1;55,7,4, 1,15,33,45	1,16,41	1,50,49	3	1,20, 0
1; 53,10,29,32,52,16	3,31,49	5, 9, 1	4	3,45, 0
1; 48,54,1,40	1, 5	1,37	5	1,12
1; 47,6,41,40	5,19	8, 1	6	6, 0
1; 43,11,56,28,26,40	38,11	59, 1	7	45, 0
1; 41,33,33,59,3,45	13,19	20,49	8	16, 0
1; 38,33,36,36	8, 1*	12,49	9	10, 0
1; 35,10,2,28,27,24,26,40	1,22,41	2,16, 1	10	1, 48, 0
1; 33,45	45	1,15	11	1, 0
1; 29,24,54,2,15	27,59	48,49	12	40, 0
1; 27,0,3,45	2,41*	4,49	13	4, 0
1; 25,48,51,35,6,40	29,31	53,49	14	45, 0
1; 23,13,46,40	56	1,46*	15	1,30

*Correct values substituted for incorrect ones on the tablet.

Figure 4.4

what was it? We can only hazard a guess, helped by a possible clue contained in Column 1.

Let us first assume that the height is normalized to 1 (i.e. that $(d/h)^2 - (b/h)^2 = 1$). Now, if $\alpha = d/h$ and $\beta = b/h$, then

$$\alpha^2 - \beta^2 = (\alpha - \beta)(\alpha + \beta) = 1$$

Let $\alpha + \beta = m/n$ and $\alpha - \beta = n/m$, where m and n are positive integers such that $m > n$. Then

$$\alpha = \tfrac{1}{2}[(m/n) + (n/m)] \quad \text{and} \quad \beta = \tfrac{1}{2}[(m/n) - (n/m)]$$

or

$$\alpha = [(m^2 + n^2)/2mn] \quad \text{and} \quad \beta = [(m^2 - n^2)/2mn]$$

But $b = \beta h$ and $d = \alpha h$. And if we put $h = 2mn$ so as to obtain a solution in integers, then

$$h = 2mn, \qquad b = m^2 - n^2, \qquad d = m^2 + n^2 \qquad (4.12)$$

This method of generating integral Pythagorean triples is usually attributed to Diophantus (*c.* AD 250), who, as we have seen, may be thought of as working in the Babylonian algebraic tradition and introducing it into Greek mathematics. It is worth noting that to arrive at these formulae we need nothing more than the ability to add and subtract fractions, and a knowledge of the algebraic identity $\alpha^2 - \beta^2 = (\alpha + \beta)(\alpha - \beta)$.

We can use the formulae (4.12) to generate the first three triples on the Plimpton Tablet, as shown in Table 4.4. With one exception, the integers chosen for m and n in the complete table are all products of prime factors of 60. For example, in the first row $m = (2)(2)(3)$ and $n = 5$.

One of the difficulties with this explanation of how the Babylonians generated Pythagorean triples is the lack of an underlying rationale for

Table 4.4 Generating the first three Pythagorean triples from the Plimpton Tablet.

m	n	$h = 2mn$	$b = m^2 - n^2$	$d = m^2 + n^2$
12	5	120	119	169
64	27	3456	3367	4825
75	32	4800	4601	6649

the choice of the particular values of (m, n), which seem to vary in an erratic fashion – there is certainly no discernible pattern to the first three sets of values in Table 4.4. It is possible that the rows may have been ordered so as to ensure an approximately linear increase in the values in Column 1, which contains the square of the ratio of two sides of the triangle shown in Figure 4.4 (i.e. $(d/h)^2$).

But there is another explanation, one which does seem quite convincing. It can be shown[*] there are exactly sixteen Pythagorean triples (of which fifteen are found in the Plimpton Tablet) that result if the following conditions hold: (i) $n \leq 60$; (ii) $30° \leq \theta \leq 45°$; and (iii) $\tan^2\theta = \dfrac{h^2}{b^2}$ has a finite sexagesimal expansion which implies that m, n and b are all products of prime factors of 60 (i.e. 2, 3 and 5). If this is so, the Plimpton Tablet contains a striking example of a classification exercise in early number theory.

Irrespective of which of these explanations, or any other, is valid or not, there can be little doubt that the Babylonians knew and used the Pythagorean theorem. This is confirmed by a problem from a tablet found at Susa, a couple of hundred miles from Babylon, belonging to the Old Babylonian period. It is one of the oldest examples of the use of the theorem in the history of mathematics:

EXAMPLE 4.9 *Find the circum-radius of a triangle whose sides are 50, 50 and 1,0.*
Solution In terms of Figure 4.5, the problem is to calculate the radius, r. The solution proceeds thus:

$$AD = \sqrt{50^2 - 30^2} = 40$$
$$OC^2 = r^2 = 30^2 + (40 - r)^2$$

---

[*]I am grateful to Professor Christopher Zeeman for this explanation.

Figure 4.5

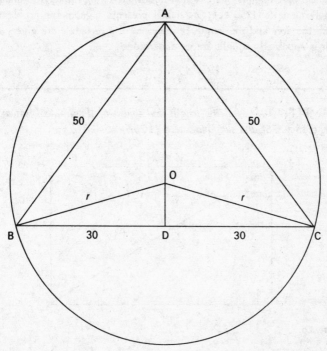

Therefore

$$r^2 = 30^2 + 40^2 - 80r + r^2$$
$$80r = 2500$$

or

$$r = 2500/80 = 31 \cdot 25 \quad (31;15)$$

There is another example of the application of the Pythagorean theorem which is notable in that it uses a rather long-winded method of solution which has a stronger geometric rationale than a shorter method based on algebraic techniques. The source of this evidence is interesting. In 1962, archaeologists working at Tell Dhibayi, near Baghdad, unearthed about 500 clay tablets. Most of them dealt with the commercial transactions and administrative matters of a city that flourished during the reign of Ibalpiel II of Eshunna (*c.* 1750 BC). One tablet presents a geometric problem in which the area and the length of diagonal of a rectangle are given, and what is apparently sought are its dimensions:

EXAMPLE 4.10 *Find the length and width of [Figure 4.6], given its area, 0;45 [0·75] and diagonal, 1;15 [1·25].*

$A = 0;45$

y

$d = 1;15$

x

Figure 4.6

Suggested solution   The tablet gives the following steps. The results at each step are given here both in sexagesimals and in decimals.
 1 Multiply the area by 2: Result 1;30 (1·5).
 2 Square the diagonal: Result 1;33,45 (1·5625).
 3 Subtract (1) from (2): Result 0;3,45 (0·0625).
 4 Find the square root of (3): Result 0;15 (0·25).
 5 Halve (4): Result 0;7,30 (0·125).
 6 Find one-quarter of (3): Result 0;0,56,15 (0·015 625).
 7 Add area to (6): Result 0;45,56,15 (0·765 625).
 8 Find square root of (7): Result 0;52,30 (0·875).
 9 Length = Result in (5) + Result in (8) = 1.
10 Width = Result in (8) − Result in (5) = 0;45 (0·75).

The procedure followed above is quite baffling at first sight. We might well have expected to see a solution along the following lines: Let *x* be the length, *y* the width, *d* the diagonal and *A* the area. Then

120

$$xy = A = 0.75 \quad \text{(area of a rectangle)} \tag{4.13}$$

$$x^2 + y^2 = d^2 = (1.25)^2 \quad \text{(Pythagorean theorem)} \tag{4.14}$$

We solve equation (4.13) for x (or y) and then substitute into equation (4.14) to solve for y (or x) after reducing the resulting biquadratic (i.e. quartic) equation to a quadratic one. Thus, substituting $y = (0.75)(1/x)$ into equation (4.14) and simplifying gives

$$16x^4 + 9 = 25x^2 \tag{4.15}$$

Setting $x^2 = z$ in equation (4.15) yields

$$16z^2 - 25z + 9 = 0$$

which has the solution $z = 1$ or $9/16$, and so $x = 1$, $y = 3/4$ (or $x = 3/4$, $y = 1$).

The Babylonian solution was quite ingenious and follows from the recognition that

$$d^2 - 2A = x^2 + y^2 - 2xy = (x - y)^2 \quad \text{(Steps (1) to (3))}$$
$$d^2 + 2A = x^2 + y^2 + 2xy = (x + y)^2 \quad \text{(the sum of Steps (1) and (2))}$$

so that

$$\sqrt{d^2 - 2A} = \text{length} - \text{width}$$
$$\sqrt{d^2 + 2A} = \text{length} + \text{width}$$

Hence the suggested solution for Example 4.10 is largely a procedure for forming these two relationships: the result from Step (5) is

$$\tfrac{1}{2}\sqrt{d^2 - 2A} = \tfrac{1}{2}(\text{length} - \text{width})$$

and the result from Step (8) is

$$\sqrt{\tfrac{1}{4}(d^2 - 2A) + A} = \tfrac{1}{2}\sqrt{d^2 + 2A}$$
$$= \tfrac{1}{2}(\text{length} + \text{width})$$

So the result from (5) plus the result from (8) gives the length, and the result from (8) minus the result from (5) gives the width.

This example epitomizes the versatility of Babylonian mathematics. Here was a group of people who for the first time combined arithmetic, algebra and geometry in tackling problems – a remarkable feat.

Similar triangles

One of the clay tablets (Figure 4.7(a)) excavated at Tell Harmal in Iraq is thought to date back to about 2000 BC, making it one of the earliest problem

Figure 4.7 Similar triangles: (a) the original tablet (Baqir, 1950, opp. p. 54); (b) the modern 'translation'.

texts we know of. The problem may be stated thus:

EXAMPLE 4.11   *Given the sides of* △ABC *and areas of* △s BAD, ADE, DEF *and* EFC, *as shown in Figure 4.7(b), find the lengths* BD, DF, AE *and* AD.

(It is possible to infer from the lengths of the sides of △ABC that it is a right-angled triangle. The original diagram and the procedures would indicate that AD and EF are perpendicular to BC, and that DE is perpendicular to AC, as shown in Figure 4.7(b). So what we have is a series of similar right-angled triangles.)

Solution   The Babylonian procedure will be given in both its rhetorical and symbolic forms. In its rhetorical form, the steps are:

1 Take the reciprocal of 1,0 and multiply it by 45: Result 0;45.
2 Multiply the result by 2: Result 1;30.
3 Multiply the result by the area of △ABD: Result $(8,6)(1;30) = 12,9$.
4 Find the square root of 12,9: Result BD = 27.

Now that BD is known, the Pythagorean theorem can be used to show that the length of AD is 36. If the area and hypotenuse of △ADE are known, the application of the above procedure would give the length of AE. This is followed by a further application of the Pythagorean theorem to evaluate ED. And this process may be continued *ad infinitum* to work out the required dimensions of an infinite series of similar right-angled triangles. (The reader is invited to check that AE and ED are 21;36 and 28;48 respectively.)

In symbolic terms, Steps 1 to 3 of the Babylonian procedure are as follows.

1 (1/AC)AB.
2 (AB/AC).
3 (2AB/AC) (area of △ABD) = BD^2.

To make any sense of these steps, it is necessary to introduce two results which must have been known to the Babylonians:

(a) If △ABC is similar to △ABD, then AB/AC = BD/AD.
(b) Area of △ABD = $\frac{1}{2}$(BD . AD).

Applying (a) and (b) to Step 3, we get

$$(2BD/AD) . \tfrac{1}{2}(BD . AD) = BD^2$$

It is clear from this example that the Babylonians had some knowledge of the properties of similar triangles (though we know of no explicit contemporary statement), in particular that they were familiar with one of the Euclidean theorems:

> In a right-angled triangle, if a perpendicular is drawn from the right angle to the hypotenuse, the triangles on each side are similar to the whole triangle and to one another.

It is reasonable to infer that Euclid took the kernel of his ideas about similar triangles either directly or indirectly from the Babylonians, and then imbued it with that peculiarly Greek contribution to mathematics, the method of axiomatic deductive logic, whose importance to the future development of the subject cannot be overestimated.

Before we attempt, in the next chapter, a final assessment of both Egyptian and Babylonian mathematics, it would be useful to summarize the character and achievements of Babylonian mathematics. The evidence presented in this chapter provides a full testimony to the quality and range of the mathematical achievements of this ancient civilization. It is clear that, with their numerical and algebraic skills, the Babylonian mathematicians produced work that stands comparison favourably with what was being done in sixteenth-century Europe before the advent of modern mathematics. Yet in emphasizing the quality of their arithmetic and algebra, we should not ignore their achievements in other areas. Their work on Pythagorean triples and similar triangles provides fine examples of their interest in and contributions to geometry.

There is a tendency to label all mathematics before the Greeks as utilitarian and pre-scientific. This view should be critically re-evaluated in the light of the content as well as the spirit of the work examined in this chapter. Even after four thousand years, some of the Babylonian contributions to mathematics remain quite awe-inspiring.

5 Egyptian and Babylonian Mathematics: An Assessment

◇◇

A widely held view of Egyptian and Babylonian mathematics is crystallized in the writings of Morris Kline, a contemporary American historian of mathematics. In his book *Mathematics, A Cultural Approach*, published in 1962, he devotes just three pages out of seven hundred to the contributions of these two civilizations. Dismissing all the evidence to the contrary marshalled by both ancient Greeks and modern scholars, he considers that the Egyptian and Babylonian contributions to mathematics were 'almost insignificant'. This is followed by his astonishing statement that, compared with what the Greeks achieved, 'the mathematics of the Egyptians and Babylonians is the scrawling of children just learning to write, as opposed to great literature'. In any case, Kline continues, these civilizations 'barely recognized mathematics as a distinct discipline,' so that 'over a period of 4000 years hardly any progress was made in the subject'.

I have quoted quite extensively from a single page (p. 14) of Kline's book because his views represent a concise summary of what we labelled in Chapter 1 'Eurocentric scholarship'. In Chapter 1 we identified the main characteristics of this Eurocentric outlook, the chief of which is a tendency to ignore new findings that go against deeply entrenched views about the origins of mathematics. The last two chapters have spelt out in great detail some of the contributions to early mathematics made by these most ancient of civilizations. Evidence of their contributions is not all hidden away in obscure journals or expressed in languages which tend to be ignored by many Western scholars: much is published in English in 'respectable' journals and books, brought out by major publishers on both sides of the Atlantic. The reason for the neglect is not that the relevant literature is inaccessible or 'unrespectable', but something deeper – a serious flaw in Western attitudes to historical scholarship (one not confined to histories of mathematics or science). An excessive enthusiasm for everything Greek, arising from the belief that much that is desirable and worthy of emulation in Western civilization originated in ancient Greece, has led to a reluctance to allow other ancient civilizations any share in the historical heritage of mathematical discovery. The belief in a 'Greek miracle' and the way of attributing any significant mathematical discoveries to Greek influences are all part of this syndrome. This view of history is a symptom of the intellectual arrogance that is often below the surface in Eurocentric scholarship. In Chapter 1 we mentioned Martin Bernal's recent book *Black Athena* (1987). It is notable for a searching analysis of how in the last two centuries Eurocentric scholarship has ignored, rejected or even suppressed knowledge of the Afro-Asian contributions to Classical Greek civilization.

A more substantial reason for neglecting Egyptian and Babylonian mathematics is contained in Kline's view that these civilizations 'barely recognized mathematics as a distinct discipline'. Behind views such as this can be discerned a number of assertions. The mathematics of Egypt and Babylonia, it is argued, (1) had no general rules, (2) contained no 'proofs', (3) lacked abstraction, (4) failed to distinguish clearly between exact and approximate results; and generally (5) there was no clearly discernible activity which we may label 'mathematics' and which was studied for its own sake. Let us examine each of these alleged shortcomings in detail.

1 No explicit general statements of algebraic rules and their appropriateness are found in the mathematical sources of either civilization. But this is hardly surprising, given both the nature of the mathematical evidence that has come down to us and the lack of symbolic notation. Also, it must not be forgotten that there was no general deductive algebra before the emergence of modern mathematics. Now, a rule can be general without being deductive. Consider two notable mathematical achievements, one from Egypt and the other from Babylonia, which we discussed earlier: the Egyptian discovery of the rule for finding the volume of a truncated pyramid, as shown in the Moscow Papyrus, and the Babylonian calculation of Pythagorean triples, contained in the Plimpton Tablet. It cannot be argued that these were merely empirical rules arrived at through a painful process of trial and error for specific problems, without any awareness of their general application. In any case, the very fact that problems requiring specific algorithms for their solutions are grouped together in both the Ahmes Papyrus and the Babylonian tablets would indicate that there was some understanding of the generality of the underlying rules. It has also been pointed out that in a number of Greek geometric solutions each step is identical to the corresponding step in the algebraic solutions of the Babylonians. For example, the Babylonian 'take square root of a certain number A' would correspond to the Greek 'take the side of a square whose area is A'.

There is sometimes a tendency to devalue the role of algorithms in the development of mathematics. From the practical concerns of society have arisen a number of rules which should be judged for both their effectiveness and their intrinsic qualities. A 'good' algorithm should have three properties:

(a) it should be clear and simple, laying out step by step the procedures to be followed,

(b) it should emphasize the general character of its applications by pointing out its appropriateness, not to a single problem but to a group of similar problems, and

(c) it should show clearly the answer obtained after the prescribed set of operations is completed.

It will be left to the reader to judge whether the mathematical sources that we have examined in the last two chapters contain algorithms which satisfy these requirements.

2 There is a hardly a trace, according to the next argument, in any of the mathematical sources that we have examined, of what is commonly recognized as 'proof'; this implies a non-scientific approach to the subject. However, what constitutes 'proof' is a difficult question. Today, a rigorous mathematical proof that is not symbolic is inconceivable. A modern proof is a procedure, based on axiomatic deduction, which follows a chain of reasoning from the initial assumptions to the final conclusion. But is this not taking a highly restrictive view of what is proof? Could we not expand our definition to include, as suggested by Imre Lakatos (1976), explanations, justifications and elaborations of a conjecture constantly subjected to counter-examples? Is it not possible for an argument or proof to be expressed in rhetoric rather than symbolic terms, and still be quite rigorous? As Gillings (1972, p. 233) states:

> A non-symbolic argument or proof can be quite rigorous when given for a particular value of the variable; the conditions for rigor are that the particular value of the variable should be typical, and that a further generalization to any value should be immediate.

It is possible to distinguish between logically deductive and axiomatically deductive algebraic reasoning. Once David Hilbert (1862–1943) and Bertrand Russell (1872–1970) had laid the foundations of mathematical logic, it became possible to construct an algebra from a limited set of axioms. Previously, what great mathematicians such as Euler, Gauss and Lagrange had considered as proof was logically deductive proof.

These questions are relevant not just to Egyptian and Babylonian mathematics, but to other mathematical traditions that we shall be examining in subsequent chapters. By posing the questions here, I am stressing how important it is not to be blinded by present-day preconceptions of what constitutes mathematical demonstration and proof when studying the mathematics of the past.

The mathematical papyri and tablets from Egypt and Babylonia show a considerable technical facility in computation, and also a recognition of the applicability of certain procedures to a similar set of problems, and of the importance of verifying the correctness of a procedure by checking, say, a division by multiplication or the solution of an equation by substitution of the calculated value of the unknown into the original equation. These procedures and checks in the mathematics of these early civilizations must be regarded as a form of 'proof' in the broader sense.

3 It is the supposed absence of abstraction in the Egyptian and Babylonian sources that sways many critics in their judgement of whether these civilizations produced mathematics or merely some form of applied arithmetic. In a number of Babylonian examples discussed in the last chapter, we found close parallels between the steps of the Babylonian procedure and the steps of the corresponding modern analysis in algebraic symbols. The Babylonian symbols *ush* and *sag* for length and width, respectively, served the same purpose as our algebraic symbols x and y. The transition from the specific to the abstract generalization is not absent. How else are we to interpret meaningfully the addition of length (*ush*) and area (*asha*)?

4 An implicit acceptance that π is a finite quantity would be an illustration of the failure to distinguish between exact and approximate results. It is often argued that the ancient civilizations did not make such a distinction, and therefore that what they practised was not mathematics, only something that resembled mathematics. We have seen that the Babylonians, in their evaluation of the reciprocal of 7 and their calculation of the square root of 2, were aware of the fact that their results were approximations and not the true values. Indeed, their omission of irregular sexagesimals from mathematical tables implied an uncertainty as to whether they would ever obtain accurate results. Balanced against this was the need for practical or numerical precision in solving real-life problems. But were the Babylonians aware of the important distinction between mathematical precision and numerical precision? On the existing evidence, it is impossible to tell.

5 A point often made about Egyptian and Babylonian mathematics is that it was more a practical tool than an intellectual pursuit. This implied criticism is symptomatic of the attitude, often attributed to the Greeks, that mathematics devoid of all utilitarian purpose is in some sense a nobler or better mathematics.

An important distinction running right through Greek thought is between *arithmetica*, the study of the properties of pure numbers, and *logistica*, the use of numbers in practical applications. The cultivation of the latter discipline was left mainly to the slaves. Legend has it that when Euclid was asked what was to be gained from studying geometry, he disdainfully told his slave to toss a coin at the inquirer. That the Egyptian and Babylonian cultures were entirely utilitarian, with little or no interest in mathematics for its own sake, is not borne out by the nature of some of the problems we have examined, which appear to have no practical implications. In any case, the pursuit of mathematics as an aesthetic activity for its own sake presupposes the existence of a leisured class, freed from the concerns of survival, including the need to make a livelihood. Greek civilization, with its substantial slave population, allowed a small elite the freedom to pursue activities which had no practical significance. Both the character of the Greeks' mathematics and their conception of mathematics as a deductive science was to some extent influenced by this form of social stratification. In civilizations where such a luxury was not possible, mathematics would have had little chance of transcending its utilitarian origins.

Ultimately, whether one characterizes the activities of the Egyptians and Babylonians as mathematics will depend on how one perceives the long algorithmic phase that preceded the development of modern algebra. Was this phase an early stage in the emergence of 'true' algebra, or did algebra begin only with the introduction of algebraic symbolism? The long period in which material was collected in the form of problems and valid methods invented for their solution was algebra's gestation period; the rearrangement of old procedures into new deductive structures marks its birth.

6 Ancient Chinese Mathematics

◇◇◇

BACKGROUND AND SOURCES

To understand the history of Chinese mathematics requires some familiarity with Chinese history. The history of China is a vast subject, as befits a country that can trace the continuity of its civilization through four thousand five hundred years. The period we are concerned with runs from the dawn of civilization in China to the end of the Ming dynasty (AD 1260–1644) and the beginnings of European contact. It will help if we divide this long time span into five shorter periods. The reader may find it useful to refer to the map of China, Figure 6.1, for places mentioned in the following account.

The first period began with the civilization that developed along the banks of the Yangtze and Huang Ho (or Yellow) Rivers during the legendary Hsia kingdom in the third millennium BC. It continued through the Shang rule, which began around 1500 BC and lasted for five hundred years. The earliest evidence of numeration in China is from this period and consists of oracle bones, so named because inscriptions on them indicate that they were used for divination or fortune-telling. The social organization during the Shang dynasty was a primitive form of feudalism, with extensive use of bronze for weaponry and armour, as well as the practical arts. Cowrie shells were widely used as money.

The Chou invaders completed their conquest of the Shang around 1030 BC. They extended their territorial control and instituted a more developed form of feudalism, reminiscent in its structure of the feudal system that would emerge in parts of Europe some two thousand years later. The development of a written language that had come into use during the Shang period was well under way. But from around 700 BC the Chou dynasty came under increasing attack from groups of insurgents. The empire disintegrated, and for a period of two hundred years from about 400 BC there existed a number of independent states virtually at war with one another for most of the time.

This period, usually known as the period of the Warring States, is notable

Figure 6.1 Map of eastern China.

from a mathematical viewpoint because from it comes the oldest source of Chinese mathematics, the *Chou Pei Suan Ching* (The Arithmetical Classic of the Gnomon and the Circular Paths of Heaven). It is believed that the earliest version of this book dates from the Shang dynasty. The first part contains a dialogue between Duke Chou Kung and a notable called Shang Kao about the properties of right-angled triangles, during which the Pythagorean theorem is stated and a geometric demonstration given. While it is no longer believed that this treatise predates Pythagoras by five centuries, it is still thought likely that it was composed before the time of the Greek mathematician. Apart from the consideration of the right-angled triangle and a brief discussion of simple arithmetic operations, the *Chou Pei* is primarily an archaic astronomical text. (This and other Chinese mathematical sources are listed in Table 6.1, together with their authorship, date and important points of coverage.)

Table 6.1 Major Chinese mathematical sources up to the seventeenth century.

Title	Author	Date	Notable subjects covered
Chou Pei Suan Ching, The Arithmetic Classic of the Gnomon and the Circular Paths of Heaven	Unknown	c. 500–200 BC	Pythagorean theorem; simple rules of fractions and arithmetic operations
Suan Shu Shu, A Book on Arithmetic	Unknown	c. 300–150 BC	Operations with fractions; areas of rectangular fields; fair taxes
* *Chiu Chang Suan Shu*, Nine Chapters on the Mathematical Arts	Unknown	c. 300 BC to AD 200	Root extraction; ratios (including the rule of three and the rule of false position); solution of simultaneous equations; areas and volumes of various geometrical figures and solids; right-angled triangles
Ta Tai Li Chi, Records of Rites Compiled by Tai the Elder	Unknown	AD 80	Magic square of order 3
Commentary on *Chiu Chang*	Chang Heng	130	$\pi \simeq$ square root of 10
Shu Shu Chi Yi, Manual on the Traditions of the Mathematical Arts	Hsu Yue	c. 200	Theory of large numbers; magic squares; first mention of the abacus
Commentary on *Chou Pei*	Chao Chun Chin	c. AD 200–300	Solution of quadratic equations of the type $x^2 + ax = b^2$

Work	Author	Date	Description
Hai Tao Suan Ching, Sea Island Mathematical Manual	Liu Hui	263	Extensions of problems in geometry and algebra from the Nine Chapters
Sun Tsu Suan Ching, Master Sun's Mathematical Manual	Sun Tsu	400	A problem in indeterminate analysis; square root extraction. Operations with rod numerals
Sui Shu (Official History of the Sui Dynasty)	Tsu Chung Chih	450	Evaluation of π; method of finite differences
Chiku Suan Ching, Continuation of Ancient Mathematics	Wang Hsiao Tung	625	Solution of third-degree equations; practical problems for engineers, architects and surveyors
* *Suan Ching Shih Shu*, The Ten Mathematical Manuals	Li Chun Feng	656	An encyclopedia of mathematical classics of the past
Meng Chi Pi Than, Dream Pool Essays	Shen Kua	1086	Summation of series by piling up a number of kegs in a space shaped like a dissected pyramid
* *Shu Shu Chiu Chang*, Nine Sections of Mathematics	Chin Chiu Shao	1247	Numerical solutions of equations of high degree; indeterminate analysis

table continues

*Works most influential in the development of Chinese mathematics.

Table 6.1 *continued*

Title	Author	Date	Notable subjects covered
* *Tshe Yuan Hai Ching,* The Sea Mirror of the Circle Measurements	Li Yeh	1248	Solutions of high-degree equations; applications of the Pythagorean theorem to practical problems; use of a diagonal line across a digit to indicate a minus quantity
* *Hsiang Chieh Chiu Chang Suan Fa Tsuan Lei,* Detailed Analysis of the Nine Chapters	Yang Hui	1261	Arithmetic progressions; decimal fractions; quadratic equations with negative coefficients of x
Yuan Shih (Official History of the Yuan Dynasty)	Kuo Shou Ching	1280	Foundations of spherical trigonometry; cubic interpolation formula; biquadratic equations
* *Szu Yuen Yu Chien,* The Precious Mirror of the Four Elements	Chu Shih Chieh	1303	Pascal's triangle; solution of simultaneous equations with five unknowns by matrix methods
Suan Fa Tung Tsung, A Systematic Treatise on Arithmetic	Cheng Tai We	1593	Magic squares; introduction to abacus
Chi Ho Yuan Pen, Elements of Geometry	Ricci and Hsu	1607	Six books of Euclid's *Elements* translated into Chinese

*Works most influential in the development of Chinese mathematics.

The time of the Warring States was also a period when 'a hundred schools of philosophers' flourished: feudal lords, faced with uncertain times of popular unrest and technological innovation (iron was introduced into China at this time), employed itinerant philosophers as advisers. One of these philosophers was Confucius, but his deep and abiding loyalty to the Chou dynasty permitted him only a short spell as an adviser to the local ruler of Lu. The emphasis Confucius placed on unity and stability in his philosophy may have been a reaction to the troubled times he lived in.

It would be useful to put this period in a global context. It is one of the more remarkable coincidences of history that the middle of the first millennium BC saw the emergence of many of the great religious and ethical leaders whose influence is felt even today. Between 650 and 450 BC lived Confucius, Gautama Buddha, Mahavira and, probably, Zoroaster. The same period saw Babylon fall to the Persians (538 BC), India invaded by the Persian Emperor Darius (512 BC), and the Persian advance to the west halted by the Greeks (480 BC). And the last two centuries of the Warring States saw the conquests of Alexander (c. 330 BC), the foundation of the Mauryan empire in India, of which Asoka (273–232 BC) was the most illustrious ruler, and the protracted Punic Wars in the Mediterranean (c. 250–150 BC).

The Second Punic War was contemporaneous with the successful reunification of China by the Chin emperor Shih Huang Ti, who was master of all China between 221 and 210 BC. He rebuilt the Great Wall; a more dubious claim to fame is his order to burn all books. Since its recent excavation, the famous army of life-size ceramic soldiers buried with him has attracted considerable attention outside China, and may be seen as one of the more positive manifestations of his egomania.

During the subsequent period, which saw the emergence of the Han dynasty (200 BC to AD 220), scholars devoted a considerable amount of their time to transcribing from memory literary and scientific texts and seeking out manuscripts that had escaped destruction. This was when the most influential of all Chinese mathematical texts was compiled: the *Chiu Chang Suan Shu* (Nine Chapters on the Mathematical Arts), which occupies a similar position in Chinese mathematics to that of Euclid's *Elements* in Western mathematics.* As

---

*In 1984, a collection of bamboo strips was excavated from three Han tombs. One of them bears a title which shows this collection to be a mathematical work: *Suan Shu Shu* (A Book on Arithmetic). It is probably as old as *Chou Pei* and may even have been a direct antecedent of the Nine Chapters. Further analysis of the text may show the level of advance reached by the Chinese in arithmetic around the turn of the millennium.

a product of the late Chin and early Han dynasties, it was written at a time when the Roman empire was at its height and Buddhism was starting to have an impact in China. It was reputed to have been arranged and commented upon by Chang Tsang (c. 150 BC) and later Keng Shou Chang (c. 50 BC), from some earlier texts which have not survived. Despite its crucial importance in Chinese mathematics, there is as yet no complete English translation.*

The Han period is noted for significant developments in Chinese science and technology. Much was achieved in astronomy and calendar construction. Hsu Yue (c. AD 200) wrote a treatise entitled *Shu Shu Chi Yi* (Manual on the Traditions of the Mathematical Arts) which discusses calendar construction and gives an early description of magic squares. Foundations were laid for the systematic study and classification of plants and animals. One of the greatest technological inventions of mankind – paper – was a product of this period. An extensive bibliography was compiled by experts in medicine, history, military science, philosophy, astronomy, magic and divination; some of the volumes, recorded on wooden or bamboo tablets or on silk, have survived. Palace revolutions, peasant rebellions and religious uprisings weakened the dynasty until it was overthrown in AD 220, when a united China split into the Three Kingdoms.

The next period saw considerable divisions and upheavals, lasting for about a hundred years, which were brought to an end by a second unification of China under the Chin dynasty. However, the troubled times apparently did not disrupt mathematical activity, for during this period lived Liu Hui (c. 263), the great commentator on the Nine Chapters, on whose work we are so dependent for information on early Chinese mathematics. Foreign contacts through the spread of Buddhism, which began during the last decades of the Han dynasty, continued – in art, sculpture, medicine and the sciences as well as religion. Fa Hsien, the great Buddhist pilgrim, travelled for fifteen years the length and breadth of northern India, and over Central Asia, after setting out for India in 399.

This period produced two great mathematicians: Sun Tsu (c. 400), in whose work we find the beginnings of indeterminate analysis, and Tsu Chung Chih (c. 450), who takes us into the succeeding Liu dynasty (420–479). In 656 there appeared an encyclopedia of Chinese mathematical classics entitled *Suan Ching*

---

*It is a sobering thought, as Needham (1959) pointed out, that modern European histories of mathematics were being written without access or reference to about one thousand original Chinese mathematical texts to be found in the libraries of Beijing.

Shih Shu (The Ten Mathematical Manuals), which was to remain an influential text for several centuries.

A number of northern and southern dynasties followed in relatively quick succession after the second partitioning of China, until the short-lived Sui dynasty (581–618) reunified the country once more. It was a period of construction of large-scale waterworks, of which the most notable was the Grand Canal linking the Huang Ho and the Yangstze Rivers. The labour requirement was enormous: at times over five million workers were needed, which in certain districts meant all adult males between the ages of 15 and 50. The benefit to posterity was immense, for a transport system had been developed which would link the productive agriculture of southern China with the politically and demographically more influential north.

The immense cost borne by ordinary people under the Sui dynasty bore fruit during the rule of the Tang dynasty, which many would consider one of the most productive periods of Chinese history. For about three hundred years (618–906) the Tang emperors ruled over a China whose territorial boundaries and cultural dominance had never been so extensive. This was a period characterized by a remarkable openness to foreign influences. Just as Baghdad, the centre of intellectual activity to the west, welcomed scholars from all lands, the great Tang capital Chang'an numbered Arabs, Koreans, Japanese and Indians among a population estimated to have reached a million. This was a period of literary and artistic renaissance, while major technological innovations included printing and gunpowder. Surprisingly, no important mathematical work from this period has been discovered, but it could well have been the fertilization through foreign contacts and the scrutiny of past Chinese texts that took place in this period which led to the upsurge in Chinese mathematics of a few centuries later.

The Tang dynasty was followed in relatively quick succession by the years of the Five Dynasties and the Ten Independent States (907–60). Although these were chaotic times there were great advances in printing; what began initially as a means of disseminating religious texts slowly spread into secular fields. A great desire for unity elevated Chao Khuang Yin to the throne in 960, marking the beginning of one of longest dynasties in Chinese history, the Sung (900–1279), in which may be seen the culmination of the developments of the previous two centuries.

The scientific and technological achievements of the Sung period are too numerous to list here; we shall concentrate on the mathematics. The Sung period produced some of the greatest mathematicians of China, especially in the thirteenth century. In 1247 Chin Chiu Shao wrote *Shu Shu Chiu Chang* (Nine Sections of Mathematics, not to be confused with the Nine Chapters of a thousand years before). In this work Chin explained the numerical solution of equations of all degrees, and extended the work on indeterminate analysis begun by Sun Tsu. A year later appeared Li Yeh's *Tshe Yuan Hai Ching* (The Sea Mirror of the Circle Measurements), which explained how to construct equations of various degrees from a given set of data, thus complementing Chin's methods of solving these equations. Between 1261 and 1275 appeared a series of works by Yang Hui, the most influential of which was the *Hsiang Chieh Chiu Chang Suan Fa Tsuan Lei* (Detailed Analysis of the Mathematical Methods in the Nine Chapters). It starts as a commentary on the *Chiu Chang*, and goes on to present some remarkable extensions of the original work in a number of areas including mathematical series, quadratic equations with negative coefficients of x, and higher-order numerical equations. The fourth name in this grand quartet of mathematicians is Chu Shih Chieh, who lifted Chinese algebra to the highest level it was ever to attain. In the two treatises he wrote, *Suan Shu Chi Meng* (Introduction to Mathematical Studies, 1299) and *Szu Yuen Yu Chien* (The Precious Mirror of the Four Elements, 1303), are to be found a treatment of 'Pascal's' triangle three hundred and fifty years before Pascal, the solution of simultaneous equations with five unknowns using the 'Method of Rectangular Arrays' (or what we would now call matrix methods) and detailed applications of what was known as the 'celestial element method' for solving equations of high degree. This list of notable mathematicians would be incomplete without mention of Kuo Shou Ching, who lived in the second half of the thirteenth century. While there is no extant treatise on mathematics which can be traced to Kou, there are records from the Ming period (1368–1648) which show his influence on astronomy and calendar construction. The first Chinese work on spherical trigonometry, probably based on Arab sources, is directly attributed to Kou. By the closing stages of the Sung dynasty, then, the Chinese algebraists had forged so far ahead that it was only during the eighteenth century that the gap between Chinese and European algebra, particularly with respect to the solution of equations, was finally closed.

The last period of this short historical survey of China and Chinese mathematics stretches over four hundred years, and takes in the Yuan (Mongol) and the Ming dynasties, which ruled over the whole of China. The Yuan dynasty began with Shih Tsu (better known in Europe as Kublai Khan, and the grandson of Genghis Khan) in 1280 and ended with Shun Ti (Toghan Timur) in 1367. This was not a period of great creative activity in mathematics, but it was a time when contacts between China and Europe were at their height.

Mongol control extended right across the Asian heartland, from the Yellow Sea in the east to the Black Sea in the west. Appointment of non-Chinese to positions of authority was state policy, as an important means of preserving control. Ideas and technology were transmitted via the trade routes across Central Asia. Marco Polo was one of many travellers who came to China and wondered at its marvels; he later served at the court of the Khan. It was also during this period and the subsequent two centuries that four technological innovations from China, which, in the words of Francis Bacon (1561–1626), 'changed the whole face and state of things throughout the world', started making their way slowly westwards to Europe. These were printing, gunpowder, paper-making and the magnetic compass. It is an interesting reflection that while there was a time-lag of ten centuries between use of paper in China and in Western Europe, the time-lags for gunpowder and the magnetic compass were four and two centuries respectively. The last two would later be used by Europeans, to dramatic and devastating effect, on the rest of the world.

The Ming period, which began in 1368, saw the restoration of indigenous culture and values after a century of foreign rule, as well as expanding Chinese influence abroad. Great maritime expeditions were sent to the southern parts of India, Sri Lanka, the eastern coast of Africa and the Persian Gulf. Exotic items, including African animals such as giraffes, zebras and ostriches, were brought back from these areas. However, this interest in things foreign was not matched by a desire to learn from foreigners. Chauvinism and intolerance of other people prevailed, and led to a stagnation of science which would only be briefly relieved with the arrival of the Jesuits during the later part of the sixteenth century. It is not surprising, given the spirit of the age, that the first translation of Euclid's *Elements* into Chinese had little impact in China. It was felt more in Japan, which had until then been very much under the influence of its larger neighbour. At the beginning of the seventeenth century, the Manchus from Manchuria mounted an invasion which led to the establishment in 1644 of a dynasty that would survive until its overthrow in 1911. Foreign interventions became increasingly common, particularly in the nineteenth century; and Chinese mathematics became influenced by the West once more, but this time in a way that was to change its character completely. But this takes us far beyond the period covered by this book.

The following discussion of Chinese mathematics takes a thematic approach rather than examining in detail the various sources of Chinese mathematics listed in Table 6.1. However, because of the great importance of the *Chiu Chang Suan Shu* (hereafter referred to simply as the *Chiu Chang*) in the development of Chinese mathematics, its contents will be subjected to a detailed analysis.

In this chapter we consider developments in Chinese mathematics up to the end of the first millennium AD; subsequent progress, particularly during the thirteenth century, will be examined in the next chapter.

THE DEVELOPMENT OF CHINESE NUMERALS

Types of Chinese numerals

It is possible to distinguish four main types of Chinese numerals, all based on the decimal system.

1 The 'standard' or 'modern' numerals may have originated from common number-words in use from about the third century BC.

2 The 'official' numerals are merely a highly decorative version of the standard numerals, used on legal documents and bank notes which need to be protected from forgery or unauthorized alterations.

3 The 'commercial' numerals, again based on the standard ones, were devised for writing quickly and were widely used in trade and commerce. They are of more recent origin, dating from the sixteenth century, and are found most commonly today on price-tags or bills in Chinese shops and restaurants.

4 The 'stick' or 'rod' numerals served until recently as the principal instrument for mathematical and scientific work, and had been in use from at least the second century BC. The name derives from their origins in arrangements of sticks or rods on Chinese counting-boards. We shall be concentrating on these numerals.

However, the earliest known Chinese numerals appear in the form of Shang oracle bones, from some time between 1500 and 1200 BC. A group of farmers, tilling their fields near Anyang at the end of the last century, came across a collection of tortoise shells and animal bones with inscriptions on them. They were eventually sold to an apothecary, who believed them to be the bones of a dragon, endowed with medicinal properties. Fortunately they were rescued before being consumed, and attracted the interest of several Chinese scholars, who were instrumental in deciphering the inscriptions they bore. It turned out that the bones had belonged to Shang nobles who were in the habit of appealing to the spirits of their ancestors for advice on the best times for travelling, harvesting, celebrating feasts and other activities. Such questions, together with answers recorded after the prophecies had been fulfilled, were inscribed on the bones.

The numbers one to ten as found on these bones were represented by the following symbols:

1	2	3	4	5	6	7	8	9	10

This numeral system was more advanced than all contemporary systems except the Babylonian, since it enabled any number, however large, to be expressed by the use of ten basic symbols (1 to 10), and a selected number of additional symbols to represent 20's, 100's and 1000's. Thus, the numbers 537 and 1348 were written as

(five hundreds, three tens, seven)

(one thousand, three hundreds, four tens, eight)

where the symbol represents 1000's, 100's and 20's.

In the centuries that followed, different variants of the above characters were used, as can be seen on artefacts such as coins and bronze vessels. All were used in a decimal system of notation, with extra symbols to denote different orders of magnitude. The standard number system is a direct descendant of the ancient Shang system: its symbols for the numbers from one to ten are

1	2	3	4	5	6	7	8	9	10
一	二	三	四	五	六	七	八	九	十

The number 842 may be written from left to right as:

八百四十二

where 百 and 十 represent hundred and ten respectively.

However, over the years the demands of commerce, administration and science led to the development of a distinctive Chinese place-value number system which involved the use of counting-rods. These rods, originally of ivory or bamboo, were arranged in columns from right to left representing increasing powers of ten. Positive and negative numbers were represented by red and black rods, respectively. Our information on their use goes back only to the Chin and early Han dynasties, though the system was probably invented earlier.

By the third century AD these rod numerals were being described as *heng* and *tsung* numerals. A later notation for these two variants of rod numerals took the form shown in Table 6.2. The *tsungs* represent units, hundreds, tens of

Table 6.2 Chinese *heng* and *tsung* rod numerals. The *tsungs* represent units, hundreds, tens of thousands, etc., and the *hengs* tens, thousands, hundreds of thousands, etc.

thousands, etc., and the *hengs* tens, thousands, hundreds of thousands, etc.*
Thus the number 3614 would be written as

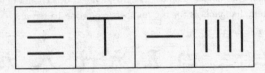

Note that the columns of numbers need not be demarcated since the type of numeral used defines the column value. Thus, in common with our present-day number system, to show 3614, a reckoner places rods to represent 4 in the extreme right column, 1 in the second, 6 in the third and 3 in the fourth column. The alternating *tsung* and *heng* representation would help to distinguish units of different orders in odd-numbered and even-numbered columns of the counting-board. This provided an in-built means of checking whether the digits were correctly represented before undertaking arithmetic operations. And as computation could be carried out by placing the rods on any flat

---

*In Sun Tsu's Arithmetic Manual (AD 280), the following verse explains this method of recording numbers:

Units are vertical, tens are horizontal,
Hundreds stand, thousands lie down,
Thus thousands and tens look the same,
Ten thousands and hundreds look alike.

surface, Chinese reckoners needed no materials other than their bundle of counting-rods.

During the Han period, the counting-rods were round bamboo sticks about 2·5 mm in diameter and 140 mm long, tied together in a hexagonal bundle which could be conveniently carried by hand. But from the sixth century AD they became shorter and rectangular in shape. Besides bamboo, counting-rods were also made from wood, cast iron, jade or ivory. Counting-rods of other shapes and sizes were also used in Korea and Japan, which came within the Chinese sphere of influence in mathematics.

Operations with rod numerals

Elementary operations with counting-rods were carried out as one would on an abacus. If numbers were to be added or subtracted, then the rods were repositioned column by column. For example, the addition of 8 and 7 would be shown thus:

where the first and second rows represent 8 and 7, respectively, while the third row shows the sum, 15. The subtraction of 6 from 12 would proceed as shown on the next page:

where 12 and 6 are first laid down in the first and second rows. The horizontal rod in the first row is then converted into an additional set of ten vertical rods before subtraction. The third row shows the answer, 6.

For multiplication the top row is the multiplier, the middle row is left blank for the intermediate steps to be entered as multiplication proceeds, and the bottom row shows the multiplicand. As an illustration we take the multiplication of 387 by 147, using modern numerals for clarity. The product of the two numbers is found via the five sets of arrangements of counting-rods shown in Figure 6.2. As a first step, the rods are arranged as in (a). In (b), 441 is obtained by multiplying 147 by 3. In the next step, shown in (c), 147 is multiplied by 8 and the result added to 4410 to give 5586. Then, in (d), 147 is multiplied by 7 and the result added to 55 860 to give the final product, 56 889. The multiplication process is complete in (e). At each stage the rods are rearranged so that the numerals fall in the appropriate columns, 147 being moved one place to the right at each step. Both cancelling and correcting mistakes are easier to do with counting-rods than with paper and pencil. With sufficient practice, the rods could be manipulated with such speed that a writer in the eleventh century comments on the rods 'flying so quickly that the eye loses track of their movements'.

The method of division begins with the divisor (*fa*) in the bottom row and the dividend (*shih*) in the middle row, with the top row left blank for the

Figure 6.2 Multiplication with counting-rods.

Figure 6.3 Division with counting-rods.

	(a)						(b)						(c)						(d)			
Shang								3						3	8					3	8	7
Shih	5	6	8	8	9		5	6	8	8	9		1	2	7	8	9		1	0	2	9
Fa			1	4	7		1	4	7					1	4	7				1	4	7

quotient (*shang*). As an illustration, we shall divide 56 889 by 147; the process takes four steps, as shown in Figure 6.3. In (a) the rods are set up for division. The number to be divided, 56 889, is inserted in the middle row and the divisor, 147, in the bottom row. The divisor is moved left to the point where it is exactly below 568. Dividing 568 by 147 gives 3, which is inserted in the top row in (b). The divisor 147 is multiplied by the quotient and subtracted from 568 to give 127, which is placed in the middle row in (c) followed by the last two digits, 8 and 9, of 56 889. The divisor is moved one place to the right and divided into the number above it, 1278. The result, 8, is inserted in the top row in (c). Then 147 is multiplied by 8 and the result subtracted from 1278, giving 102, which is inserted in the middle row followed by the last digit, 9, of 56 889, as in (d). The division of 1029 by 147 gives 7, with no remainder, which is inserted in the top row in (d). The answer is 387.

The representation of zero in this system poses no problem: a space is simply left blank. But there was a difficulty, as with the Babylonian notation, when the rod numerals were written down. A circular sign for zero makes its first appearance in Chin Chiu Shao's work in 1247, probably an influence from India. However, the blank space standing for what we now call zero in Chinese numerals was conceptually different from the blank space in Babylonian numerals. In the Chinese system the blank space is itself a numeral, whereas in the Babylonian system it represents the absence of a digit – a place-holder. This is immediately evident if we consider what is implied by the absence of a blank space as a right-hand terminal indicator in each of the two systems. In the Babylonian sexagesimal place-value notation, the symbols $\top$ $\langle$ could have stood for seventy, or three thousand six hundred and sixty, or some other number. Such ambiguities did not occur with the Chinese rod numerals since the reckoner would always have been aware of the positional values of the digits in the numbers operated with. Also, the ingenious device of alternating the orientation of the rods in successive place values meant that it was easy to check that the numerals were correctly positioned relative to one another. Thus

$$102: \quad | \quad \quad || \quad \quad 12: \quad \text{---} \quad ||$$

An important advantage of a place-value number system, as we saw earlier, is that the fractional and integral parts of a mixed number can be represented as economically as possible. In the rod numeral system the integral and fractional parts are separated by a line. Thus, the number 48·125 is represented as

The counting-rods were not merely a calculating device: they eventually became a vehicle for expressing certain mathematical concepts. Apart from its notational facility, the rod numeral system was helpful in suggesting new approaches to algebraic problems, and also ways of operating with negative numbers. It may even be argued that geometric algebra, which began with the Babylonians and was extended by the Greeks, was given an arithmetic dimension by the Chinese counting-rods. It was merely a matter of time before the positions of the counting-rods came to stand for algebraic symbols, and operations with the rods for algebraic operations. Let us now look at how the rods were used as a representational device (their use for algebraic operations will be examined in a later section).

To represent a system of equations, the counting-rods were arranged in such a way that one column was assigned to each equation of the system and one row to the coefficients of each unknown in the equations. The elements of the last row consisted of the entries on the right-hand side of each equation. Red rods were used to represent positive (*cheng*) coefficients and black rods for negative (*fu*) coefficients.

Figure 6.4 Representing simultaneous equations with counting-rods.

Figure 6.4 shows the representation of the following system of equations in three unknowns:

$$2x - 3y + 8z = 32$$
$$-6x - 2y - z = 62$$
$$3x + 21y - 3z = 0$$

It has been argued that because equations were represented in this form, it was only a matter of time before matrix methods of solving systems of linear simultaneous equations or numerical methods of solving higher-order equations would logically suggest themselves. Whether or not this happened, it is clear that these methods became an integral part of Chinese mathematics, probably as early as the third century BC (they made their first appearance in the *Chiu Chang*), and remained unique to Chinese mathematics until the eighteenth century AD. We shall examine these methods more thoroughly later. The early development in China of an algebra of negative numbers (*cheng fu shu*, or 'positive and negative' operations) is another by-product of the rod numeral system. With later (written) rod numerals, a negative number was shown by placing a diagonal rod through its last column, for example

$$— \text{\Large\textcancel{||||}} \quad : -12$$

It is easy to be deceived by the simplicity of our present-day decimal number system into overlooking the advantages of other systems. We have seen how the Babylonians had developed a sexagesimal place-value system whose lack of a symbol for zero hindered both number representation and computation. The distinctive feature of the decimal system, which it shares with the Chinese rod numeral system, is an economy both in the number of symbols used and in the space occupied by a written number. In the Chinese system, each of the first nine numbers is represented by no more than five rods. By letting a single rod represent the number five, the numbers from six to nine became much simpler to represent than in, say, the Egyptian system, and arithmetical operations were easier to carry out. Alternating the orientation of the rods provided a clear indication of the place positions of different digits of a number, and all that was needed to carry out computations was a bundle of rods, together with a flat surface to place them on.

However, as well as these virtues the rod system had its shortcomings. In long and complicated calculations the rods took up much space. Errors might be made when rods were moved rapidly, and there was no possibility of detecting errors once a sequence of manipulations was complete. It may be that the remarkable success of these devices for calculation in China inhibited the development of alternative mechanical devices such as the abacus, which did not become widespread until the time of the Ming dynasty, and delayed the adoption of the decimal number system until recent times.

CHINESE MAGIC SQUARES (AND OTHER DESIGNS)

A magic square is a square array of numbers arranged in such a way that the numbers along any row, column or principal diagonal add up to the same total. In most magic squares of n rows and n columns the n^2 'cells' are occupied by the natural numbers from 1 to n^2. For example, a magic square of four rows and four columns (i.e. of order 4) would contain all the integers from 1 to 16.

Magic squares have some interesting mathematical properties. If s is the constant sum of the numbers in each row, column or principal diagonal, and S is the grand total of the numbers in the n^2 cells, then

$$S = \tfrac{1}{2}n^2(n^2 + 1) \quad \text{and} \quad s = S/n$$

If n is odd, the number of the central cell is given by S/n^2, which is also the arithmatic mean of the series $1 + 2 + \ldots + n^2$. (For magic squares of even order, for which there is no single central cell, S/n^2 is not a whole number.) This is a key number for all odd-order magic squares, since from this number and the value of n it is possible to work out the partial sum s and total sum S. For example, if $n = 3$ and the middle number is 5,

$$S = 9 \times 5 = 45 \quad \text{and} \quad s = 45/3 = 15$$

A magic square of odd order in which every pair of numbers on opposite sides of the central cell add up to twice the middle number, is known as a regular magic square. For example, in Figure 6.5 in each of the pairs (4, 6), (2, 8), (3, 7) and (1, 9), the two numbers lie on opposite sides of the middle number (5) and add up to 10, which is twice 5. Figure 6.5 represents the most famous of the regular Chinese magic squares, known as the *Lo shu*. It was seen as a symbol of the universe itself (but more of this later).

4	9	2
3	5	7
8	1	6

Figure 6.5 A regular magic square of order three.

Magic squares are only of marginal interest today, forming part of a peripheral area known as 'recreational mathematics'. Yet until four hundred years ago, in almost all mathematical traditions, magic squares engaged the interests of notable mathematicians as a challenging object of study. Their attractions were heightened not only by their aesthetic appeal, but also by their association with divination and the occult; they were engraved on ornaments worn as talismans.

Magic squares in China

In China, where the earliest recorded magic squares have been found, there was a long-established fascination with number patterns and the associated combinatorial analysis. The Chinese shared with the Greeks an interest in numerology and number mysticism (for example odd numbers were thought to be lucky, even numbers unlucky), but there is nothing in Chinese mathematics that resembles the Pythagorean fascination for figurate numbers (i.e. triangular, square, pentagonal numbers) or for types of numbers such as perfect or amicable numbers.* Neither is there anything in Greek mathematics that suggests even the slightest interest in magic squares. These differences may exemplify two different views of numbers: the Chinese having a marked preference for the concrete, the Greeks being more interested in the abstract and the metaphysical.

The first record of a magic square in China goes back to the time of the semi-mythical Emperor Yu, who was reputed to have lived in the third millennium BC. There is a legend that Yu acquired two diagrams, the first one (*Ho thu*, meaning the River Chart) from a magical dragon-horse which rose from the waters of the Huang Ho (Yellow River), and the second (*Lo shu*, meaning Lo River Writing) copied from the design on the back of a sacred turtle found in the Lo, a tributary of the Huang Ho. Figure 6.6 shows these gifts: *Ho thu*, a cruciform array of numbers from 1 to 10, and *Lo shu*, a regular magic square of order 3. (The number 10 is shown in the *Ho thu* as a square of black beads surrounding the 5 central white beads.) The *Ho thu* is so arranged that, disregarding the central 5 and 10, both the odd and even sequences of numbers add up to 20. The *Lo shu*, which we discussed earlier, is a magic square in which the figures in any diagonal, row or column add up to 15, a remarkable balance

---

*Figurate numbers can be represented by a geometrical array of dots; the first four triangular numbers, for example, are 1, 3, 6 and 10. The meanings of perfect and amicable numbers will be made clear in Chapter 10, on Arab mathematics.

Figure 6.6 (a) *Ho thu* and (b) *Lo shu* (Needham, 1959, p. 57).

(a)   (b)

being maintained between the odd (white beads) and even (black beads) numbers around the central number, which is again 5. Both diagrams represent an important principle of Chinese philosophy – balancing the two complementary forces of yin (female) and yang (male) in nature, represented here by odd and even numbers respectively.

There can be no doubt about the antiquity of this story, and aspects of it suggest that it cannot have originated any later than the second century BC. Certain passages in the ever-popular manual of divination, *I Ching* (The Book of Changes), written at about that time, emphasize not only the magical/divinatory nature of these diagrams, but also their numerical properties. Gradually an extensive folklore grew up around the magical properties of the *Lo shu*, and more generally of magic squares of order 3. It came to be described as the nine rooms or halls of the cosmic temple, the Ming Tang, and later writers would refer to the construction of this square as the 'nine halls calculation (*chiu kung suan*). The belief in the magical powers of this square spread into neighbouring areas, among the Tibetans, Koreans and Mongolians, who depicted it as an arrangement of black and white knots or beads on short lengths of cord, as shown in Figure 6.6(b).

Given the early start the ancient Chinese had in the development of magic squares, one would have expected them to progress to squares of order higher than 3. However, there is no evidence that they did, though brief quotations from existing texts show that commentaries had been written on the *Lo shu*, especially during the periods of disorder and unrest that occurred between the fourth and sixth centuries AD. It can only be assumed that this great interest in magic and divination was but a sign of the times — a desperate search for a better tomorrow.

The long hiatus ended with the emergence of Yang Hui, who in 1275 published his *Hsu Ku Chai Chi Suan Fa* (Continuation of Ancient Mathematical Methods for Elucidating the Strange Properties of Numbers). In the preface to his book, Yang Hui pointed out that he was merely passing on the works of earlier scholars and would make no claim to originality. Some of the magic squares he constructed were very complicated, and instructions for building them were either absent or cryptic to the point of obscurity. Let us consider briefly the methods he outlined for constructing magic squares of orders 3, 4 and 5.

Construction of the *Lo shu*

Yang Hui's instructions may be expressed as follows:

1 Arrange the numbers 1 to 3, 4 to 6 and 7 to 9 (as shown in Figure 6.7(a)) so that they slant downwards to the right.
2 Replace 1 on the 'head' with 9 on the 'shoe', and vice versa.
3 Interchange 7 and 3, other numbers remaining in their old positions. (The new positions are shown in Figure 6.7(b).)
4 Lower 9 to fill the slot between 4 and 2, and raise 1 to fill the slot between 8 and 2. (Figure 6.7(c) shows the *Lo shu* that results, after similarly moving 7 and 3 inwards.)

Figure 6.7 Construction of the *Lo shu*.

Construction of a magic square of order 4

To construct a magic square of order 4, place the numbers 1 to 16 in an array of four columns and four rows, as shown in Figure 6.8(a). Then interchange the numbers at the corners of the outer square (16 and 1, 4 and 13), to give the arrangement in Figure 6.8(b). Finally, interchange the numbers at the corners of the inner square (6 and 11, 7 and 10). This will produce a magic square in which all the columns, rows and diagonals add up to 34, as shown in Figure 6.8(c).

As Yang Hui points out, different variants of this 'method of interchange' can produce different magic squares. For example, Figure 6.9 was obtained by arranging the sixteen numbers in four columns beginning at the top left-hand corner. Figure 6.10 was obtained by first listing the sixteen numbers beginning at the top right-hand corner and going down the four columns from right to left. This arrangement for a magic square was referred to as the yin (female) square. 'The diagram of the sixteen flowers', shown in Figure 6.11, was constructed from an initial arrangement of the numbers 1 to 16 in four rows running from right to left.

1	2	3	4
5	6	7	8
9	10	11	12
13	14	15	16

(a)

16	2	3	13
5	6	7	8
9	10	11	12
4	14	15	1

(b)

16	2	3	13
5	11	10	8
9	7	6	12
4	14	15	1

(c)

Figure 6.8

16	5	9	4
2	11	7	14
3	10	6	15
13	8	12	1

Figure 6.9

4	9	5	16
14	7	11	2
15	6	10	3
1	12	8	13

Figure 6.10

2	16	13	3
11	5	8	10
7	9	12	6
14	4	1	15

Figure 6.11

Figure 6.12

21	16	11	6	1
22	17	12	7	2
23	18	13	8	3
24	19	14	9	4
25	20	15	10	5

(a)

```
        7
   14      18
17      13      9
   8       12
       19
```

(b)

1	23	16	4	21
15	14	7	18	11
24	17	13	9	2
20	8	19	12	6
5	3	10	22	25

(c)

Construction of a magic square of order 5

Yang Hui provides no explanations of how he constructed magic squares of order 5 onwards. However, by examining the two squares of order 5 that he included in his book we can get an idea of this method. The first (Figure 6.12(c)) is a magic square within a magic square, the inner one of order 3 having a constant of 39. The central number is 13, which is also the middle number of the sequence 1 to 25. The numbers in this inner square are (in ascending order) 7, 8, 9; 12, 13, 14; and 17, 18, 19; this suggests that to construct the complete magic square, we should proceed as follows.

First write the numbers from 1 to 25 in columns starting from the top right (Figure 6.12(a)). Then proceed to make the inner square into a magic square of order 3 by following the method of forming a diamond-shaped pattern, as for the construction of the *Lo shu*, discussed earlier, interchanging the numbers 8 and 18 and rearranging the sequence 12, 13, 14 to 14, 13, 12, to produce Figure 6.12(b). Around the inner magic square arrange the rest of the numbers in complementary pairs such as (1, 25), (2, 24) and (3, 23), such that the rows, columns and diagonals of the outer square sum to the constant of the outer square, 65 (Figure 6.12(c)). This arrangement of the numbers around the inner magic square follows no particular pattern, and is done by trial and error.

Yang Hui constructs magic squares up to order 10, although the squares of order 10 are 'incomplete'. Both the ingenuity and 'number sense' represented by these constructions are remarkable, and provide yet further evidence of the Chinese knack in computation, surely a product of their facility with rod numeral operations. The other striking feature of their work on magic squares is the crucial importance of the *Lo shu* in all their constructions.

Figure 6.13

46	8	16	20	29	7	49
3	40	12	14	18	41	47
44	37	33	23	19	13	6
28	15	11	25	39	35	22
5	24	31	27	17	26	45
48	9	38	36	32	10	2
1	42	34	30	21	43	4

Figure 6.13 shows one of Yang Hui's magic squares of order 7. It encompasses three magic squares, all consistent with the *Lo shu* principles of associated pairs and the yin–yang balance, but Yang Hui does not explain how he derived it. The name given to this magic square is *yen shu tu*, which means the 'diagram of the abundant number'. This must be a reference to 50, which is twice the central number 25 and also the sum of each associated pair: $46 + 4 = 50$, $3 + 47 = 50$, $14 + 36 = 50$, and so on.

Yang Hui also provides six magic circles of varying complexity. The simplest of them, shown in Figure 6.14, consists of a total of seven circles arranged in such a manner that each has four numbers on its circumference, with one other number at the centre of the diagram. Each of the numbers on the central circle lies on one of the four circles that touch it. Thus, associated with each of the seven circles is a group of five numbers, and each of these groups adds up to 65. If we compare this magic circle with the pattern of numbers in one of Yang Hui's magic squares of order 5 (Figure 6.12(c)), we can see that there is a correspondence between some circles and rows.

Yang Hui's work continued to arouse interest among later mathematicians. But apart from Chang Chao (*c.* 1650), who produced the first complete magic square of order 10, and Pao Chi Shou (*c.* 1880), who constructed three-dimensional magic cubes, spheres and tetrahedrons, there were hardly any

Figure 6.14 Magic circles.

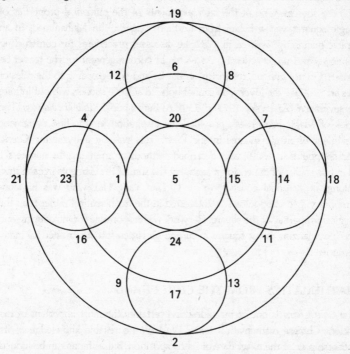

innovations after Yang Hui. Ever more elaborate magic circles continued to be constructed, though the decline in the hold of the *Lo shu* principle may have contributed to the emergence of incomplete designs. But China's influence spread abroad to produce interesting developments in Japan and India. In Japan, Isomura's (1660) interest in magic shapes was directly derived from his work in geometry; and Seki Kowa (1683), who devoted one of his seven books to the theory of magic squares and magic circles, used his algebraic talents to produce one of the first scientific treatises on the subject. These developments in Japan are well covered by Mikami (1913).

West of China, the subject of magic squares was first discussed by the Arabs, towards the end of the ninth century AD. Possible influences from China through trade links cannot be ruled out. In the works of Thabit ibn Qurra (*c.* 850), al-Ghazzali (*c.* 1075) and al-Buni (*c.* 1225), and also of the medieval Indian mathematician Narayana Pandit (*c.* 1350), one finds more or less the same ingredients of occult, numerology and combinatorial analysis as in the Chinese sources. One of the later exponents of the numerical properties of magic squares was a Fulani from northern Nigeria, ibn Muhammad. In an Arabic manuscript written in 1732, he discusses procedures for constructing magic squares up to order 11. As words of encouragement for the reader he writes: 'Do not give up, for that is ignorance and not according to the rules of this art . . . Like the lover, you cannot hope to achieve success without infinite perseverance' (Zaslavsky 1973, p. 139). Al-Buni's book, *Kitab al-Khawass* (The Book of Magic Properties), provided the inspiration for the first systematic treatment of magic squares in the West. This was by a Byzantine Greek, Moschopoulos (c. 1300), who described methods of arranging the numbers 1 to n^2 in a square of dimension n such that the sum of the elements in each row, column or diagonal equals $\frac{1}{2}n(n^2 + 1)$. Like Yang Hui, who was his contemporary, Moschopoulos was interested in the mathematical rather than the magical properties of the square. His work was responsible for introducing and then popularizing magic squares in Europe – an interest that remained for many centuries.

MATHEMATICS FROM THE *CHIU CHANG*

The *Chiu Chang* is one of the oldest and certainly the most important of the ancient Chinese mathematical texts. We have no reliable knowledge of its authorship or of the exact date of its composition, but as far as can be judged it is a product of the late Chin or early Han dynasty, which places it near the beginning of the first century AD. It presents a detailed summary of contemporary Chinese mathematical knowledge, and attracted in subsequent generations a line of distinguished commentators, including Liu Hui (third century) and Yang Hui (thirteenth century), who elaborated and extended its contents.

The *Chiu Chang* consists of nine sections, or chapters, with a total of two hundred and forty-six problems. Each chapter deals with a topic in mathematics relevant to the Chinese society of the time. Information is provided through the statement of a specific problem and the answer, followed by the rule for solution, which is often terse and occasionally obscure – hence the invaluable role of the later commentators. While there is nothing in the way of algebraic notation or proofs as we understand them today, an examination of the general context in which the problems were solved firmly places the book within an algebraic/arithmetic tradition similar to that of the Babylonian mathematics we examined in Chapter 4.

Field measurement and operation with fractions

The first chapter is entitled *Fang thien* (Land Surveying). Its central theme is the calculation of areas of fields (*thien*) of different shapes, and the basic unit of measurement is the *fang*, 'square unit'. Correct rules are given for finding the areas of rectangles, triangles, trapeziums and circles (π having an implicit value of 3). Liu Hui was at pains to point out in his commentary that 3 is the ratio, not of the circumference, but of the perimeter of an inscribed regular hexagon to the diameter, and that the more sides the inscribed regular polygon had, the closer its perimeter would approach the circumference of the circle. However, an approximate value of 3 was sufficient for most practical purposes for which such calculations were required.

The chapter also contains a discussion of methods for adding, subtracting and multiplying fractions. And here appears a rule for simplifying which is identical to that of the 'repeated subtraction' algorithm found in the work of the Hellenistic mathematician Nicomachus of Gerasa, who lived around the first century AD. Consider a reducible fraction of the form m/n. The rule given in the text is:

> If the two numbers [m and n in our notation] can be halved, then halve them. Otherwise set down the denominator below the numerator, and subtract the smaller number from the greater number. Continue this process until the common divisor, *teng*, is obtained. Simplify the original fraction by dividing both numbers by *teng*.

An example from the text illustrates the procedure.

EXAMPLE 6.1   *Simplify the fraction 49/91.*
Solution   Following the text, lay out the solution as follows:

49	49	7	7	7	7	7	7
91	42	42	35	28	21	14	7

So the common divisor (or *teng*) is 7, and the simplified fraction is 7/13.

The *Chiu Chang* contains rules for adding and subtracting fractions identical to the ones we would use today if we were operating without a lowest common multiple – multiply the numerator of each fraction by the denominator of the other, and add or subtract the product before dividing the result by the product of the denominators; in modern notation,

$$\frac{a}{b} + \frac{c}{d} = \frac{ad + cb}{bd}$$

If necessary, the resulting fraction can be simplified by using the algorithm explained above. Multiplication would proceed in the same way as today, but without cancellations. The algorithm may again be used to simplify the fraction that results.

Proportions and rule of false position

The second chapter, entitled *Su mi* (Millet and Rice), deals with questions of simple percentages and proportions relating to these commodities. The third chapter, 'Distributions by proportions' (*Cui fen*), is concerned with the distribution of property and money according to prescribed rules which lead, in some cases, to arithmetic and geometric progressions. Solutions often use the 'rule of three' for determining proportions. This rule, according to our present knowledge, was first applied in China. Here is an example from the text:

EXAMPLE 6.2 *Two and one half piculs [a measure of weight carried by a man on his back, approximately 65 kg] of rice are purchased for 3/7 of a taiel of silver. How many [piculs of rice] can be bought for 9 taiels?*
Solution   The suggested solution, expressed in modern terms, is to let x be number of *piculs* of rice bought for 9 *taiels* of silver; applying the 'rule of three' then gives

$$\frac{2\frac{1}{2}}{3/7} = \frac{x}{9}$$

$$x = 52\frac{1}{2} \ piculs$$

Our next example illustrates how the Chinese tackled simple problems involving series by using the rule of false position (or assumption), which we came across in the section on Egyptian algebra in Chapter 3.

EXAMPLE 6.3 *A weaver, improving her skills daily, continues to double her previous day's output. In five days she produces five chih of cloth [1 chih is about 23 cm]. How much did she produce each day?*

Solution   A modern solution would start by letting the output of the first day be x, so that the output on successive days is $2x$, $4x$, $8x$ and $16x$. Then

$$x + 2x + 4x + 8x + 16x = 5$$
$$31x = 5$$
$$x = 5/31$$

Successive doubling will give the output for the next four days.

The argument of false position would proceed as follows: If the total output is 1, then the weaver would have produced only 1/31 of it on the first day. But since the total output is 5, the output on the first day would be 5/31. The outputs of successive days until the fifth would therefore be 10/31, 20/31, 40/31 and 80/31, which when added to 5/31 give 155/31, or 5 *chih*.

Extraction of square and cube roots

The fourth chapter, *Shao kuang*, contains twenty-four problems on land mensuration (*shao* means 'how much', and *kuang* means 'width'). An important objective was to parcel out squares of land, given the area and one of the sides. This chapter is notable for the first occurrence of an important topic in the development of Chinese mathematics – how to find square and cube roots. Although the original text is very vague, later commentators on the *Chiu Chang*, notably Liu Hui and Yang Hui, have no doubts about the geometric basis of the method as used for both square and cube roots. To illustrate the calculation of square roots, here is a problem from the chapter:

EXAMPLE 6.4   *There is a [square] field of area 71 824 [square] pu [or paces]. What is the side of the square?* Answer: *268 pu.*
Solution   In the text only the answer is given. Fortunately, a detailed description of how the problem was solved is given in a fifteenth-century encyclopedia, *Yung Lo Ta Tien*, reproduced from Yang Hui's commentary on the *Chiu Chang*. (The Encyclopedia of Yung Lo's Reign, as its title translates, originally consisted of 11 095 volumes, of which only about 370 survive. It covered almost every field of human knowledge, and was compiled by over three thousand scholars under the supervision of Hsieh Chin.) Here we give both the algorithmic and geometric approaches so that the correspondence between the two can be easily established.

159

Figure 6.15 illustrates the steps of the algorithm for finding the square root of 71 824. On the left-hand side of each diagram is the algebraic rationale for the numerical calculations. In the diagrams N is a number whose square root is a three-digit integer, and α, β and γ are the digits standing in the 'hundreds', 'tens' and 'units' places, respectively. Thus, if the square root of N is the three-digit number abc, then $\alpha = 100a$, $\beta = 10b$ and $\gamma = c$. Therefore

$$N = (100a + 10b + c)^2 = (\alpha + \beta + \gamma)^2$$
$$= \alpha^2 + (2\alpha + \beta)\beta + [2(\alpha + \beta) + \gamma]\gamma \qquad (6.1)$$

It is a simple matter to extend this formula to numbers with more than three digits by expanding $(\alpha + \beta + \gamma + \delta + \ldots)^2$.

The Chinese method uses the above relationship but reverses the procedure and hence the ensuing calculations. The procedure is begun by finding an appropriate value for α by 'inspection'. It is, for example, easily deduced that $\alpha = 200$ (i.e. $\alpha = 100a$, where $a = 2$) if we are seeking the square root of $N = 71\,824$. The procedure continues with a calculation of α^2, and this quantity is subtracted from N. We next estimate β ($= 10b$), the second place of the square root, and form $(2\alpha + \beta)\beta$. We can now work out

$$N - \alpha^2 - (2\alpha + \beta)\beta = N - (\alpha + \beta)^2 \qquad (6.2)$$

and the procedure continues along similar lines until the third component on the right-hand side of equation (6.1) is calculated. If N is a perfect square, the final subtraction of this component from equation (6.2) would leave a remainder of 0.

How did the Chinese apply this method in calculating the square root of 71 824? They began by laying out counting-rods in four rows, as shown in Figure 6.15(a). The top row (*shang*) shows the result obtained at each stage of the rod operations. The second row (*shih*) contains the number on which further

Result (shang)						
Given number (shih)	N	7	1	8	2	4
Square element (fang fa)						
Carrying rod (chieh suan)	1					1

Figure 6.15(a)

Result (*shang*)	α			2		
Given number (*shih*)	$N - \alpha^2$	3	1	8	2	4
Square element (*fang fa*)	α	2				
Carrying rod (*chieh suan*)	1	1				

Figure 6.15(b)

operations are carried out. The third row, known as the 'square element' (or *fang fa*), shows the adjustments made to the element in the previous row in the process of extracting the square root. The final row has two different interpretations, depending on the context. In this context it is called a 'carrying rod' row and used to fix the positions of the digits as calculation proceeded.

This method of extracting square roots was eventually extended to the solution of quadratic equations. Indeed, a clear connection was established between the extraction of roots of any degree and the solution of equations of the same degree – at the time a unique feature of Chinese mathematics. It was eventually adopted with interesting variations by both the Koreans and the Japanese. The interested reader may wish to consult Smith and Mikami (1914) for further details.

The procedure used by the Chinese may be summarized in the following steps:

Step 1   Lay out the counting-rods (shown as modern decimal numbers for clarity) as shown in Figure 6.15(a). This is the initial configuration before calculations begin. Empty cells represent zero.

Step 2   Find the value of the 'hundreds' component of the square root (i.e. the value $\alpha = 100a$). In this example $\alpha = 200$ is entered in the first, 'result' row. Move the 'carrying rod' in the fourth row to the 'tens of thousands' position. Multiply the value of the 'carrying rod' by $a = 2$, and place the result, 2, in the 'square element' row (*fang fa*). Multiply the 'square element' row by $a = 2$ and subtract the result from the 'given number': $71\,824 - 2(20\,000) = 31\,824$. The new entries are shown in Figure 6.15(b).

Step 3   Double the value α to get 2α (400) and enter it as the new 'square element' in the third row of Figure 6.15(c) after moving the entry one step

Result (*shang*)	$\alpha + \beta$			2	6	
Given number (*shih*)	$N - \alpha^2$	3	1	8	2	4
Square element (*fang fa*)	2α		4			
Carrying rod (*chieh suan*)	1			1		

Figure 6.15(c)

Result (*shang*)	$\alpha + \beta$		2	6	
Given number (*shih*)	$N - \alpha^2 - \beta(2\alpha + \beta)$	4	2	2	4
Square element (*fang fa*)	$2\alpha + \beta$	4	6		
Carrying rod (*chieh suan*)	1		1		

Figure 6.15(d)

backwards. Make the necessary adjustment to the 'carrying rod' by moving its entry two spaces forwards. The 'tens' value is then obtained by estimating that the square root of the 'square element' (4000) is approximately 60 ($\beta = 60$, $b = 6$). The resulting entries are shown in Figure 6.15(c).

Step 4 The product of $b = 6$ and the 'carrying rod', when added to the 'square element' in the third row of Figure 6.15(c), gives $2\alpha + \beta$. Then β is multiplied by the new 'square element' ($2\alpha + \beta$) and subtracted from the given number ($N - \alpha^2$). The new entries are shown in Figure 6.15(d).

By continuing this line of reasoning, passing through the stages shown in Figures 6.15(e) and (f), the digit in the 'units' position ($\gamma = c$) is found to be 8, and thus the square root of 71 824 is obtained as 268, as shown in the top row of Figure 6.15(g).

A correspondence is easily established between this algorithmic approach and the geometric approach shown in Figure 6.16. We begin by constructing a square with side (α) 200 *pu* (paces), and therefore of area $A = 40\,000$ square

Result (*shang*)	$\alpha + \beta$		2	6	
Given number (*shih*)	$N - (\alpha + \beta)^2$	4	2	2	4
Square element (*fang fa*)	$2(\alpha + \beta)$	5	2		
Carrying rod (*chieh suan*)	1		1		

Figure 6.15(e)

Result (*shang*)	$\alpha + \beta + \gamma$		2	6	8
Given number (*shih*)	$N - (\alpha + \beta)^2$	4	2	2	4
Square element (*fang fa*)	$2(\alpha + \beta)$		5	2	
Carrying rod (*chieh suan*)	1				1

Figure 6.15(f)

Result (*shang*)	$\alpha + \beta + \gamma$	2	6	8
Given number (*shih*)	$N - (\alpha + \beta)^2 - \gamma[2(\alpha + \beta) + \gamma]$			
Square element (*fang fa*)	$2(\alpha + \beta) + \gamma$	5	2	8
Carrying rod (*chieh suan*)	1			1

Figure 6.15(g)

pu. Two additional rectangular sections, each of dimensions 200 by 60 ($B = \alpha\beta$, $C = \alpha\beta$), have a combined area of 24 000. To complete a square figure, we add a smaller square, of side 60 and area 3600 ($D = \beta^2$). The area

163

Figure 6.16

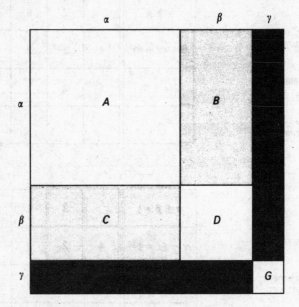

of the larger square, $A + B + C + D$, is $40\,000 + 24\,000 + 3600 = 67\,600$. This falls short of $71\,824$, the number whose square root we are seeking, by 4224. It is seen that this is equal to the area of two rectangular strips of dimensions 260 by 8 ($E = F = (\alpha + \beta)\gamma$), and a small square of side 8 ($G = \gamma^2$): $2(260 \times 8) + 8^2 = 4224$. Thus the geometric representation of the procedure for extracting the square root of $71\,824$ is equivalent to finding the length of the side of a square of area $71\,824$ square *pu*. Figure 6.16 indicates that the side has length $\alpha + \beta + \gamma = 200 + 60 + 8 = 268\,pu$.

The method of extracting cube roots is based on the following expansion:

$$\begin{aligned}
(\alpha + \beta + \gamma)^3 &= \alpha^3 + (3\alpha^2 + 3\alpha\beta + \beta^2)\beta + \{[(3\alpha^2 + 3\alpha\beta + \beta^2) \\
&\quad + (3\alpha\beta + 2\beta^2)] + [3(\alpha + \beta)\gamma + \gamma^2]\}\gamma \\
&= (\alpha + \beta)^3 + [3(\alpha + \beta)^2 + 3(\alpha + \beta)\gamma + \gamma^2]\gamma \quad (6.3)
\end{aligned}$$

The following example from the *Chiu Chang* illustrates the method.

EXAMPLE 6.5 *Find the cube root of 1 860 867 [or, solve the cubic equation $x^3 = 1\,860\,867$].*

Solution The solution algorithm is begun by laying out the counting-rods in five rows, as shown in Figure 6.17(a). The arrangement is similar to the one used for extracting square roots, except that the third and fourth rows, now known as the 'upper' and 'lower' elements, respectively, are for numbers obtained during the operations. For simplicity, we ignore operations with the 'carrying rod'.

Result (*shang*)							
Given number (*shih*)	N	1	8	6		8	6	7
Upper element (*shang fa*)							
Lower element (*chia fa*)							
Carrying rod (*chieh suan*)	1						1

Figure 6.17(a)

Result (*shang*)	α			1		
Given number (*shih*)	$N - \alpha^3$	8	6		8	6	7
Upper element (*shang fa*)	$3\alpha^2$		3			
Lower element (*chia fa*)	3α			3		
Carrying rod (*chieh suan*)	1					1

Figure 6.17(b)

Result (*shang*)	$\alpha + \beta$				1	2	
Given number (*shih*)	$N - \alpha^3 - [3\alpha^2 + (3\alpha + \beta)\beta]\beta$	1	3	2	8	6	7
Upper element (*shang fa*)	$3\alpha^2 + (3\alpha + \beta)\beta$		3	6	4		
Lower element (*chia fa*)	$3\alpha + \beta$				3	2	
Carrying rod (*chieh suan*)	1						1

Figure 6.17(c)

Result (*shang*)	$\alpha + \beta$				1	2	
Given number (*shih*)	$N - (\alpha + \beta)^3$	1	3	2	8	6	7
Upper element (*shang fa*)	$3(\alpha + \beta)^2$		4	3	2		
Lower element (*chia fa*)	$3(\alpha + \beta)$				3	6	
Carrying rod (*chieh suan*)	1						1

Figure 6.17(d)

It is easily deduced from the problem that the cube root of N is a three-digit number *abc*, where the 'hundreds' value is $\alpha = 100a$, the 'tens' value is $\beta = 10b$ and the 'units' value is $\gamma = c$. Thus the cube root of N is $\alpha + \beta + \gamma$. We start by reversing the calculation implied in the identity (6.3), and arrange the rods as in Figure 6.17(b). The procedure continues with the calculation of $\beta = 20$, giving the configuration shown in Figures 6.17(c) and (d). Then we identify $\gamma = 3$ and obtain the final rod arrangement shown in Figure 6.17(e). Here the 'given number' is zero, so the calculation is complete: the cube root of 1 860 867 is 123.

Result (*shang*)	$\alpha + \beta + \gamma$			1	2	3
Given number (*shih*)	$N - (\alpha + \beta)^3 -$ $[3(\alpha + \beta)^2 + 3(\alpha + \beta)\gamma + \gamma^2]\gamma$					
Upper element (*shang fa*)	$3(\alpha + \beta)^2 + [3(\alpha + \beta) + \gamma]\gamma$	4	4	3	4	
Lower element (*chia fa*)	$3(\alpha + \beta)$			3	6	
Carrying rod (*chieh suan*)	1					1

Figure 6.17(e)

The difficulties of presenting all the stages of the solution here should not lead us to suppose that the Chinese method was particularly laborious or time-consuming. Indeed, contemporary records suggest that solving problems of this nature was a matter of a few minutes, and made use of mechanical routines (and also perhaps an auxiliary set of counting-rods where necessary) which were second nature to the reckoners.

As with the procedure for square roots, it is possible to provide a geometric interpretation of this method of extracting cube roots. The resulting geometric figure is three dimensional, and its analysis, although more complex than that for the plane region shown in Fig. 6.16, is along similar lines (it is omitted here).

Engineering mathematics

The fifth chapter of the *Chiu Chang*, entitled *Shang kung* (a reference text for engineers), contains rules for computing the volumes of three-dimensional shapes that would be familiar to the builders of castles, houses and canals. They include the correct formulae for several solids, which were referred to by the names of common objects (see Figure 6.18). The formula for the truncated triangular prism was later to appear in Adrien Marie Legendre's *Éléments de géométrie* (1794), and in the West he is usually credited with its discovery.

Particularly noteworthy is the rule for the volume of a tetrahedron (or *bienao*) whose opposite edges, of lengths w and g, share a common perpendicular, a third edge of length h. This problem is seen in modern mathematics as forming part of the theory relating to volumes of polyhedra. The *Chiu Chang*'s rule, in modern notation, is

$$V = \tfrac{1}{6}wgh$$

In his commentary on the *Chiu Chang*, Liu Hui explains why the volume so measured is half that of a rectangular pyramid (a *yangma*) with base wg and height h. A cuboid of sides w, g and h can, he says, be divided into three congruent *yangmas* so that the volume of each of them is one-third that of the cube ($\tfrac{1}{3}wgh$). He proceeds to show that a *yangma* and a *bienao* can be slotted

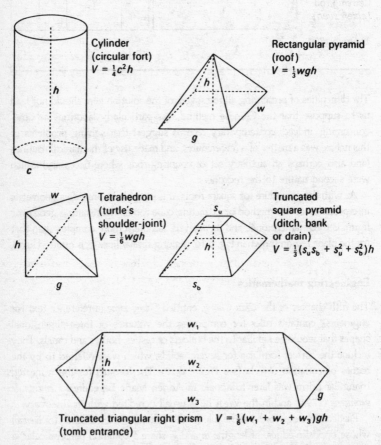

Cylinder
(circular fort)
$V = \tfrac{1}{4}c^2 h$

Rectangular pyramid
(roof)
$V = \tfrac{1}{3}wgh$

Tetrahedron
(turtle's
shoulder-joint)
$V = \tfrac{1}{6}wgh$

Truncated
square pyramid
(ditch, bank
or drain)
$V = \tfrac{1}{3}(s_u s_b + s_u^2 + s_b^2)h$

Truncated triangular right prism $V = \tfrac{1}{6}(w_1 + w_2 + w_3)gh$
(tomb entrance)

Figure 6.18 Formulae for the volume V of various solids from the *Chiu Chang*.

together to produce a right triangular prism (a *qiandu*) whose volume is $\frac{1}{2}wgh$. So the volume V of the *bienao* is

$$V = \tfrac{1}{2}wgh - \tfrac{1}{3}wgh = \tfrac{1}{6}wgh$$

Liu shows that this formula holds not only for a prism cut from a cube, but for any right triangular prism. Liu's principle may be expressed in modern terms as follows:

> If any rectangular parallelepiped is cut diagonally into two prisms, and the prisms are further cut into pyramids and tetrahedra, the ratio between the volumes of the pyramid and tetrahedron so produced is always $2:1$.

The method of 'proof' used by Liu is based on a principle which had wide applications in traditional Chinese geometry, including the famous *kou ku* (or Pythagorean) theorem, to be discussed in the next chapter. The essence of this principle, sometimes referred to as the 'out–in complementarity principle' in Chinese texts (or what we would call today 'dissection and reassembly'), follows from two common-sense assumptions:

1 both the area of a plane figure and the volume of a solid remain the same under rigid translation to another place, and
2 if a plane figure or solid is cut into several sections, the sum of the areas or volumes of the sections is equal to the area or volume of the original figure.

The reasoning behind this approach was very different from that behind Euclidean geometry, but the method was often just as effective, as we shall see when we come to look at how it was applied in obtaining a proof of the Pythagorean theorem. However, it is wrong to dismiss it as a trial-and-error method, where the rule is merely a result of experimenting with concrete structures: the rules are stated far too precisely for that. (We have discussed this point in relation to Egyptian geometry in Chapter 3.)

Fair taxes

The sixth chapter, *Chun shu* (literally, 'fair taxes'), contains a medley of problems – some, relating to the title of the chapter, are about the distribution of the tax burden among different sections of the population, while others are about the time required to transport taxes (paid in grain) from outlying towns to the capital. Among the latter are 'pursuit' problems, which were introduced into Europe by the Arabs and enjoyed considerable popularity between the twelfth and fifteenth centuries. Most of them have a hound chasing a hare, as in the next example from the chapter.

EXAMPLE 6.6   *A hare runs 50 pu [paces] ahead of a dog. The dog pursues the hare for 125 pu but the hare is still 30 pu ahead. In how many pu will the dog overtake the hare?*

Solution   The text gives the following working:

$$\frac{125 \times 30}{50 - 30} = 187\frac{1}{2} pu$$

Excess and deficiency

The seventh chapter is entitled *Ying buzu*, which may be translated as 'Too much and not enough'. The origin of the phrase may lie in the Chinese description of the phases of the Moon, from full (too much) to new (too little). There are twenty problems in this chapter and, as Yang Hui showed, most of them can be tackled by alternative methods which are less cumbersome than the ones proposed in the text. An illustration is Problem 2, which may be restated as follows:

EXAMPLE 6.7   *A group of people buy hens together. If each person gave 9 wen, there will be 11 wen of money left after the purchase. If, however, each person contributed only 6 wen, there would be a shortfall of 16 wen. How many persons are there in the group and what is the total cost of hens?*

Suggested solution

It is worth stating the rule for solution as given in the text (with minor modifications for clarity): Arrange the two types of contribution made by members of the group towards the purchase of the hens along the first row. The excess and deficiency that result are arranged as a row below the first row, which contains the members' contributions. Cross-multiply diagonally. Add the products together and label the sum as *shih*. Add the excess and deficiency and label the sum as *fa*. If a fraction occurs in either *shih* or *fa*, make them have the same denominator. Divide *shih* by the difference between the two contributions to get the total cost of hens. Divide *fa* by the difference between the contributions to get the number of persons in the group.

Expressed in algebraic terms, the application of the rule is simple: Let the two contributions be a and a', and let the excess and deficiency be b and b', respectively; then

$$\begin{pmatrix} a & a' \\ b & b' \end{pmatrix} = \begin{pmatrix} 9 & 6 \\ 11 & 16 \end{pmatrix}$$

$$\begin{pmatrix} ab' & a'b \\ b & b' \end{pmatrix} = \begin{pmatrix} 144 & 66 \\ 11 & 16 \end{pmatrix}$$

$$\begin{pmatrix} ab' + a'b \\ b + b' \end{pmatrix} = \begin{pmatrix} 210 \\ 27 \end{pmatrix}$$

Therefore, the total cost of the hens is

$$\frac{\text{shih}}{a - a'} = \frac{ab' + a'b}{a - a'} = \frac{210}{3} = 70$$

and the number of persons in the group is

$$\frac{\text{fa}}{a - a'} = \frac{b + b'}{a - a'} = \frac{27}{3} = 9$$

The above problem may be restated in terms of a system of equations in two unknowns x and y, where x is the number of persons and y is the cost:

$$ax - cy = b, \qquad a'x - c'y = -b' \qquad (6.4)$$

If $a = 9$, $a' = 6$, $b = 11$, $b' = 16$, $c = 1$ and $c' = 1$, equations (6.4) become

$$9x - y = 11, \qquad 6x - y = -16 \qquad (6.5)$$

Equations (6.4) and (6.5) may be expressed in matrix form as

$$\begin{pmatrix} a & -c \\ a' & -c' \end{pmatrix}\begin{pmatrix} x \\ y \end{pmatrix} = \begin{pmatrix} b \\ -b' \end{pmatrix}, \qquad \begin{pmatrix} 9 & -1 \\ 6 & -1 \end{pmatrix}\begin{pmatrix} x \\ y \end{pmatrix} = \begin{pmatrix} 11 \\ -16 \end{pmatrix}$$

Applying the rule attributed to the Swiss mathematician Gabriel Cramer (1750) for solving simultaneous equations by determinants gives

$$x = \frac{bc' + b'c}{ac' - a'c} \qquad y = \frac{ab' + a'b}{ac' - a'c}$$

so that

$$x = \frac{11 + 16}{9 - 6} = 9, \qquad y = \frac{144 + 66}{9 - 6} = 70$$

For $c = c' = 1$, the result is identical to the solution given in the *Chiu Chang*.

So, as early as the beginning of the Christian era a variant of Cramer's rule for solving two equations in two unknowns was known in China, though there is nothing in either this treatise or any of the subsequent commentaries that hints at an awareness of the rule for three equations in three unknowns, or of the general rule for p equations in p unknowns. However, we can detect in this method the concept of a determinant. Apparently this escaped the notice of Chinese mathematicians, but it was taken up by the Japanese mathematician Seki Kowa in 1683 – ten years before Leibniz, to whom historians of mathematics usually attribute the discovery of determinants.

There was another method for tackling problems involving 'excess and deficiency'. The 'rule of double false position' was particularly popular during the period when the lack of a suitable symbolic notation made the solution of even simple linear equations a difficult undertaking. This rule was brought to Europe by the Arabs, and is found in the works of the ninth-century Arab mathematician al-Khwarizmi under the name *hisab al-khataayn*. The name is quite revealing, for from the Arabic word *khataayn* came the word 'Cathay', referring to China.

The method is best explained in present-day notation. We want to find an unknown quantity x in a linear equation of the form

$$ax + b = 0$$

Let g_1 and g_2 be two preliminary guesses for the value of x, and let f_1 and f_2 be the errors arising from these guesses; then

$$ag_1 + b = f_1, \qquad ag_2 + b = f_2 \tag{6.6}$$

Hence

$$a(g_1 - g_2) = f_1 - f_2 \tag{6.7}$$

The first of equations (6.6) is multiplied by g_2 and the second by g_1; subtracting the two resulting equations one from the other gives

$$b(g_2 - g_1) = f_1 g_2 - f_2 g_1 \tag{6.8}$$

Equation (6.8) is divided by equation (6.7) to give, since $x = -b/a$,

$$x = \frac{f_1 g_2 - f_2 g_1}{f_1 - f_2} \tag{6.9}$$

This rule is illustrated by Problem 9 from the seventh chapter of the *Chiu Chang*. The 'excess and deficiency' is not explicitly stated, but has to be inferred from the question.

EXAMPLE 6.8   *A tub of full capacity 10 tou contains a certain quantity of coarse [i.e. husked] rice. Grains [i.e. unhusked rice] are added to fill up the tub. When the grains are husked it is found that the tub contains 7 tou of coarse rice altogether. Find the original amount of rice in the tub.*

(Assume that 1 *tou* of grains yields 6 *sheng* of coarse rice, where 1 *tou* is equal to 10 *sheng*.)

Suggested solution   If the original amount of rice in the tub is 2 *tou*, a shortage of 2 *sheng* occurs; if the original amount of rice is 3 *tou*, there is an excess of 2 *sheng*. Cross-multiply 2 *tou* by the surplus 2 *sheng*, and then 3 *tou* by the deficiency of 2 *sheng*, and add the two products to give 10 *tou*. Divide this sum (i.e. 10) by the sum of the surplus and deficiency (i.e. 4) to obtain the answer: 2 *tou* and 5 *sheng*.

This method gives the same answer as the rule of double false position: equation (6.9) is used, with $g_1 = 2$, $g_2 = 3$, $f_1 = -2$ and $f_2 = 2$, to give

$$x = \frac{f_1 g_2 - f_2 g_1}{f_1 - f_2} = \frac{-10}{-4} = 2\tfrac{1}{2} \text{ tou}$$

Solution of simultaneous equations

Chapter Eight is called *Fang cheng* (Method of Rectangular Arrays) and deals with the solution of simultaneous equations with up to five unknowns by placing them in a table and then operating on the columns in a way that is identical to the row transformations of the modern matrix method of solution. A notable feature of this method is that it is just as easy to use with negative as with positive numbers. The best way to explain it is by examples, and we begin with Problem 13, from Chapter Seven:

EXAMPLE 6.9   *5 large containers and 1 small container have a total capacity of 3 hu. 1 large container and 5 small containers have a capacity of 2 hu. Find the capacities of 1 large container and 1 small container. [1 hu = 10 tou].*

Solution   The method of tables starts by setting up the information contained in the problem in the form of a matrix:

$$\begin{array}{c} \text{Large containers} \\ \text{Small containers} \\ \text{Total capacity} \end{array} \begin{pmatrix} 1 & 5 \\ 5 & 1 \\ 2 & 3 \end{pmatrix}$$

Step 1

Multiply the first column by 5 and then subtract the second column from the result. Put this down as the first column of the next matrix:

$$\begin{array}{c} \text{Large containers} \\ \text{Small containers} \\ \text{Total capacity} \end{array} \begin{pmatrix} 0 & 5 \\ 24 & 1 \\ 7 & 3 \end{pmatrix}$$

Step 2

Multiply the second column by 24 and then subtract the first column from the result. Put this down as the second column of the next matrix:

$$\begin{array}{c} \text{Large containers} \\ \text{Small containers} \\ \text{Total capacity} \end{array} \begin{pmatrix} 0 & 120 \\ 24 & 0 \\ 7 & 65 \end{pmatrix}$$

Thus a small container has a capacity of $7/24\,hu$, and a large container has a capacity of $65/120$ or $13/24\,hu$.

There are in all 18 problems requiring the method of rectangular arrays to be used to solve systems of simultaneous linear equations with up to five unknowns. The text explains how counting-rods can be set up for column operations which would finally yield the solutions. Negative numbers that appear in the course of such operations do not pose any problem since they could be represented (as we saw earlier in this chapter) by black rods, red rods being used for positive numbers. Here is an example where there are three unknowns, and negative quantities appear in the course of solution:

EXAMPLE 6.10   *The yield of 2 sheaves of superior grain, 3 sheaves of medium grain and 4 sheaves of inferior grain is each less than 1 tou. But if one sheaf of medium grain is added to the superior grain or if one sheaf of inferior grain is added to the medium, or if 1 sheaf of superior grain is added to the inferior, then in each case the yield is exactly 1 tou. What is the yield of one sheaf of each grade of grain?*

Solution   In modern notation, the solution begins by letting x, y and z be the yields from 1 sheaf of superior, medium and inferior grain respectively. The question may then be posed as follows: given $2x \leqslant 1$, $3y \leqslant 1$ and $4z \leqslant 1$, solve

$$2x + y = 1$$
$$3y + z = 1$$
$$4z + x = 1$$

The information contained in the problem is first arranged on a counting-board so that it resembles a matrix:

$$\begin{pmatrix} 1 & 0 & 2 \\ 0 & 3 & 1 \\ 4 & 1 & 0 \\ 1 & 1 & 1 \end{pmatrix}$$

The column on the extreme right contains the coefficients and constant of the first equation, $2x + 1y + 0z = 1$, and similarly for the other columns. The first column is multiplied by 2, and column 3 subtracted from the result; the first column is replaced by the new column. We then have

$$\begin{pmatrix} 0 & 0 & 2 \\ -1 & 3 & 1 \\ 8 & 1 & 0 \\ 1 & 1 & 1 \end{pmatrix}$$

The first column is multiplied by 3, and the second column is added to the result:

$$\begin{pmatrix} 0 & 0 & 2 \\ 0 & 3 & 1 \\ 25 & 1 & 0 \\ 4 & 1 & 1 \end{pmatrix}$$

Hence:

25 sheaves of inferior grain cost 4 *tou*, so 1 sheaf of inferior grain costs 4/25 *tou*

3 sheaves of medium grain and 1 sheaf of inferior grain cost 1 *tou*, so 1 sheaf of medium grain costs 7/25 *tou*

2 sheaves of superior grain and 1 sheaf of inferior grain cost 1 *tou*, so 1 sheaf of superior grain costs 9/25 *tou*

This method of solving simultaneous linear equations is essentially the same as the one we use today, the development of which is attributed in the West to the famous German mathematician Karl Friedrich Gauss (1777–1855). But, over one thousand five hundred years before Gauss, Chinese mathematicians were using a variant of one of Gauss's methods. This innovation, together with the use of a special case of Cramer's rule for tackling certain problems in the previous chapter of the *Chiu Chang*, raises some interesting questions about the subsequent treatment of these subjects in Chinese mathematics. First, despite the early promise, there was no work on determinants until the Japanese mathematician Seki Kowa, working very much within the Chinese mathematical tradition, developed the concept of a determinant in his book *Kai Fukadai No Ho* (1683).*

Second, this way of tackling simultaneous equations is not found in any other mathematical tradition until the advent of modern mathematics. We are therefore driven to the conclusion that the method is a logical outcome of rod numeral computational techniques. However, the reliance on these very techniques inhibited the development of abstract algebra – a prerequisite for more advanced work on matrices and determinants.

The last chapter of the *Chiu Chiang* is called *Kou ku* (Right-angled Triangles). In the Chinese mathematical literature the shortest side of a right-angled triangle was called the *kou*, the longer side the *ku* and the hypotenuse the *hsian*. Twenty-four problems on right-angled triangles are presented in this chapter. The demands of accurate surveying and astronomical observation must have required an understanding and application of the Pythagorean theorem well before the *Chiu Chang* was written, and this would explain its presence in the earliest text, the *Chou Pei*. We shall examine this subject in Chapter 7.

We have devoted many pages to the *Chiu Chang*. The range of topics it covers is impressive, and indicates the level of sophistication reached by Chinese mathematicians at the beginning of the Christian era. It is one of the oldest mathematical texts in the world, with problems more varied and richer than in any Egyptian or Babylonian text. But although it was to have a powerful influence on the course of Chinese mathematics, with a number of notable mathematicians writing commentaries on it, by becoming a classic it also acted as an impediment to progress. Its influence on the Sung mathematicians of the

---

*Seki Kowa's contribution will not be examined in this book, as the level of mathematics required is too high, and this book covers developments only up to 1600. A short discussion of the life and work of Seki Kowa can be found on pp. 392–4.

thirteenth century was perhaps even counter-productive since they were obliged to refer to it, just as some of today's academics routinely cite standard authorities to make their work 'respectable'. (There is a parallel too with generations of students in the West being taught from Euclid's *Elements*, a practice abandoned only in the present century.) When the *Chiu Chang* was written the status of the mathematician was high. This status was gradually eroded as mathematics came to be perceived as a diligent and unquestioning application of ancient wisdom rather than a process of building on the solid foundations that the early texts had laid.

7   Special Topics in Chinese Mathematics

◇◇◇

The last half of the thirteenth century and the early fourteenth marked the culmination of over a thousand years of development of Chinese mathematics, built on the solid foundation of the *Chiu Chang*. In the historical introduction to the last chapter we saw that the Sung period produced outstanding scientific and technological achievements. Four of the greatest Chinese mathematicians – Chin Chiu Shao, Li Yeh, Yang Hui and Chu Shih Chieh – lived during this period, and there were more than thirty mathematical schools scattered across the country. Mathematics was a compulsory subject for national public service examinations, for which the recommended text was *Suan Ching Shih Shu* (The Ten Mathematical Manuals), a compendium of major mathematical works of earlier years. And mathematics was finding more practical applications in a widening number of disciplines – calendar-making, surveying, chronology, architecture and meteorology, as well as areas relating to trade and barter, the payment of wages and taxes, and simple mensuration, as we have already seen in the *Chiu Chang*.

The most striking feature of the Chinese mathematics of this period is its essentially algebraic character; there was little in contemporary geometry, particularly mensuration, that was not to be found in the *Chiu Chang* or its early commentaries. Many of the algebraic innovations of the period were extensions of previous work. They fall into three main categories:

1 *Numerical equations of higher order*   Although the procedure had its origins in the method of extracting square and cube roots found in the *Chiu Chang*, discussed in the last chapter, its fullest development is contained in the works of Chin Chiu Shao (*c.* 1250). The breakthrough came with the use of what we know as Pascal's triangle for the extraction of roots. Not until the beginning of the nineteenth century would European mathematicians, notably Horner and Ruffini, make any substantial progress in this area.

2 *Pascal's triangle*   Although this triangular array of numbers is named after the seventeenth-century French mathematician Blaise Pascal, it received detailed treatment in the hands of Yang Hui and Chu Shih Chieh, some 350 years previously. And Yang reported that his discussion of Pascal's triangle was derived from an earlier work by Chia Hsien (*c.* 1050) which has not survived. Chia had differentiated between two methods of extracting square and cube roots. The first, known as *tseng cheng fang fa*, the method of 'extraction by adding and multiplying', is similar to the method examined in the previous chapter. The second was known as *li cheng shih shuo*, 'unlocking the coefficients by means of a chart', which uses binomial coefficients taken from Pascal's triangle to solve numerical equations of higher order.

3 *Indeterminate analysis* Interest in this subject arose both in China and India in connection with calculations in calendar-making and astronomy. The basic problem was one of finding a procedure for solving a system of n equations with more than n unknowns. At the simplest level, how does one go about solving an equation in two unknowns,

$$3x + 8y = 100$$

where the solutions for x and y are either real numbers or positive integers? Such problems were successfully tackled in Europe only in the eighteenth century, by Euler and Gauss.

There were other specific developments worthy of note, though they will not be examined in any detail in this book. They include:

1 the derivation of a cubic interpolation formula popularized by Kuo Shou Ching (*c.* 1275), which later came to be known in the West as the Newton–Stirling formula;

2 the *tien yuan* notation for non-linear equations, first used by Li Yeh and Chu Shih Chieh, which made possible the development of algorithms for solving different types of equation; and

3 the use of geometric methods to study mathematical series by a number of mathematicians of the period. (We shall examine Indian work in this area in Chapter 9.)

There were also two subjects of a geometric character to which the Chinese made significant contributions: they investigated the properties of right-angled triangles, and continued the age-old search for more accurate estimates of π. Chinese interest in both these subjects goes back to well before AD 1000, in the former case to the earliest source of Chinese mathematics, the *Chou Pei*.

THE 'PILING-UP OF RECTANGLES': THE PYTHAGOREAN THEOREM IN CHINA

The Pythagorean theorem is generally held to be one of the most important results in the early history of mathematics. From it came important discoveries in theoretical geometry as well as practical mensuration. We saw in Chapter 4 how the Babylonians' understanding of geometry, based on similar triangles and circles, was enhanced by the discovery of the Pythagorean result, and how their algorithmic procedure for extracting square roots of 'irregular' (irrational) numbers was also based on this result. In China too, the study of the right-angled triangle had a considerable impact on mathematics.

The earliest extant Chinese text on astronomy and mathematics, the *Chou Pei*, is notable for a diagrammatic demonstration of the Pythagorean (or *kou ku*) theorem. Needham's translation of the relevant passage is illustrated by Figure 7.1(a), drawn from the original text. The passage reads:

> Let us cut a rectangle (diagonally), and make the width 3 (units) wide, and the length 4 (units) long. The diagonal between the (two) corners will then be 5 (units) long. Now, after drawing a square on this diagonal, circumscribe it by half-rectangles like that which has been left outside, so as to form a (square) plate. Thus the (four) outer half-rectangles, of width 3, length 4 and diagonal 5, together make two rectangles (of area 24); then (when this is subtracted from the square plate of area 49) the remainder is of area 25. This (process) is called 'piling up the rectangles'. (Needham, 1959, pp. 22–3)

Figure 7.1 The *kou ku* (Pythagorean) theorem: (a) the original illustration from the *Chou Pei* (Needham, 1959, p. 22); (b) the modern 'translation'.

In terms of Figure 7.1(b), the larger square ABCD has side $3 + 4 = 7$ and thus area 49. If, from this large square, four triangles (AHE, BEF, CFG and DGH), making together two rectangles each of area $3 \times 4 = 12$, are removed, this leaves the smaller square HEFG. And implicitly,

$$(3 + 4)^2 - 2(3 \times 4) = 3^2 + 4^2 = 5^2$$

The extension of this 'proof' to a general case was achieved in different ways by Chao Chun Ching and Liu Hui, two commentators living in the third century AD. In modern notation, Chao's extension may be stated thus: if the shorter (*kou*) and longer (*ku*) sides of one of the rectangles are a and b respectively, and its diagonal (*hsian*) is c, then the above reasoning would produce

$$c^2 = (b - a)^2 + 2ab = \text{square IJKL} + \text{rect DGIH} + \text{rect CFLG}$$
$$= \text{square AMIH} + \text{square MBFL}$$
$$= a^2 + b^2$$

An alternative explanation is based on the identity

$$(a + b)^2 = a^2 + 2ab + b^2$$
$$c^2 = (a + b)^2 - 2ab = a^2 + b^2$$
$$= \text{square ABCD} - 4 \triangle \text{DGH}$$

A geometric interpretation of this identity is fairly easily established, and was certainly known to the authors of the *Sulbasutras* (*c.* 500 BC) of Vedic India, which we shall examine in the next chapter. There is also the possibility that it was also known to the Babylonians of the Hammurabi dynasty. Later, geometric proofs of this identity are found in Euclid's *Elements* (*c.* 300 BC) and al-Khwarizmi's *Algebra* (*c.* AD 800). The Chinese may have deduced the identity from Figure 7.1(a) itself, where two squares of areas $a^2 = 4^2$ and $b^2 = 3^2$, together with two rectangles of area ab, together make up the large square of area $(a + b)^2$.

However, there is a third explanation, found in Liu's commentary, which does not refer to the diagram in the *Chou Pei* (i.e. Figure 7.1(a)), but is based on a principle of Chinese geometry well known at the time – the 'out–in complementarity' (dissection and reassembly) principle, which was discussed in the last chapter. We saw that this principle is based on two common-sense assumptions:

1 both the area of a plane figure and the volume of a solid remain the same under rigid translation to another place, and

2 if a plane figure or solid is cut into several sections, the sum of the areas or volumes of the sections is equal to the area or volume of the original figure.

If these conditions hold, it is possible to infer simple arithmetic relations between the areas or volumes of various sections of the plane or solid figures resulting from dissection or reassembly. It was this principle that Liu used to 'prove' the Pythagorean theorem. Lam and Shen's (1984, p. 95) translation of the relevant passage from Liu reads:

> Let the square on *kou* (*a*) be red and the square on *ku* (*b*) be blue. Use the [principle] of mutual subtraction and addition of like kinds to fit into the remainders, so that there is no change in [area on the completion of] a square on the hypotenuse (*c*).

The principle referred to is the 'out–in complementarity' principle. There have been a number of attempts to construct the missing diagram that accompanied this statement. One of the more plausible is shown in Figure 7.2. Here ABC is a space for a right-angled triangle. BCIS is the red square on the *kou* (or *a*), while AQDC (=FERI) is the blue square on the *ku* (or *b*). From SBCFER cut off the triangle GBS and put it in the space ABC. Next, cut off the triangle EGR and place it over the triangle EAF. What we have now is the square AEGB, which is the square on the hypotenuse (or *c*). This completes the 'proof' of the *kou ku* theorem.

It is important to recognize a basic difference between this Chinese proof

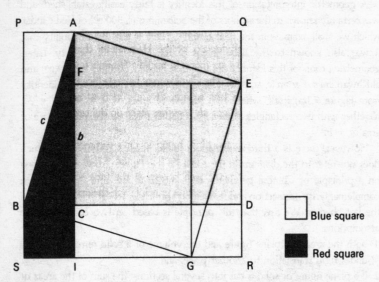

Figure 7.2 A reconstruction of Liu's proof of the *kou ku* theorem.

and the Euclidean proof of the Pythagorean theorem. Considerable geometric knowledge of properties relating to identical triangles and areas is required to understand the Euclidean proof, which probably explains why the theorem does not appear in Euclid's *Elements* until the end of Book I. The Chinese proof is a matter of common sense, enabling the theorem to be applied to many practical problems with relative ease. We now examine a few of these applications.

Applications of the *kou ku* theorem

The *kou ku* theorem is applied via various permutations built on the relation $kou^2 + ku^2 = shian^2$, or $a^2 + b^2 = c^2$. The reader is invited to solve the following problems, which are mainly from the ninth chapter of the *Chiu Chang*, before consulting the solutions given below them.

EXAMPLE 7.1   *Under a tree 20 chih high and 3 chih in circumference, there grows an arrow-root vine which winds seven times round the stem of the tree and just reaches its top. How long is the vine? [1 chih is about 23 cm]*

The problem is of the form: Given a and b, find c (see Figure 7.3).

Suggested solution   Take $7 \times 3 = 21$ as one of the sides (a) of a right-angled triangle. Take the height of the tree as a second side (b). Find the hypotenuse (c), which is then the length of the vine:

$$\text{length of vine} = \sqrt{(21^2 + 20^2)} = 29 \ chih$$

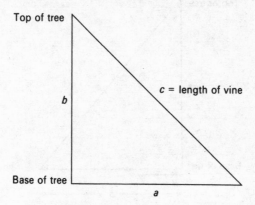

Figure 7.3

EXAMPLE 7.2 *There is a rope hanging from the top of a tree with 3 chih of it lying on the ground. When it is tightly stretched, so that its end just touches the ground, it reaches a point 8 chih from the base of the tree. How long is the rope?*

The problem is of the form: Given b and $c - a$, find c.

Suggested solution

$$\text{length of rope} = \tfrac{1}{2}(64/3 + 3)$$
$$= 73/6$$
$$= 12\tfrac{1}{6} \ chih$$

This is an interesting solution, displaying a considerable degree of algebraic sophistication. With modern notation and Figure 7.4, we can attempt to reconstruct the route that may have been followed to arrive at the above solution. So, let

$c =$ length of rope

$d = c - a =$ length of rope on the ground

$b =$ distance of end of rope, if tightly stretched, from base of tree

The following relationship may be deduced from Figure 7.4:

$$c^2 = (c - d)^2 + b^2$$

Therefore

$$0 = -2cd + d^2 + b^2$$
$$c = \tfrac{1}{2}(b^2/d + d)$$

and it was this expression for c that was probably used to solve the problem.

Figure 7.4

EXAMPLE 7.3   *The height of a door is 6 chih 8 tsun larger than its width. The diagonal is 10 chih. What are the dimensions of the door? [1 chih = 10 tsun].*

The problem is of the form: Given $b - a$ and c, find a and b (see Figure 7.5).

Suggested solution   From the square of 10 *chih*, subtract twice the square of half the given difference (6 *chih* 8 *tsun*). Take half this result, and find its square root. Subtract half the given difference from this root to obtain the width of the door, and add the root to half the given difference to obtain the height:

$$\text{width} = 2 \text{ chih } 8 \text{ tsun}$$

$$\text{height} = 9 \text{ chih } 6 \text{ tsun}$$

The solution is derived from the equation (in modern notation)

$$(a + b)^2 = 2c^2 - (b - a)^2 \tag{7.1}$$

where a is the width of the door, b the height of the door and c the diagonal. This is easily seen since

$$(a + b)^2 + (b - a)^2 = 2a^2 + 2b^2 = 2c^2$$

However, Shao gave a geometric demonstration (which is discussed by Lam and Shen (1984)).

From the numerical values given in the problem, it is easily established that $b - a = 6 \cdot 8$ *chih* and $c = 10$ *chih*. Substituting these values into equation (7.1) and taking the square root gives

$$a + b = \sqrt{2(10)^2 - (6 \cdot 8)^2} \tag{7.2}$$

Adding $b - a = 6.8$ to equation (7.2) and halving gives the value of b; subtracting the same quantity and halving gives the value of a.

Top of door

b c

a

Figure 7.5

The Crest of the Peacock

EXAMPLE 7.4   *There is a bamboo 10 chih high, the upper end of which, being broken, touches the ground 3 chih from the foot of the stem. What is the height of the break?*

This is a famous problem in the history of mathematics. Figure 7.6(a) shows the problem as illustrated in the original text. It kept re-appearing in the works of Indian mathematicians, from Mahavira in the ninth century to Bhaskaracharya in the twelfth century, and eventually in European works, probably thus charting a westward migration of Chinese mathematics via India and the Arab world.

The problem is of the form 'Given a and $b + c$, find b'.

Suggested solution   Take the square of the distance from the foot of the bamboo to the point at which its top touches the ground, and divide this

(a)   (b)

Figure 7.6 The 'broken bamboo' problem: (a) as illustrated in the *Hsiang Chieh Chiu Chang Suan Fa Tsuan Lei* (Needham, 1959, p. 28); (b) the modern 'translation'.

quantity by the length of the bamboo. Subtract the result from the length of the bamboo, and halve the resulting difference. This gives the height of the break.

These instructions may be expressed in modern notation, with reference to Figure 7.6(b), as follows. Let

$$a = \text{distance from foot of bamboo}$$
$$b + c = \text{length of bamboo}$$
$$b = \text{height of erect section of bamboo}$$

Then the above rule is equivalent to

$$b = \frac{1}{2}\left(b + c - \frac{a^2}{b + c}\right)$$

which yields

$$b = \tfrac{1}{2}(10 - 9/10) = 91/20 \; chih.$$

We could continue this discussion of applications of the *kou ku* theorem in Chinese mathematics almost indefinitely. The applications are remarkable for their range and ingenuity. The reader who wishes to know more will find Ang (1978), Gillon (1977), Lam and Shen (1984), and Swetz and Kao (1977) informative. The title of Swetz and Kao's book is particularly appropriate: *Was Pythagoras Chinese? An Examination of Right Triangle Theory in Ancient China.*

Our detailed treatment of the properties and problems of right-angled triangles is justifiable for a number of reasons. Of all the ancient mathematical traditions, the Chinese contained the most extensive, sustained and ingenious treatments. In contrast to the mathematics of ancient Greece, the corpus of knowledge built up was primarily for the purpose of practical applications in height and distance mensuration. But the discovery of the *kou ku* theorem gave rise to work on the proportionality of sides in similar triangles, the extraction of square and cubic roots, methods of solving different types of quadratic equation, and the numerical solution of higher-order equations (to be discussed in a later section).

The importance of the *kou ku* theorem in establishing algebraic geometry and in its contribution to the broader development of Chinese algebra cannot be overestimated. It founded a tradition in geometric reasoning which belies the notion that all mathematical traditions not influenced by the Greeks were essentially algebraic and empirical. The commentaries by Liu and Chao clearly show a deductive geometry that is moving beyond the mere numerical relations connecting the sides of a right-angled triangle in an active search for general proofs of the relationship. The reasoning in the proofs was based on geometry, with the basic concept being that figures of dissimilar shape can have the same area, and the basic procedure being the 'out–in complementarity' principle. Terms such as *ji ju* ('piling up rectangles') were widely used underlining the importance of pictorial representation in Chinese mathematics. But the work on the right-angled triangle also highlighted one of the negative aspects of the development of Chinese mathematics – the excessive reverence accorded the *Chiu Chang*. The above examples, chosen to show the wide range of applications, were all either taken directly from this text or derived from it. So why were mathematicians of the calibre of Liu and Chao, and later the brilliant thirteenth century quartet, unable to free themselves from the constricting influence of the *Chiu Chang*? Was the astonishing continuity and stability of Chinese civilization responsible, coupled perhaps with a great reverence for the past? Most likely we shall never know.

ESTIMATION OF π

The Greek symbol π was first used to denote the ratio of the circumference of a circle to its diameter in 1706, by the Welshman William Jones. The same symbol had previously been used for just the circumference, at a time when the idea that a ratio could be a number was quite a novel one. In his major work *Introductio in analysin infinitorum* (1748), Leonhard Euler gave his personal approval to this use of π, thereby popularizing it. It is now perhaps the most widely known mathematical symbol.

If π represented merely the ratio of the circumference of a circle to its diameter, the determination of its numerical value would have been of limited mathematical interest. However, there are other reasons why the evaluation of π has been a continuing quest, for some four thousand years from 1800 BC to the present day:

1 a practical requirement for increasingly accurate determinations of π, in fields as diverse as building and astronomy,
2 the perennial fascination with the problem of 'squaring the circle', and
3 a growing interest in the nature of the constant represented by π.

The problem of squaring the circle was finally resolved in 1882 when it was shown to be impossible.* However, the search for more accurate estimates for π, which began with the Egyptians, continues even today as powerful

---

*Stated simply, the problem of squaring the circle (or, more formally, the quadrature of the circle) is: can one construct a square whose area is exactly equal to that of a circle of given diameter

computers are used to calculate its value to millions of decimal places. Table 7.1 contains some of the historical highlights of this search, before the advent of modern mathematics. Two ways of calculating π are represented here, though only one example (Madhava's calculation) of the second method was in use before AD 1600. They are:

1 *The 'classical' (or geometric/empirical) method* This consists of calculating the perimeters of regular polygons inscribed in and/or circumscribed about a circle of given radius whose circumference lies between these perimeters. This method or a variant of it was used in all but one of the calculations listed in Table 7.1.

2 *The 'modern' (or analytical) method* This consists of evaluating π from a converging infinite series. It was first used in Kerala, India, in about the fifteenth century. It is discussed in Chapter 9, on medieval Indian mathematics.

For a long period the Chinese were content with taking the ratio of circumference to diameter as 3 in their calculations. One of the earliest attempts to get a better estimate was by an astronomer and calendar-maker called Liu Hsing, who lived just before the beginning of the Christian era. Instructed by his ruler, Wang Mang, to construct a standard measure of capacity, Liu Hsing produced a cylindrical vessel made out of bronze which became the prototype for hundreds of such vessels produced and distributed throughout China. From an examination of the dimensions of one of these vessels (now kept in a museum in Beijing), some commentators have inferred that Liu Hsing used the value $\pi \simeq 3.1547$. This view is supported by an entry in *Sui Shu* (Official History of the Sui Dynasty) which states that Liu Hsing had found a new value for π to replace the old one of 3. Another piece of conjectural evidence comes from a stray remark by Liu Hui that Chang Heng, a court astrologer who lived in the first century AD, had made an implicit estimate of π as the square root of 10, a value also found in the Jaina mathematics of India and reported by Umasvati a few centuries earlier.

The first systematic treatment of this topic is contained in Liu Hui's notable commentary on the *Chiu Chang*, written in the third century AD. He began by examining the underlying assumption in Problems 31 and 32 of the first chapter of the *Chiu Chang*, in which the area of a circle was calculated by taking the product of half the circumference and half the diameter. After explaining

---

using only a straightedge and a compass? Only in the nineteenth century was it demonstrated that, since squaring the circle is equivalent to constructing a line segment whose length is equal to the product of the square root of π (which is not a constructible quantity) and the radius of the given circle, it cannot be done.

Table 7.1 Estimates of π before AD 1600.

Date	Source/mathematician	Method and value
c. 1650 BC	Ahmes Papyrus (Egypt)	Equating a circular field of 9 units to a square of side 8 units implies $\pi \simeq 3 \cdot 16$
c. 1600 BC	Susa Tablet (Babylonia)	Equating a regular hexagon to a circle; the ratio of the perimeter of the hexagon to the circumference of the circle, given as 0;57,36 implies $\pi \simeq 3;7,30 \ (3 \cdot 125)$
c. 800– 500 BC	*Sulbasutras* (India)	Baudhayana gave the following rule: (s = side of a square, d = diameter of a circle): $s = d[1 - 28/(8 \times 29)$ $- 1/(6 \times 8 \times 29)$ $+ 1/(6 \times 8 \times 29 \times 8)]$ which implies, if the areas are equal, that $\pi \simeq 3 \cdot 09$
c. 250 BC	Archimedes (Greece)	Calculating the perimeters of inscribed and circumscribed regular polygons with 12, 24, 48 and 96 sides within and around a circle, to obtain $223/71 < \pi < 22/7$, where both limits give $\pi \simeq 3 \cdot 14$, correct to 2 decimal places
c. 150 BC	Umasvati (India)	Inscribing a regular hexagon and then a 12-sided polygon, and applying the Pythagorean result gives a value equal to the square root of 10: $\pi \simeq 3 \cdot 16$

Date	Source/mathematician	Method and value
c. AD 260	Liu Hui (China)	Inscribing a regular hexagon within a circle and calculating by successive applications of the Pythagorean theorem the perimeters of polygons with 12, 24, . . . , 96 sides. By considering the last of these polygons, he arrived at $\pi \simeq 3 \cdot 1416$ (The method is examined later in this chapter)
c. 480	Tsu Chung Chih (China)	Similar method to Liu's, except for the successive applications of the Pythagorean theorem to polygons with up to 24 576 sides! The result was $3 \cdot 141\,592\,6 < \pi < 3 \cdot 141\,592\,7$
c. 500	Aryabhata (India)	Probably by calculating the perimeter of a regular inscribed polygon of 384 sides: $\pi \simeq 3 \cdot 1416$
c. 1400	Madhava (India)	Using an infinite-series expansion for π (to be discussed in Chapter 9): $\pi \simeq 3 \cdot 141\,592\,653\,59$, correct to 11 decimal places
1429	Al-Kashi (Persia)	Probably by calculating the perimeter of a regular polygon with 3×2^{28} sides. Expressed in decimals: $\pi = 3 \cdot 141\,592\,653\,589\,793\,2$, correct to 16 decimal places
1579	François Viète (France)	Calculating the perimeter of a regular polygon with 393 216 sides gave $\pi = 3 \cdot 141\,592\,654$, correct to 9 decimal places

the rationale for this rule and the use of the inaccurate value of 3 for the ratio of circumference to diameter, Liu proceeded to work out a method for obtaining a more accurate value for π. Since the method bears some similarity to Archimedes' innovative approach four hundred years earlier, it is interesting to compare them.

Archimedes based his method on the simple observation that if a circle is enclosed between two polygons of n sides, then, as n increases, the gap between the circumference of the circle and the perimeters of the inscribed and circumscribed polygons diminishes, so that eventually the perimeters of the polygons and the circle would become identical. Or in other words, as n increases, the difference in area between the polygons and the circle would be gradually exhausted. (This approach is an example of a Greek method known as 'exhaustion'.)

Figure 7.7 shows a circle enclosed between an inscribed regular hexagon (i.e. a six-sided polygon with equal sides) ABCDEF and a circumscribed regular hexagon RSTUVW. It is convenient to begin with a regular hexagon, since its construction with straightedge and compass is a fairly simple matter. First draw the circle. Then, with the compass still set to the radius of the circle, start at any point on the circumference and mark off the vertices A, B, C, D, E and F. Draw tangents to the circle at each of these points, and join them to give the circumscribed hexagon RSTUVW. Given the inscribed polygon ABCDEF, it is possible next to construct a 12-sided polygon by bisecting the arc subtended on each of the sides of ABCDEF and joining each of the additional six points to the two original vertices that are adjacent.

As an illustration, if G is the midpoint of the arc AF, then joining F to G, and G to A, will give two of the sides of the 12-sided inscribed polygon. And if, at each of the 12 vertices of the new polygon, tangents are drawn to the circle, we obtain the corresponding circumscribed 12-sided polygon. This process can be carried on to obtain regular polygons of 24, 48, 96, . . . sides. Archimedes stopped at 96-sided polygons.

Now, if the perimeters of the inscribed and circumscribed polygons of n sides are denoted by p_n and P_n respectively, and C is the circumference of the circle, it follows that

$$p_6 < p_{12} < p_{24} < p_{48} < p_{96} < \cdots$$
$$\cdots < p_{n/2} < p_n < C < P_n < P_{n/2} < \cdots$$
$$\cdots < P_{96} < P_{48} < P_{24} < P_{12} < P_6$$

It can be shown that p_n and P_n converge to C as n tends to infinity. Next, it is easily established that the perimeter of the inscribed regular hexagon, p_6, is

Figure 7.7 Finding the area of a circle by the method of exhaustion.

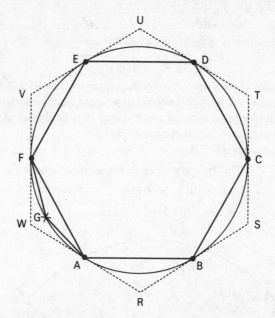

$3d$, where d is the diameter, and also that the perimeter of the circumscribed regular hexagon, P_6, is $2\sqrt{3}d$. Starting with p_6, Archimedes found close approximations to p_{12}, p_{24}, p_{48} and p_{96}. Then from P_6 he approximated P_{12} to P_{96}. Since $p_{96} < C < P_{96}$, he concluded that

$$3 + 10/71 < \pi < 3 + 1/7$$

In his computations of the square root of 3 and other calculations, Archimedes appears to have followed the Babylonian methods we discussed in Chapter 4. But the idea of finding approximate numerical values of π by establishing narrower and narrower limits, between which the value must lie, turned out to be a peculiarly Greek innovation.

Liu Hui's method of evaluating π required only inscribed regular polygons (see Figure 7.8). Starting with the known perimeter of a regular polygon of n sides inscribed in a circle, the perimeter of the inscribed regular polygon of $2n$ sides was calculated by applying the Pythagorean theorem twice. The circle in Figure 7.8 has its centre at O and radius r. Let $PQ = s$ be the side of a regular inscribed polygon of n sides having a known perimeter. Then, given s and n, the Pythagorean theorem is used to calculate, in turn, that

$$OM = u = \sqrt{r^2 - (\tfrac{1}{2}s)^2}$$
$$MR = m = r - u$$
$$RQ = w = \sqrt{m^2 + (\tfrac{1}{2}s)^2}$$

where w is one of the sides of the regular polygon with $2n$ sides. Repetition of this procedure will produce closer and closer approximations to the circumference of the circle, in terms of which π may then be defined. A similar procedure can be used with circumscribed polygons.

In Liu's own example, $OR = r = 10$, $n = 6$ and $PQ = s = 10$, and so

$$u = \sqrt{100 - 25} \simeq 8.660\,254$$
$$m = 10 - 8.660\,254 = 1.339\,746$$
$$w_1 \simeq 5.176\,381$$

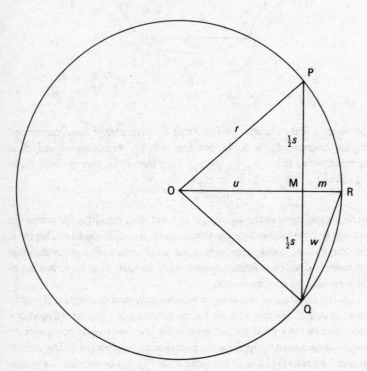

Figure 7.8 Liu's method of finding π.

The first iteration produces $w_1 \simeq 5 \cdot 176\,381$ as the length of the side of a 12-sided regular polygon. Repeating the same process for a 12-sided polygon with $s = 5 \cdot 176\,381$ gives

$$w_2 \simeq 2 \cdot 610\,524$$

The iterative process is continued until we find the length of the side of a 96-sided regular inscribed polygon. Now, the area of a regular inscribed polygon of $2n$ sides is $\frac{1}{2}nsr$, where s is the length of the side of a regular polygon with n sides and r is the radius of the circle. For the dimensions given above,

$$A_{12} = \tfrac{1}{2}(6 \times 10 \times 10) = 300 \text{ is the area of a 12-sided polygon}$$

Similarly,

$$A_{24} = \tfrac{1}{2}nsr = \tfrac{1}{2}(12 \times 5 \cdot 176\,381 \times 10) = 310 \cdot 5829$$

is the area of a 24-sided polygon

Continuing in this way, Liu Hui found

$$A_{96} = 313 \cdot 9344 \quad \text{and} \quad A_{192} = 314 \cdot 1024$$

By a geometrical argument, he showed that

$$A_{192} < A < 2A_{192} - A_{96}$$

where A is the area of the circle. This leads to

$$314 \cdot 1024 < A < 314 \cdot 2704$$

and by an interpolation Liu Hui then arrived at

$$\pi = 3 \cdot 1416$$

Like Archimedes, he enclosed his result between two limits.

The Chinese fascination with the value of π reached its climax in the work of Tsu Chung Chih and his son Tsu Keng Chih, who established new boundaries for π. In his text *Su Shu* (Method of Interpolation), written in AD 479 but unavailable today, the elder Tsu may have explained his method for approximating π. All we have are passages in the Official History of the Sui Dynasty, recording the efforts made by Tsu (who is described as a historian) to improve the accuracy of the value of π. Using a circle of diameter 10 *chih*, subdivided into 10^8 units, he found an upper limit of $3 \cdot 141\,592\,7$ and a lower limit of $3 \cdot 141\,592\,6$ for π. The same official history states that Tsu gave the Archimedean value of 22/7 as an inaccurate approximation, and 355/113 as an accurate one. Tsu's book was adopted by the Tang government as a text for their civil service examinations.

We can only offer conjectures about how Tsu achieved his highly accurate estimate of π. Most probably, Tsu extended Liu's method to polygons of 24 576 sides, as extending the above procedure to A_{24576} yields an approximation for π which is correct to the seventh decimal place. An accurate approximation for π is 355/113, correct up to the sixth decimal place, which appears in the work of medieval Indian mathematicians of the fifteenth century. In 1585 a Dutch mathematician, Adriaen Anthoniszoon, apparently discovered this value by subtracting the numerators and denominators of the Ptolemaic and Archimedean values, 377/120 and 22/7! More likely is that the Indian and European knowledge of this highly accurate value for π came from China, as Tsu's discovery spread southwards and westwards over a number of centuries. In China itself the work of Liu and Tsu was soon forgotten, and various inaccurate approximations – including 3 – continued to be used. In 1247 Chin Chiu Shao stated that 3, 22/7 and $\sqrt{10}$ were all in use! In 1275, Yang Hui gave five formulae for finding the area of a circle. In one of them π was 3, two had π as 22/7, and in the other two π was 3·14. There was no guidance on choosing which value to use. One is left with the impression that the innovative work of Liu and Tsu was ignored by the mathematicians who came after them.

There are certain similarities between the work of Archimedes and of Liu and Tsu. All three used the method of exhaustion where, by increasing the number of sides of a polygon inscribed in a circle, the sides became so short that it eventually became possible to approximate the circle by the polygon. Archimedes was mainly concerned with evaluating the circumference of the circle. He drew no conclusions about π; this was done by later commentators. Liu's interest, though, was clearly focused on π. By showing that the area of the circle is half the product of its circumference and its radius, he proceeded to establish an equality between the ratio of the area of a circle to the square of its radius, and the ratio of the circumference to the diameter. However, Archimedes' achievement seems the more remarkable when weighed against the limited scope offered by the numerals and computational techniques available to him, and the strongly ingrained aversion to experimentation and computation that marked Greek mathematics until the Egyptian and Babylonian empirical influences subtly altered its character. However, Liu and Tsu's highly elaborate calculations were products of a mathematical culture sympathetic to computation and offering methods which made its mathematicians far better equipped to carry out complicated calculations. We shall return to this point, about how a penchant for numerical work determined the character of Chinese mathematical achievements, particularly in the solution of higher-order and indeterminate equations.

SOLUTION OF HIGHER-ORDER EQUATIONS
AND PASCAL'S TRIANGLE

The Chinese development of Pascal's triangle

The invention of the 'arithmetical' triangle named after Blaise Pascal cannot be traced or credited to any individual. Its earliest use was among the Indians and the Chinese. In India it was an outgrowth of interest in a subject called *vikalpa*, or what is known today as permutations and combinations. A discussion of the Indian work on this subject will be found in Chapter 8.

During the first half of the eleventh century there appeared in China the earliest reference to a basic diagram for solving equations. In a book unfortunately no longer extant, Chia Hsien constructed a table of binomial coefficients up to the sixth power, the exponents being positive integers. In modern notation, Chia was selecting the coefficients of the binomial expansion:

$$(a + b)^n = {}_nC_0a^nb^0 + {}_nC_1a^{n-1}b^1 + {}_nC_2a^{n-2}b^2 + \cdots + {}_nC_na^0b^n,$$

$$\text{for } r = 1, 2, \ldots, n$$

Here

$$ {}_nC_r = \frac{n!}{r!(n - r)!} $$

where $n!$ is the usual notation for 'n factorial',

$$ n! = n \times (n - 1) \times (n - 2) \times \cdots \times 2 \times 1 $$

with $0!$ taken to be 1.

For $n = 0$, the coefficient is ${}_0C_0 = 1$

For $n = 1$, the coefficients of $(a + b)^1$ are ${}_1C_0$ and ${}_1C_1$, or 1, 1

For $n = 2$, the coefficients of $(a + b)^2$ are ${}_2C_0$, ${}_2C_1$ and ${}_2C_2$, or 1, 2, 1

For $n = 3$, the coefficients of $(a + b)^3$ are ${}_3C_0$, ${}_3C_1$, ${}_3C_2$ and ${}_3C_3$, or 1, 3, 3, 1

For $n = 4$, the coefficients of $(a + b)^4$ are ${}_4C_0$, ${}_4C_1$, ${}_4C_2$, ${}_4C_3$ and ${}_4C_4$,

$$\text{or } 1, 4, 6, 4, 1$$

In a similar fashion, the coefficients can be determined for any value of n. The Pascal's triangle for $n = 0, 1, 2, \ldots, 8$ depicted at the beginning of Chu Shih Chieh's book 'The Precious Mirror of the Four Elements' is shown in Figure 7.9(a), with a transliteration in our numerals given in Figure 7.9(b). Below Chu's representation of the triangle is a comment that provides both an explanation of how it was constructed and an indication of the uses to which it might be put. Recast in present-day terminology, it reads:

The numbers in the $(n + 1)$th row show the coefficients of the binomial

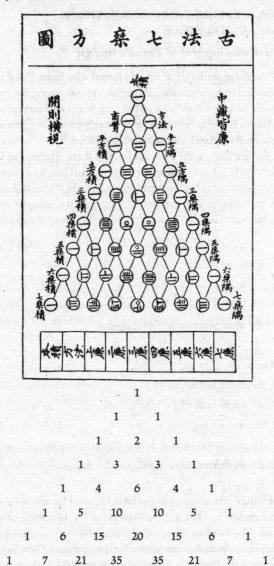

(a)

(b)

```
                        1
                    1       1
                1       2       1
            1       3       3       1
        1       4       6       4       1
    1       5      10      10       5       1
  1     6      15      20      15       6       1
1   7      21      35      35      21       7       1
1  8    28    56    70    56    28    8    1
```

Figure 7.9 Pascal's triangle: (a) as depicted by Chu Shih Chieh (Needham, 1959, p. 135); (b) in modern numerals.

expansion of $(a + b)^n$, n being a positive integer. The unit coefficients along the extreme left slanting line (the *chi shu*) and the extreme right slanting line (the *yu suan*) are the coefficients of the first and last term respectively in each expansion. The inner numbers, '2', '3, 3', '4, 6, 4', . . . on the third, fourth and fifth, etc., rows (the *lien*) are the inner terms of the binomial equations of the second, third, fourth, etc., degree.

Chu continues by indicating the close relationship between the construction of the triangle and the solution of numerical equations of higher order:

> Multiply the coefficients of the $(n + 1)$th row by a suggested value for the root; then subtract the nth power of the suggested row from *shi* (i.e. the constant whose root is to be extracted); and divide the difference by the product of the suggested value and the coefficient to obtain a new value for the root.

The Chinese origin of Horner's method*

In our earlier discussion of the extraction of square and cube roots in the *Chiu Chang*, we found that the basic procedure had a strong geometric rationale: the method consisted of adding and subtracting sections from a given geometric figure. But the limitation of such a geometric approach is obvious: it cannot be used to solve equations beyond the third degree. Indeed, the geometric approach, even when used in the context of cubic equations, is highly cumbersome, which was why we avoided it in solving the cubic equation. What Pascal's triangle did for Chinese mathematics was to help break the geometric mould by establishing a clear link between the patterns of the coefficients of the triangle and the derivation of transformed equations.

A few examples will help. Let us begin with a simple one, expressed in symbolic notation. Suppose we want to find the square root of N, i.e. to solve the quadratic

$$x^2 = N \tag{7.3}$$

---

*A method named after an English schoolteacher and mathematician, William George Horner, who in 1819 published a numerical method of finding approximate values of the root of equations of the type

$$f(x) = a_0 x^n + a_1 x^{n-1} + \cdots + a_{n-1} x + b_n = 0$$

The procedure that Horner rediscovered is identical to the computational scheme used by the Chinese over five hundred years earlier.

We take the following steps:

1 Let $x = h + y$. Equation (7.3) is then transformed as

$$N = x^2 = (h + y)^2 = h^2 + 2hy + y^2 = h^2 + (2h + y)y \qquad (7.4)$$

where h is a 'guesstimate' of the root x, and the equation is formed with the coefficients from the third row of Pascal's triangle shown in Figure 7.9.

2 Obtain y by dividing $N - h^2$ in equation (7.4) by $2h$ to get

$$\frac{N - h^2}{2h} = \frac{2hy}{2h} + \frac{y^2}{2h}$$

or

$$N^* = \frac{N - h^2}{2h} = y + ay^2, \quad \text{where } a = 1/2h$$

3 Using equation (7.4) again, add y to $2h$, multiply the sum by y and subtract from $N - h^2$. If the result is zero, the answer is correct, and the square root of N is $h + y$.

A more difficult example is from the fourth chapter of Chin Chiu Shao's *Su Shu Chiu Chang*, which requires the solution of the equation

$$-40\,642\,560\,000 + 763\,200x^2 - x^4 = 0 \qquad (7.5)$$

The first step in the solution is to 'guesstimate' the number of digits in the answer and also the value of the first digit; this was done by trial and error. Here it is easily established that the final answer will be a three-digit number starting with 8. The approach used by Chin is analogous to the present-day method of synthetic division to factor out $y = x - 800$ from equation (7.5). The first division gives

$$-40\,642\,560\,000 + 763\,200x^2 - x^4$$
$$= (x - 800)(98\,560\,000 + 123\,200x - 800x^2 - x^3) + 38\,205\,440\,000$$

or

$$-40\,642\,560\,000 + 763\,200x^2 - x^4$$
$$= y(98\,560\,000 + 123\,200x - 800x^2 - x^3) + 38\,205\,440\,000$$

Four repetitions of this procedure, each giving a new remainder, finally produce

$$38\,205\,440\,000 - 826\,880\,000y - 3\,076\,800y^2 - 3200y^3 - y^4 = 0 \qquad (7.6)$$

from which we can estimate that the first digit of y, a two-digit number, is 4. The process of synthetic division is repeated, with equation (7.6) divided by $y - 40$. The first division leaves no remainder, so the solution to equation (7.5) is $x = 840$.

Clearly, Chin had neither the notation nor the technique to follow the synthetic division used above. Instead, he proceeded by setting up a row of numbers corresponding to the coefficients in equation (7.5) and then applying a procedure resembling synthetic division. We shall examine Chin's procedure for solving equation (7.5) by using the counting-rods format, but with decimal rather than rod numerals. Each row in Figures 7.10(a)–(c) represents a certain quantity. The first is reserved for the root, the eventual result, and is labelled R. The next five rows represent the coefficients of the zeroth, first, second, third and fourth powers of the unknown x in equation (7.5) and are labelled accordingly. If counting-rods were used, the negative constant and the coefficient of the fourth power of x would be shown by black rods, while the positive coefficient of the second power of x would be shown by red rods. Further, instead of leaving blank spaces for zeros, as the Chinese did, here we insert the symbol 0 for clarity.

After arranging the numbers as in Figure 7.10(a), the calculation begins by advancing the elements in the rows labelled 1st to 4th by one to four places respectively, into the positions shown in Figure 7.10(b). We can now deduce that the solution will be a three-digit number, and that the 'hundreds' digit is 8. Next the 'hundreds' digit is multiplied by -1 from the 4th row, and added to 0 from the 3rd row. This gives the new entry for the 3rd row in Figure 7.10(c), which is -800. This -800 is multiplied by the root 800 and added to 763 200, which is the entry in the 2nd row of Figure 7.10(b). This gives the new entry in the 2nd row of Figure 7.10(c) as

$$(-800)(800) + 763\,200 = 123\,200$$

Next 123 200 is multiplied by the root 800 and added to the quantity in the 1st row in Figure 7.10(b), which is 0. This gives the new entry for the 1st row of Figure 7.10(c), 9 856 000. Finally, this quantity is multiplied by the root 800 and added to $-40\,642\,560$ to give

$$(9\,856\,000)(800) + (-40\,642\,560) = 38\,205\,440\,000$$

which is the new entry in the 0th row of Figure 7.10(c).

These calculations are equivalent to taking $h = 800$ as the first approximation, so that $x = h + x_1$, and then carrying out the steps (1) to (3) outlined above to obtain the new equation

$$38\,205\,440\,000 + 9\,856\,000x + 123\,200x^2 - 800x^3 - x^4 = 0 \quad (7.7)$$

Subsequent iterations follow the same procedure as above. Some iterations it is possible to skip by making use of the property that the coefficient of each term of the transformed equation will involve the numbers of Pascal's triangle

R											
0th	-4	0	6	4	2	5	6	0	0	0	0
1st											
2nd						7	6	3	2	0	0
3rd											
4th											-1

Figure 7.10(a)

R											
0th	-4	0	6	4	2	5	6	0	0	0	0
1st											
2nd				7	6	3	2	0	0		
3rd											
4th						-1					

Figure 7.10(b)

R									8	0	0
0th	3	8	2	0	5	4	4	0	0	0	0
1st					9	8	5	6	0	0	0
2nd						1	2	3	2	0	0
3rd									-8	0	0
4th							-1				

Figure 7.10(c)

lying on a line slanting across the triangle. Thus for $n = 4$ we use the first five rows of the triangle, and the coefficients of the general equation of the fourth degree:

$$\begin{array}{cccccc}
& & & 1 & & R_0 = a_4 = -1 \\
& & 1 & & 1 & R_1 = 4a_4h + a_3 = -3200 \\
& 1 & & 2 & & 1 \quad R_2 = 6a_4h^2 + 3a_3h + a_2 = -3\,076\,800 \\
1 & & 3 & & 3 & 1 \quad R_3 = 4a_4h^3 + 3a_3h^2 + 2a_2h + a_1 = -826\,880\,000
\end{array}$$

$$1 \quad 4 \quad 6 \quad 4 \quad 1 \quad R_4 = a_4h^4 + a_3h^3 + a_2h^2 + a_1h + a_0 = 38\,205\,440\,000$$

where R_0 to R_4 are the last five rows of Figures 7.10(a)–(c) written from top to bottom. They also show how these quantities are calculated. The transformed equation in y (at the end of the fourth iteration, which is the same as equation (7.6) above) can then be written, using the coefficients of Pascal's triangle, as

$$-y^4 - 3200y^3 - 3\,076\,800y^2 - 826\,880\,000y + 38\,205\,440\,000 = 0 \quad (7.8)$$

Now, taking equation (7.8) and setting $y = h' + z$ or $x = h + h' + z$, where $h' = 40$ is taken as the next approximation, proceed exactly in the same manner as before. The next transformed equation, in z, is

$$z(-z^3 - 3240z^2 - 3\,206\,400z - 955\,136\,000) = 0, \quad \text{so } z = 0$$

Since $x = h + h' + z$, and $h = 800$, $h' = 40$ and $z = 0$, it would follow that the solution is $x = 840$. We are advised by Chin Chiu Shao to check the correctness of this solution by substituting $x = 840$ in equation (7.5).

There are certain general features of this method which need further elaboration:

1 The representation of an equation on the counting-board conformed to what has been described as the *tian yuan* notation. For example, the equation

$$x^5 + 4x^3 - 5x^2 + 8x - 65 = 0$$

would be represented as

Constant	-65
x	8
x^2	-5
x^3	4
x^4	0
x^5	1

It is characteristic of this representation that the constant term was always expressed as a negative quantity, putting the equation into the form $f(x) = 0$ so that the method described above could be used. It is interesting that the idea of rearranging an equation to give a sum of terms equal to zero did not arise in European mathematics until the seventeenth century, with René Descartes.

2 The absence of symbols for 'equals' and 'minus' was unimportant in Chinese mathematics since the colour of the counting-rods indicated sign, and their arrangement indicated the relations between terms.

3 While handling negative quantities posed few problems in Chinese mathematics, negative solutions to equations were ignored. This was a reflection of the practical nature of their problems, which rendered such solutions meaningless.

4 The origins of the method described go back to the time of the *Chiu Chang*, when techniques for extracting square and cube roots were first used to solve numerical equations. The method remained unique to China until the eleventh century AD, when it appears in the works of Arab mathematicians such as al-Nasawi (*c.* 1025), who used it to extract cube roots, and later al-Kashi (*c.* 1450), who used it to extract roots of any degree. In Chapter 10 we shall examine al-Kashi's numerical solution of a cubic equation for evaluating the sine of 1°, which bears some resemblance to the method discussed above. So, a simple version of what is known as Horner's method may have made its way from China to the Arab world, and may even have been known to Fibonacci through the Arab mathematicians. If this is so, then Horner's method entered the mainstream of European mathematics, via the Arabs, as early as the thirteenth century.

The Chinese method in the West

The possibility of later Chinese mathematical influence on the West cannot be ruled out altogether. It is true that there is as yet no direct evidence of either William Horner or his Italian contemporary and rival, Paolo Ruffini, being aware of the Chinese solution. However, there is a tendency to dismiss too easily any circumstantial evidence of Chinese influence in this area. This is at least partly a consequence of ignoring the possibility that the Jesuit link established during the closing decades of the sixteenth century may have led to a two-way exchange of ideas. The emphasis has always been on how European mathematics reached China through the Jesuits.

In 1582 an Italian Jesuit, Matteo Ricci, was sent on a mission to China. He was one of the most remarkable men of his time: not only an accomplished linguist, with an extraordinary mastery of the Chinese language, but also a scientist and mathematician of note. He was received warmly by the imperial court, where he, together with a talented group of fellow Jesuits and Chinese, set out to acquaint the Chinese with scientific works from the West. To that end, they translated the first six books of Euclid; they also made detailed reports of aspects of Chinese life to their parent organization, the Society of Jesus. It is not unreasonable, given Ricci's knowledge of mathematics and his personal acquaintance with the works of algebraists of the calibre of Girolamo Cardano (1501–76) and François Viète (1540–1603), to suggest that he studied Chinese mathematics with some care and reported back his findings. But this must remain a conjecture until further research is undertaken on the mathematical content of the communications from the Jesuits in China, and on the extent of their contacts with the Italian mathematicians of the sixteenth and seventeenth centuries who were mainly responsible for the revival of algebra in Europe.

INDETERMINATE ANALYSIS IN CHINA

The following problem occurs in the third chapter of the fourth-century mathematical text *Sun Tsu Suan Ching* (Master Sun's Arithmetic Manual):

> There is an unknown number of objects. When counted in 'threes', the remainder is 2; when counted in 'fives', the remainder is 3; and when counted in 'sevens', the remainder is 2. How many objects are there? Answer: 23.

In modern notation, what we have here is the following set of simultaneous equations of the first degree:

$$N = 3x + 2$$
$$N = 5y + 3$$
$$N = 7z + 2$$

where N is the total number of objects, x the number of 'threes', y the number of 'fives' and z the number of 'sevens'. This information can be expressed even more concisely in the notation of linear congruences as

$$N \equiv 2(\mathrm{mod}\, 3) \equiv 3(\mathrm{mod}\, 5) \equiv 2(\mathrm{mod}\, 7)$$

where an integral value for N is required.*

---

*If two integers A and B are divided by a common integer m (called the modulus) and leave the same remainder r, then A is said to be 'congruent to B modulus m'. This is written as

$$A \equiv r(\mathrm{mod}\, m) \equiv B$$

Now, we know that in order that a solution to a given set of equations may be obtained, there must be as many equations as unknowns. But here there are three equations but four unknowns (i.e. N, x, y, z), so there are an infinite (or indeterminate) number of solutions. However, there is a further constraint implied by the answer given to the question: what we are seeking is the least (or minimum) integer value for N.

There are four main approaches to the solution of indeterminate equations. The most obvious is an arithmetic one in which, for the example given above, a solution set for (x, y, z) which satisfies the three equations is obtained by trial and error. By a long and laborious process it is possible to work out that the solution set $(x = 7, \; y = 4, z = 3)$ will give a least integer value of $N = 23$. The scope of such a method, apart from its tedium, is very restricted.

A second approach is the one that probably originated with the Indian mathematician Aryabhata I (c. AD 500), and was refined and extended by later mathematicians, notably Brahmagupta (c. 625), Mahavira (c. 800) and Bhaskaracharya (c. 1100). The method was referred to as *kuttaka*, and consisted of continuous divisions and substitutions. This approach is discussed in Chapter 9.

The third method of solving indeterminate equations of the first degree bears some resemblance to the Indian approach. It is the method favoured in more recent times. To illustrate it, consider the following problem:

EXAMPLE 7.5   *Solve $5x + 8y = 100$, in integers $\geqslant 0$.*
Solution   We have

$$8 = (1)(3) + 5 \tag{7.9}$$

$$5 = (1)(3) + 2 \tag{7.10}$$

$$3 = (1)(2) + 1 \tag{7.11}$$

$$2 = (2)(1) \tag{7.12}$$

The highest common factor (HCF) of 8 and 5 is 1, since this is the highest integer which divides 2, in (7.12), 3, in (7.11), 5, in (7.10), and 8, in (7.9), without leaving a remainder. Back-substituting, starting with equation (7.11), gives

$$
\begin{aligned}
1 &= 3 - (1)(2) \\
&= 3 - 1(5 - (1)(3)) && \text{from equation (7.10)} \\
&= (8 - (1)(5)) - 1[5 - 1(8 - (1)(5))] && \text{from equation (7.9)}
\end{aligned}
$$

Therefore

$$1 = (2)(8) - (3)(5) \tag{7.13}$$

The method is perfectly general, and with it we can always obtain from any pair of integers x and y their HCF h, and a relation

$$h = Mx + Ny$$

where M and N are positive or negative integers. Returning to the solution, multiplying (7.13) by 100 gives

$$100 = (200)(8) + (-300)(5)$$

and

$$100 = (200 - 5t)8 + (8t - 300)5$$

for any t chosen, and covers all solutions. Hence the solution sets are obtained from the equations

$$x = 8t - 300, \qquad y = 200 - 5t$$

The solutions in integers $\geqslant 0$ correspond to $t = 38, 39, 40$:

$$x = 4, \qquad y = 10$$
$$x = 12, \qquad y = 5$$
$$x = 20, \qquad y = 0$$

Finally, there is the Chinese procedure (or *ta yen*). Before we examine it, though, we must consider why there was such a long and sustained interest in this subject in both India and China. The answer lies in problems to do with time that arose in both countries in calendar-making and astronomical calculations. One problem in calendar-making attracted the attention of both Sun Tsu (*c.* AD 300), the originator of indeterminate analysis in China, and Chin Chiu Shao (*c.* 1250), whose statement of the *ta yen* rule for the general solution of indeterminate equations of the first degree predated the work of Euler and Gauss by five hundred years. We shall consider the role of astronomy in motivating Indian work in this area in a later chapter.

All calendars need a beginning. A calendar constructed during the Wei dynasty (220–65) took as its starting-point the last time that winter solstice coincided with the beginning of a lunar month and was also the first day of an artificial sexagenary (60 day) cycle, known as *chia tsu*. The objective was to locate exactly the number of years (measured in days) since the beginning of the calendar. To restate the problem in modern symbolic notation, let y be the number of days in a tropical year, N the number of years since the beginning of the calendar, d the number of days in a synodic month, r_1 the number of days in the 60 day cycle between the winter solstice and the last day of the preceding *chia tsu*, and r_2 the number of days between the winter solstice and the beginning of the lunar month. The number of years since the beginning of the calendar can then be calculated from

$$yN \equiv r_1 (\bmod 60) \equiv r_2 (\bmod d)$$

More complex alignments, including planetary conjunctions, were built into models for estimating the beginnings of calendars, and as early as the fifth century AD the mathematician-astronomer Tsu Chung Chih solved a set of ten linear congruences. However, the first general mathematical formulation for solving problems in indeterminate analysis of the first degree is found in the work of Chin Chiu Shao (*c.* 1250).

The early approach

Let us return to the problem from the *Sun Tsu Suan Ching*. The solution offered by Sun Tsu reads:

> If you count in 'threes' and have the remainder 2, then put 140.
>
> If you count in 'fives' and have the remainder 3, then put 63.
>
> If you count in 'sevens' and have the remainder 2, then put 30.

Add these numbers and you get 233; from this subtract 210 and you have the answer (23).

A popular folk-song of the time, entitled 'The Song of Master Sun', had the following mnemonic aid to offer for the problem:

> Not in every third person is there one aged three score and ten,
> On five plum trees only twenty-one boughs remain,
> The seven learned men meet every fifteen days,
> We get our answer by subtracting one hundred and five over and over again.

In modern algebraic notation we would say that, given

$$N = 3x + 2 = 5y + 3 = 7z + 2$$

or

$$N \equiv 2(\bmod 3) \equiv 3(\bmod 5) \equiv 2(\bmod 7)$$

then the solution is obtained as

$$70 \equiv 1(\bmod 3), \quad \text{so} \quad 140 \equiv 2(\bmod 3)$$
$$21 \equiv 1(\bmod 5), \quad \text{so} \quad 63 \equiv 3(\bmod 5)$$
$$15 \equiv 1(\bmod 7), \quad \text{so} \quad 30 \equiv 2(\bmod 7)$$

Therefore

$$N = [(2 \times 70) + (3 \times 21) + (2 \times 15)] - (2 \times 105) = 23$$

The key to understanding this method is to find out where the numbers 105, 70, 21 and 15 come from. It is immediately obvious that 105 is the lowest common multiple (LCM) of the moduli 3, 5 and 7. Now, if 70, 21 and 15 are to be expressed as $1 (\bmod m)$, the following relationship is implied:

$$70 = 2 \times \frac{3 \times 5 \times 7}{3} \equiv 1 (\bmod 3)$$

$$21 = 1 \times \frac{3 \times 5 \times 7}{5} \equiv 1 (\bmod 5)$$

$$15 = 1 \times \frac{3 \times 5 \times 7}{7} \equiv 1 (\bmod 7)$$

We shall not attempt to generalize this procedure; the interested reader may wish to consult Libbrecht's splendid monograph (1973), which contains an extensive discussion of this subject.

However, we should not leave the subject without saying a little more about the Chinese mathematician whose innovative work in this area must rank as one of the greatest contributions of Chinese mathematics. Chin Chiu Shao is generally regarded as one of the most accomplished mathematicians to come out of China. Indeed, in any list of great mathematicians who lived before the emergence of modern mathematics, Chin should figure prominently. He was one of the four brilliant Sung mathematicians of the first half of the thirteenth century who were responsible for developing algebra to a level which was far in advance of anything that would be achieved elsewhere until the middle of the seventeenth century. However, it should not be thought that in China at that time there was anyone like today's professional mathematicians. Of the four notable mathematicians of the period, Chu Shih Chieh was a wandering teacher, Yang Hui was a minor civil servant, Li Yeh was a scholarly recluse, and Chin Chiu Shao was a man of many parts, the most important of which were his work for the military and civil service.

The little we know of Chin's character is not very flattering. He was considered unprincipled, extravagant and boastful, and his penchant for sexual imbroglio made him the equal of Casanova himself. But nobody denies his remarkable versatility. He was well versed in astronomy, harmonics, mathematics and even architecture. In sports, there were few to match him in polo, archery or swordplay. He made notable contributions in two areas of mathematics: in the solution of numerical equations of higher degree (as we have discussed) and, more importantly, the derivation of the *ta yen* rule for solving indeterminate equations of the first degree.

His treatment of indeterminate analysis is found in his best-known book, *Su Shu Chiu Chang* (Nine Sections of Mathematics), written in 1247. The book contains 81 problems divided into nine sections, but there the resemblance to the *Chiu Chang* stops. Neither the examination of the subjects covered nor the illustrative problems included owe much to the revered text. In his preface, Chin notes that he intends to introduce a method of indeterminate analysis (*ta yen shu*) which, though known to calendar-makers and astronomers such as the famous Buddhist sage I-hsing (*c.* AD 700), was not found in the *Chiu Chang*.

Chin's approach is best illustrated by an example from his book; we do not, however, follow him in the details of the solution he offered. His procedure for solving this problem is summarized by Libbrecht (1973, p. 408).

EXAMPLE 7.6   *Three thieves, A, B and C, entered a rice shop and stole three vessels filled to the brim with rice but whose exact capacity was not known. When the thieves were caught and the vessels recovered, it was found that all that was left in Vessels X, Y and Z were 1 ko, 14 ko and 1 ko respectively. The captured thieves confessed that they did not know the exact quantities that they had stolen. But A said that he had used a 'horse ladle' (capacity 19 ko) and taken the rice from vessel X. B confessed to using his wooden shoe (capacity 17 ko) to take rice from vessel Y. C admitted that he had used a bowl (capacity 12 ko) to help himself to the rice from Vessel Z. What was the total amount of rice stolen?*

Solution   The problem, restated concisely in modern notation, is to find N given that

$$N \equiv 1(\mathrm{mod}\ 19) \equiv 14(\mathrm{mod}\ 17) \equiv 1(\mathrm{mod}\ 12)$$

Chin's answer is that the total amount of rice stolen (N) is 3193 *ko*.

The *Su Shu Chiu Chang* contains a number of practical problems of indeterminate analysis. They include problems in calendar calculations, engineering and military applications, and architecture. Chin worked not only with integers but also with fractions, for which he devised special procedures. Libbrecht (1973) gives a detailed technical discussion of Chin's work and fills in the social and economic background to the various problems. For its rigour, clarity and originality, Chin's work must rank as one of the outstanding pieces of mathematical literature. It is a measure of its quality that later Chinese mathematicians found it difficult to comprehend. It thus remained neglected until the eighteenth century, when it began to arouse some interest. By this time, though, work on linear congruence had already begun in Europe with seminal contributions from Euler (1743) and Gauss (1801), initially using techniques similar to the ones pioneered by Chin. European developments soon eclipsed the Chinese work that had begun with Sun Tsu some fifteen hundred years before.

CHINESE MATHEMATICS: A FINAL ASSESSMENT

From our survey in this chapter and the last, we can identify the areas where Chinese contributions were notable:

1 As early as the Shang dynasty (*c.* fourteenth century BC), there emerged a system of notation which used nine numerals and the place-value principle. This was about a thousand years before the Indian number system, and was the earliest instance of the use of the place-value principle, after Babylon.

2 The development of an algorithm for extracting square and cube roots was first explained in the *Chiu Chang*, and elaborated and refined by Sun Tsu (*c.* AD 300) and other commentators. The thirteenth-century mathematicians extended the algorithm to the extraction of roots of any order. It is possible that the Arab mathematician Al-Kashi (*c.* AD 1400) and later Europeans were influenced by the Chinese method.

3 It is in China, near the beginning of the Christian era, that the concept of negative numbers and operations with them appear for the first time.

4 In *Chou Pei Suan Ching* there appears one of the earliest 'proofs' of the *kou ku* (or Pythagorean) theorem. A detailed discussion of the applications of this theorem to practical problems is found in the *Chiu Chang* and its commentaries.

5 Probably the earliest applications of the 'rule of three' and the 'rule of double false position' are found in Chinese mathematics during the first few centuries AD. The methods may later have been transmitted through India to the Arabs, who in turn passed them to the West.

6 A notable contribution was the development of numerical methods of solving higher-order equations in the thirteenth century – methods similar to those associated with Horner and Ruffini, at the beginning of the nineteenth century.

7 From Sun Tsu (*c.* AD 300) onward, the Chinese forged ahead with the study of indeterminate equations of the first degree, and devised an approach based on continued fractions which was similar to the one used by Euler and Gauss five hundred years later.

8 Pascal's triangle of binomial coefficients was known in China as early as AD 1100. Chinese mathematicians used it to solve numerical equations of higher degree.

9 In the work of some Chinese mathematicians can be seen attempts to blend geometric and algebraic approaches which are reminiscent of the work of al-Khwarizmi (*c.* AD 800), discussed in Chapter 10.

10 The values of π estimated by Liu Hui (*c.* AD 200) and Tsu Chung Chih (*c.* AD 400) remained the most accurate values for a thousand years.

Much research needs to be done before we can be more certain about the nature and mode of the interchange of mathematical ideas that took place between China and other cultural centres. Throughout history China has been relatively isolated from other cultures, partly by sheer geographical distance. Archaeological evidence suggests that at the time of the river valley civilizations there were contacts between Egypt, Mesopotamia and the Harappan cultures; the civilization that developed along the Yellow River, however, was remote and separated from areas to the south and west by natural barriers such as the Himalayas and the Central Asian plains. But these geographical barriers were not sufficient to exclude all contact, as there is evidence to show.

The major impact of Chinese culture and mathematics (particularly the *Chiu Chang*) on Japan, Korea and other neighbouring countries is clear, but Chinese contacts with areas to her south and west are more difficult to establish.

By the second century AD trade over the Silk Routes from China to the West was at its height, and along with the goods went ideas and techniques. In the centuries to come, the Classical civilizations of both the East and the West would suffer invasions small and large, culminating in Mongol hegemony over vast stretches of the Eurasian plains, which both served as instruments for diffusion and led to the convergence of ideas and technological practices. In examining the dissemination of Chinese mathematics, one needs to look at the Indian and Arab connections.

There is only fragmentary evidence of Chinese–Indian cultural and scientific contacts before the rise of Buddhism around the fourth century AD. A number of Chinese Buddhist scholars (notable among the early travellers were Fa Hsien (*c.* AD 400) and Hsuan Chuang (*c.* AD 650)) made their pilgrimage to holy places in India, bringing back many texts for translation. Among the places they visited were monasteries such as Nalanda and Taxila which were Indian centres of scholarship, not only in religion but in medicine, astronomy and mathematics too. Few of the writings or commentaries of these Buddhist pilgrims from

China have been examined for what they reveal about Indian science; the main interest has been in their religious and sociological content.

Then there is the evidence of Chinese diplomats posted at the court of the Guptas in India in the middle of the first millennium AD. And from the seventh century, there is evidence that translations were made of Indian astronomical and mathematical texts, such as the *Brahman Suan Fa* (Brahman Arithmetical Rules) and *Brahman Suan Ching* (Brahman Arithmetical Classic) mentioned in the records of the Sui dynasty. The records of the Tang dynasty contain the names of Indian astronomers. One of the Indians, whose Chinese name was Chu Tan Hsia Ta (Gautama Siddharta), was reputed to have constructed a calendar, based on the Indian *Siddhantas*, at the orders of the first emperor of the Tang dynasty. The text contains sections on Indian numerals and operations, and sine tables. Yabbuchi (1954) located a surviving block-print text which contains Indian numerals, including the use of a dot to indicate zero; this is one of the earliest references to zero as a numeral in Chinese texts. There are also sine tables at intervals of 3°45' for a radius of 3438 units, which are the values given in the Indian astronomical texts *Aryabhatiya* and *Suryasiddhanta*.* This is the earliest record of a sine table in any Chinese text. Unfortunately, there is little evidence of Chinese science in any of the extant Indian texts.

The Arab connection is probably better documented. One of the better known *hadiths* (i.e. utterances of the Prophet Muhammad which have religious sanction) is: 'Seek learning, though it be as far away as China.' There are a number of reports of political and diplomatic links between the Arab world and China to supplement trade relations. Arab travellers, including ibn Battuta (*c.* AD 1350), reputedly the greatest traveller of medieval times, gave detailed accounts of Chinese society and science, including ship-building, the manufacture of porcelain, the use of paper money and even a comprehensive system of old age pensions. Chinese mathematics may have made specific borrowings from Arab sources: it is possible that trigonometric methods used in astronomy may have been transmitted through Arab and Indian contacts. In constructing a calendar in the fourteenth century, Kou Shou Ching used spherical

---

*The choice of a radius of value 3438 was determined by the practice of dividing the circumference (C) of a circle into $360 \times 60 = 21\,600$ equal parts. If the length of the arc of each of these equal parts is one unit, and the value of π is taken as $3\cdot1416$ (Aryabhata's value), then the radius (r) of the circle can easily be established from the formula $C = 2\pi r$ to be $r = 21\,600/6\cdot2832 = 3438$ (to the nearest integer). Furthermore, in the construction of a sine table for angles between 0 and 90°, Aryabhata divided the quadrant into 24 equal parts so that the table would give the sines of the multiples of the basic angle (90/24), or 3°45'.

trigonometric methods which seem to have Arab origins. It is possible that Euclid's geometry may have reached China at the end of the thirteenth century via the Arabs, though the lack of interest in the Euclidean method – or, more probably, a lack of sympathy – meant that this knowledge was forgotten until it was reintroduced by the Jesuits. Finally, the lattice method of multiplication, which will be examined in Chapter 10, appears in Chinese mathematical texts at the end of the sixteenth century. It is not clear whether the agents of transmission were the Arabs or the Portuguese.

Whether there was any direct transmission of mathematical knowledge from China to the West remains a matter of conjecture. However, the possibility should not be dismissed out of hand, as many historians of mathematics are inclined to do – either because they find the idea unpalatable or because there is insufficient documentary evidence. The fact remains that, as early as the third century BC, Chinese silk and fine ironware were to be found in the markets of Imperial Rome. And a few centuries later a whole range of technological innovations found their way slowly to Europe. It is not unreasonable to argue that some of China's intellectual products, including mathematical knowledge, were also carried westwards to Europe, there perhaps to remain dormant during Europe's intellectual Dark Ages, but coming to life once more with the cultural awakening of the Renaissance.

During the late seventeenth and the eighteenth centuries, Europe became aware of the Chinese intellectual heritage. The Jesuits were responsible through their translations for awakening the interest of people like Voltaire, Gottfried Leibniz and François Quesnay in Chinese thought and science. Leibniz (1646–1716), one of the founders of modern mathematics, was in the forefront of promoting a universal system of natural philosophy based on Confucian writings. He founded the Berlin Society of Science with the express purpose of 'opening up China and the interchange of civilizations between China and Europe'. While there is already some recognition of Europe's debt to China in the realms of philosophy and the arts (Edwardes, 1971), the possibility of an 'east–west passage' of scientific ideas during this period through the Jesuit connection has hardly been explored.

If these conjectures are implausible, then so too must be the attribution of Greek or European origins to so many developments in mathematics and astronomy in other cultures. However, it is my belief that if the idea of a westward transmission of mathematics were to be taken more seriously, and research were to be channelled in this direction, then it would be only a matter of time before evidence of east–west links comes to light.

8 Ancient Indian Mathematics

◇◇

A RESTATEMENT OF INTENT AND A BRIEF HISTORICAL SKETCH

Ancient Indian history raises many problems. The period before the Christian era takes on a haziness which seems to have prompted opposing reactions. There are those who make excessive claims for the antiquity of Indian mathematics, and others who go to the opposite extreme and deny the existence of any 'real' Indian mathematics before about 500 AD. The principal motive of the former is to emphasize the uniqueness of Indian mathematical achievements where, if there was any influence, it was always a one-way traffic from India to the rest of the world. The motives of the latter are more mixed. For some their Eurocentrism (or Graeco-centrism) is so deeply entrenched that they cannot bring themselves to face the idea of independent developments in early Indian mathematics, even as a remote possibility.

A good illustration of this blinkered vision is provided by a widely respected historian of mathematics at the turn of this century, Paul Tannery. Confronted with the evidence from Arab sources that the Indians were the first to use the sine function as we know it today, Tannery devoted himself to seeking ways in which the Indians could have acquired the concept from the Greeks. For Tannery, the very fact that the Indians knew and used sines in their astronomical calculations was sufficient evidence that they must have had it from the Greeks. But why this tunnel vision? The following quotation from G. R. Kaye (1915) is illuminating:

> The achievements of the Greeks in mathematics and art form the most wonderful chapters in the history of civilisation, and these achievements are the admiration of western scholars. It is therefore natural that western investigators in the history of knowledge should seek for traces of Greek influence in later manifestations of art, and mathematics in particular.

It is particularly unfortunate that Kaye is still quoted as an authority on Indian mathematics. Not only did he devote much attention to showing the derivative nature of Indian mathematics,* usually on dubious linguistic grounds (his knowledge of Sanskrit was such that he depended largely on indigenous 'pandits' for translations of primary sources), but he was prepared to neglect the weight of contemporary evidence and scholarship to promote his own viewpoint. So, while everyone else claimed that the Bakhshali Manuscript (to be discussed at the end of this chapter) was written or copied from an earlier text dating back to the first few centuries of the Christian era, Kaye insisted that it was no older than the twelfth century AD. Again, while the Arab sources unanimously attributed the origin of our present-day numerals to the Indians, Kaye was of a different opinion. And the distortions that resulted from Kaye's work have to be taken seriously because of his influence on Western historians of mathematics, many of whom remained immune to findings which refuted Kaye's inferences and which established the strength of the alternative position much more effectively than is generally recognized.

This tunnel vision is not confined to mathematics alone. Surprised at the accuracy of information on the preparation of alkalis contained in an early Indian textbook on medicine (*Susruta Samhita*†) dating back to a few centuries BC, the eminent chemist and historian of the subject, M. Berthelot (1827–1909), suggested that this was a later insertion, after the Indians had come into contact with European chemistry!

While non-European chauvinism (on the part of, for example, the Arabs, Chinese and Indians) does persist, the 'arrogant ignorance' (as J. D. Bernal (1969) described the character of Eurocentric scholarship in the history of science) is the other side of the same coin. But the latter tendency has done more harm than the former because it rode upon the political domination imposed by the West, which imprinted its own version of knowledge on the rest of the world.

Table 8.1 offers a brief summary of the main events in the long history of India as a backdrop to the development of mathematics; it divides Indian history up to the beginning of the sixteenth century into six periods. The map of India

---

*Attempts to show the derivative nature of Indian science, and especially its supposed Greek roots, continue even today. For example, Pingree (1981) has prepared a chronology of Indian astronomy which is notable for the absence of any Indian presence!

†In this chapter and the next, many of the titles of Indian texts are given without translation. This is sometimes because words in the title have a number of possible interpretations, and sometimes because a literal translation would not be particularly meaningful.

in Figure 8.1 shows places mentioned in the text. The earliest evidence of mathematics is found among the ruins of the Indus Valley civilization, which goes back to 3000 BC. (It is perhaps more appropriately referred to as the Harappan civilization, since at its peak it spread far beyond the Indus Valley itself.) Around 1500 BC, according to the traditional — though increasingly contentious — view among historians, a group of people descended from the North and destroyed the Harappan culture, but not before they had absorbed some of its features. These invaders are often referred to as 'Aryans' — a term which has acquired an unfortunate connotation in modern times through its association with the Nazis.

The Aryans were a pastoral people, speaking a language which belonged to the Indo-European family. Over the years this language, Sanskrit, developed sufficiently to become a suitable medium for religious, scientific and philosophical discourse. Its potential for scientific use was greatly enhanced as a result of the thorough systematization of its grammar by Panini, about two thousand six hundred years ago. In a book entitled *Astadhyaya*, Panini offered what must be the first attempt at a structural analysis of a language. On the basis of just under 4000 *sutras* (i.e. rules expressed as aphorisms), he built virtually the whole structure of the Sanskrit language, whose general 'shape' hardly changed for the next two thousand years. Sanskrit served as a useful medium for recording early scriptural texts such as the *Vedas* and *Upanishads*, early scientific literature such as the *Vedangas* and early rules of social conduct such as the *Code of Manu*.

An indirect consequence of Panini's efforts to increase the linguistic facility of Sanskrit soon became apparent in the character of scientific and mathematical literature. This may be brought out by comparing the grammar of Sanskrit with the geometry of Euclid — a particularly apposite comparison since, whereas mathematics grew out of philosophy in ancient Greece, it was, as we shall see, partly an outcome of linguistic developments in India.

The geometry of Euclid's *Elements* starts with a few definitions, axioms and postulates and then proceeds to build up an imposing structure of closely interlinked theorems, each of which is in itself logically coherent and complete. In a similar fashion, Panini began his study of Sanskrit by taking about 1700 basic building blocks — some general concepts, vowels and consonants, nouns, pronouns and verbs, and so on — and proceeded to group them into various classes. With these roots and some appropriate suffixes and prefixes, he constructed compound words by a process not dissimilar to the way in which one specifies a function in modern mathematics. Consequently, the linguistic facility of the language came to be reflected in the character of mathematical

Table 8.1 Chronology of Indian history and mathematics.

Period	Main historical events	Mathematics	Notable mathematicians
3000–1500 BC	The Indus Valley civilization (script undeciphered) covering 1·2 million square kilometres; main urban centres Harappa, Lothal and Mohenjo-Daro	Weights, artistic designs, 'Indus scale'; brick technology probably influenced the construction of Vedic altars in the next period	
1500–500 BC	The coming of the Aryans; the formation of Hindu civilization; the emergence of the *Code of Manu*; the recording of the *Vedas* and *Upanishads*	*Vedangas* and *Sulbasutras*; problems in astronomy, arithmetical operations, Vedic geometry	Baudhayana, Apastamba, Katyayana
500–200 BC	The establishment of Indian states; the rise of Buddhism and Jainism; contacts with Persia maintained; the Mauryan empire, culminating in the reign of Asoka, who spread Buddhism abroad	Vedic mathematics continues during the earlier years but declines with ending of ritual sacrifices; beginnings of Jaina mathematics: number theory, permutations and combinations, the binomial theorem; astronomy	
200 BC to AD 400	Triple Division: Kushan dynasty (North), Pandyas (South), Bactrian–Persian (Punjab); pervading influence of Buddhism in art and sculpture	Jaina mathematics; rules of mathematical operations, decimal place notation, first use of 0; algebra including simple, simultaneous and quadratic equations; square roots; details of how to represent unknown quantities and negative signs	

Period	Main historical events	Mathematics	Notable mathematicians
400–1200	Imperial Guptas reaching their height in the reign of Harsha (606–647); flowering of Indian civilization as shown in science, philosophy, medicine, logic, grammar and literature	The Classical period of Indian mathematics; important works: the Bakhshali Manuscript, *Aryabhatiya*, *Pancha Siddhantika*, *Aryabhatiya Bhasya*, *Maha Bhaskariya*, *Brahma Sputa Siddhanta*, *Patiganita, Ganita Sara Samgraha, Ganitilaka, Lilavati, Bijaganita*	Aryabhata I, Varahamihira, Bhaskara I, Brahmagupta, Sridhara, Mahavira, Bhaskara II (also known as Bhaskaracharya)
1200–1600	Early Muslim dynasties; birth of Sikhism; the Hindu kingdom of Vijaynagar in the south	Decline of mathematics and learning in the north; the rise of the Kerala school of astronomy and mathematics; work on infinite series and analysis	Narayana Madhava Nilakantha

literature and reasoning in India. Indeed, it may even be argued that the algebraic character of ancient Indian mathematics is but a by-product of the well-established linguistic tradition of representing numbers by words.

The third period of Indian history began around 800 BC. It saw not only the establishment of two of the great religions originating in India, Buddhism and Jainism, but also the growth of independent states, a number of which were later merged to form the first of the great empires of India, the Mauryan empire. This period marked the decline of Vedic mathematics and the beginnings of the Jaina school, which was to do notable work in number theory, and permutations and combinations, as well as other abstract areas of mathematics.

The fourth period, from about 200 BC, was a period of instability and fragmentation brought about by waves of foreign invasions. But it was also a time of useful cross-cultural contacts with neighbours and with the Hellenistic world, bringing fresh ideas into Indian science and laying the foundation for great advances in the next period. The Kushan empire became an important

Figure 8.1 Map of India and (inset) South East Asia.

vehicle for spreading not only Buddhist religion and art, but also Indian science, particularly astronomy, into western Asia. This is a period in which Jains made a notable contribution to mathematics.

The fifth period, from the third to the twelfth centuries AD, is often referred to as the Classical period of Indian civilization. The earlier part of this period saw much of India ruled by the Imperial Guptas, who encouraged the study of science, philosophy, medicine and other arts. Mathematical activities reached a climax with the appearance of the famous quartet, Aryabhata, Brahmagupta, Mahavira and Bhaskaracharya. Their lives and works will be examined in the next chapter. Indian work on astronomy and mathematics spread westwards, reaching the Arabs, who in turn absorbed, refined and augmented what they received before transmitting the results to Europe.

The last period, which we may describe as the 'medieval' period of Indian history, saw the rise of great states in southern India and a migration of mathematics and astronomy from the north to the south, probably as a result of political upheavals. It was believed for a long time that mathematical development came virtually to a stop in India after Bhaskaracharya in the twelfth century. There may be some element of truth in this as far as the north was concerned, but in the south – and particularly in the south-west, in the area corresponding to the present-day state of Kerala – this was a period marked by remarkable studies of infinite series and mathematical analysis which predated similar work in Europe by about three hundred years.

The mathematics of Kerala will be presented in the next chapter. In this chapter we examine Indian mathematics from its early beginnings to just before the Classical period; in the next chapter we consider classical and medieval Indian mathematics. The development of Indian numerals is dealt with in this chapter, though there is some historical overlap, particularly when one considers the spread of the numerals into countries such as Cambodia and Java to the east, and into the Arab world to the west. The reader may wish to refer to Table 8.1 and Figure 8.1 whenever necessary to sketch in the historical and geographical background to this and the next chapter.

MATHS FROM BRICKS:
EVIDENCE FROM THE HARAPPAN CULTURE

Between 1921 and 1923 a series of archaeological excavations along the banks of the Indus uncovered the remains of two urban centres, at Harappa and Mohenjo-Daro, dating back to about 3000 BC. Subsequent searches over the last four decades have revealed further remains spread across an area of about

1·2 million square kilometres, including not only the Indus Valley but parts of East Punjab and Uttar Pradesh, nothern Rajasthan, the coastal areas of Gujerat (which contained the major port and the third of the large urban settlements of this civilization, Lothal) and northern areas near the Persian border. Over seventy sites, large and small, have been excavated to uncover this most dispersed of the early civilizations, hereafter referred to as the Harappan culture.

It was a highly organized society, with the towns supplied by surrounding agricultural communities which cultivated wheat and barley and raised live-stock. Urban development was regulated by planning and characterized by a highly standardized architecture. There is every possibility that the town-dwellers were skilled in mensuration and practical arithmetic of a kind similar to what was practised in Egypt and Mesopotamia. Alas, the Harappan script remains undeciphered, so our evaluation of the mathematical proficiency of this civilization must be based on excavated artefacts. The archaeological finds described below do provide some indication, however meagre, of the nature of the numerate culture that this civilization possessed.

1 A number of different plumb-bobs of uniform size and weight, showing little change over the five hundred years for which evidence is available, have been found throughout the vast area of the Harappan culture. This uniformity of weights over such a wide area and time is quite unusual in the history of metrology. Rao (1973), who examined the considerable finds at Lothal, showed that the weights could be classified as 'decimal': if we take the plumb-bob weighing approximately 27·584 grams as a standard, representing 1, the other weights form a series with values of 0·05, 0·1, 0·2, 0·5, 2, 5, 10, 20, 50, 100, 200 and 500. Such standardization and durability is a strong indication of a numerate culture with a well-established, centralized system of weights and measures.

2 Scales and instruments for measuring length have been discovered at Mohenjo-Daro, Harappa and Lothal. The Mohenjo-Daro scale is a fragment of shell 66·2 mm long, with nine carefully sawn, equally spaced parallel lines, on average 6·7056 mm apart. The accuracy of the graduation is remarkably high, with a mean error of only 0·075 mm. One of the lines is marked by a hollow circle, and the sixth line from the circle is indicated by a large circular dot. The distance between the two markers is 1·32 inches (33·5 mm), and has been named the 'Indus inch'.

There are a number of interesting links between this unit of measurement (if indeed this is what it was) and others found elsewhere. A Sumerian *shushi* is exactly half an Indus inch, which would support other archaeological evidence of a possible link between the two urban civilizations. In north-west India a traditional yard known as the *gaz* was in use from very early times; in the sixteenth century the Mughal Emperor Akbar even attempted (unsuccessfully) to have the *gaz* adopted as a standard measure in his kingdom. The *gaz*, which is 33 inches (840 mm) by our measurement, equals 25 Indus inches. Further-more, the *gaz* is only a fraction (0·36 inches) longer than the megalithic yard, a measure that seems to have been in use in north-west Europe around the second millennium BC. This has led to the conjecture that a decimal scale of measurement may have originated somewhere in western Asia and spread widely – as far as Britain, and to ancient Egypt, Mesopotamia and the Indus Valley (Mackie, 1977).

A notable feature of the Harappan culture was its extensive use of kiln-fired bricks and the advanced level of its brick-making technology. Chattopadhyaya (1986) has argued for a closer examination of this activity, which should give us some vital clues about the direction and character of later mathematical developments in India. So let us examine the socio-economic origins of making and using kiln-fired brick in the Harappan culture.

There is general agreement among archaeologists of the Harappan culture that, during its formative stages, farmers had to produce a substantial agri-cultural surplus to support a rapidly growing urban population. Presumably, this occurred not as a result of the introduction of any revolutionary agricul-tural technology, but because of improved knowledge about how to exploit the annual flooding and thus raise agricultural productivity. When the floods receded, principal food crops such as wheat and barley would be sown on the land which had been submerged, to be harvested in March or April. The land would need no ploughing, no manure, and no additional irrigation. With a minimum of labour and equipment, a substantial yield could be achieved. The return of the floods would mark the period when the autumn crops, such as cotton and sesame, would be sown for harvesting at the end of the autumn. By recognizing how the floods could be utilized to prepare the land for cultivation, and by following a strict sequence of sowing and harvesting spring and autumn crops, it became possible to build up a large agricultural surplus. However, before such a system of cultivation could be considered, an effective system of flood control was necessary. In areas where stone was not readily available (and this included most of the Harappan sites), there was a need for something more solid than mud-brick, which was easily destroyed by rain or floodwater. The momentous discovery was the technology for firing bricks. There is evidence, especially from Kalibanga, a pre-Harappan site, that kiln-fired bricks were already in use, but by the time the Harappan culture had matured there had been a veritable explosion in the production and use of such bricks.

The story is told of how William Brunton, a nineteenth-century railway-builder, dug up the ruins of Harappa for bricks to use as ballast for a railway line between Multan and Lahore, a distance of over a hundred miles! Despite this and other new uses for antique bricks, a massive quantity of them remain at Harappa. They are exceptionally well baked and of excellent quality, and may still be used over and over again provided some care is taken in removing them in the first place. They contain no straw or other binding material. While fifteen different sizes of Harappan bricks have been identified, the standard ratio of the three dimensions – the length, breadth and thickness – is always 4 : 2 : 1. Even today this is considered the optimal ratio for efficient bonding.

A correspondence between the Indus scales (from Harappa, Mohenjo-Daro and Lothal) and brick sizes has been noted by Mainikar (1984). Bricks of different sizes from these three urban centres were found to have dimensions which were integral multiples of the graduations of their respective scales. This apparent relationship between brick-making technology and metrology was to reappear about 1500 years later during the Vedic period, in the construction of sacrificial altars of brick. We take up the story again later in this chapter.

MATHEMATICS FROM THE *VEDAS*

The sources

Vedic literature went through four stages of development: the *Samhitas* (*c.* 1000 BC), the *Brahmanas* (*c.* 800 BC), the *Aranyakas* (*c.* 700 BC) and the *Upanishads* (*c.* 600–500 BC). The *Samhitas* were lyrical collections of hymns, prayers, incantations, and sacrificial and magical formulae. From what must have been a vast corpus of such writings, four great collections have come down to us – the oldest surviving literary efforts of mankind. In order of their age,* they are:

1 *Rigveda*, which contains hymns and prayers to be recited during the per-

---

*The dating of early Indian texts is highly uncertain. The dates given here are rough and conservative estimates of when the first versions of the texts were recorded. It is very likely that before they were written, an earlier oral tradition kept the contents alive. Copying old texts was a common pursuit of the Indian scholar and student, sanctioned by religion and custom. It is therefore important not to depend on evidence of the dates of copies of mathematical texts in assessing the true age of a particular method or technique.

formance of rituals and sacrifices,

2 *Samaveda*, which contains melodies to be sung on suitable occasions,

3 *Yajurveda*, which contains sacrificial formulae for ceremonial occasions, and

4 *Atharvaveda*, a collection of magical formulae and spells.

The *Brahmanas*, the second great division of Vedic literature, have been described as practical handbooks for those conducting sacrifices. As the Brahmana communities gradually dispersed from the north to the eastern and southern parts of the country, there arose a need for a record of ritual procedures and duties for a travelling class of priests, and a means of allocating special tasks among different priests.

For mathematics, a more important source is provided by the 'appendices' to the main *Vedas*, known as the *Vedangas*. These were classified into six branches of knowledge: (1) phonetics, the science of articulation and pronunciation, (2) grammar, (3) etymology, (4) metronomy (*chandah*), the art of prosody, (5) astronomy, and (6) rules for rituals and ceremonials (*kalpa*). In the last two *Vedangas* are found the most important sources of mathematics from the Vedic period. The evidence is usually in the form of *sutras*, a peculiar form of writing which aims at the utmost brevity and often uses a poetic style to capture the essence of an argument or result. By avoiding the use of verbs as far as possible and compounding nouns at great length, a vast body of knowledge was made easier to memorize. Condensation into *sutras* was also a way of eking out scarce writing materials. This was the form in which the contents of the *Brahmanas* were preserved, and it was adopted later not only by various philosophical and scientific schools, but also by writers of books on statecraft (*arthasastra*) and sex manuals (*kamasastra*).

We have referred to the *Kalpasutras* as an important source of Vedic mathematics. This ritual literature included *Srautasutras*, which gave directions for constructing sacrificial fires at different times of the year. Part of this literature dealt with the measurement and construction of sacrificial altars, and came to be known as the *Sulbasutras*. The term originally meant rules governing 'sacrificial rites', though later the word *sulba* came to refer to the cord of rope used in measuring altars. Most of what we know of Vedic geometry comes from these *sutras*.

The *Sulbas*: Mathematics in the service of religion?

There is a view that Indian mathematics originated in the service of religion. The proponents of this view have sought their main support in the complexity of motives behind the recording of the *Sulbasutras*. Since time immemorial, they argue, the needs of religion have determined not only the character of Indian social and political institutions, but also the development of her scientific knowledge. Astronomy was developed to help determine the auspicious day and hour for performing sacrifices. The forty-nine verses of *Jyotisutras* (the *Vedanga* containing astronomical information) gave procedures for calculating the time and position of the Sun and Moon in various *naksatras* (signs of the zodiac). Also, a strong reason in Vedic India for the study of phonetics and grammar was to ensure perfect accuracy in pronouncing every syllable in a prayer or sacrificial chant. And the construction of altars (or *vedi*) and the location of sacred fires (or *agni*) had to conform to clearly laid-down instructions about their shapes and areas if they were to be effective instruments of sacrifice.

The *Sulbasutras* provided such instructions for two types of ritual, one for worship at home and the other for communal worship. Square and circular altars were sufficient for household rituals, while more elaborate altars whose shapes were combinations of rectangles, triangles and trapeziums were required for public worship. One of the most elaborate of the public altars was shaped like a falcon just about to take flight, as shown in Figure 8.2. It was believed that offering a sacrifice on such an altar would enable the soul of the supplicant to be conveyed by a falcon straight to heaven.

Early researchers on the *Sulbasutras*, notably Thibaut in the second half of the nineteenth century, were at pains to stress the religious element of these texts, but ignored their secular side. It is worth reiterating our earlier argument that the *Sulbasutras* may well provide a connecting thread between the Harappan culture, which came to an end around 1750 BC, and the emergence of a literate Vedic culture around the beginning of the first millennium BC, for the highly developed brick-making technology of the Harappan culture was replicated in the construction of sacrifical altars during the Vedic period. According to this view, then, the instructions given in the *Sulbasutras* were mainly for the benefit of craftsmen laying out and building altars. To overemphasize the religious and ritual features of altar design and construction at the expense of the technological aspects is to diminish the role of craftsmen in ancient Indian society, at the same time buttressing the stereotypical view of a society dominated by priests and overwhelmed by ritual.

Three of the more mathematically important *Sulbasutras* were the ones recorded by Baudhayana, Apastamba and Katyayana. Little is known about these *sulbakaras* (i.e. authors of *Sulbasutras*), except that they were not just scribes but probably also priest-craftsmen performing a multitude of tasks including constructing *vedi* (sacrificial altars), maintaining *agni* (sacred fires) and

Figure 8.2 The first layer of a Vedic sacrificial altar in the shape of a falcon; the wings are each made from 60 bricks of type a, and the body from 46 type b, 6 type c, and 24 type d bricks (after Thibaut, 1875).

instructing worshippers on the appropriate choice of both sacrifices and altars. It is difficult to assign firm dates to these three texts. All we can say is that the earliest of them, the one composed by Baudhayana, was probably first recorded between 800 and 600 BC; and that the other two were recorded one or two centuries later. (From their style, they predate the Sanskrit grammarian Panini, who lived in the fourth century BC.) Baudhayana's *Sulbasutra* is complete in three chapters. It contains a general statement of the Pythagorean theorem, an approximation procedure for obtaining the square root of 2 correct to five decimal places, and a number of geometric constructions including ones for 'squaring the circle' (approximately) and constructing rectilinear shapes whose area was equal to the sum or difference of areas of other shapes. The next oldest text, by Apastamba, contains six chapters and treats in more detail the topics examined by Baudhayana. Katyayana's *Sulbasutra* adds little to the work of his predecessors.

Sulba geometry

The geometry of the *Sulbasutras* grew out of the need to ensure strict conformation of the orientation, shape and area of altars to the prescriptions laid down in the Vedic scriptures. Such accuracy was just as important for the efficacy of the ritual as was the meticulous pronunciation of Vedic chants (or *mantras*). However, while accurate geometric methods were used, the principles underlying the constructions were often not discussed. It will therefore be useful, while going through the illustrative examples given below, to bear in mind the distinction between three aspects of the geometry found in the *Sulbasutras*:

1 geometric results and theorems explicitly stated,
2 procedures for constructing different shapes of altars, and
3 algorithmic devices contained in (1) and (2).

In the first category the most notable is the Pythagorean theorem for a right-angled triangle. As we have seen, knowledge of this result was found in a number of early mathematical traditions. Here we shall just look at a specific application of it to the design of a particular type of altar. It is the second aspect of *sulba* geometry that forms a substantial part of the mathematical evidence of the period. Again, the emphasis is on designing altars with the minimum of tools. We shall briefly examine just three of the fifteen such constructions discussed in the texts. Finally, as an illustration of an algorithmic device born out of constructional needs, we consider the approximation procedure for evaluating the square root of 2.

The actual statement of the Pythagorean theorem, expressed in terms of the sides and diagonals of squares and rectangles, is found in the Baudhayana, Apastamba and Katyayana *Sulbasutras*. The Baudhayana version states:

> The rope which is stretched across the diagonal of a square produces an area double the size of the original square.

All three *Sulbasutras* gave a more general proposition:

> The rope [stretched along the length] of the diagonal of a rectangle makes an [area] which the vertical and horizontal sides make together.

Figure 8.3 illustrates how this proposition was applied in the construction of altars. It shows a drawing of the base of the *Mahavedi* altar (a 'cemetery' altar at which an intoxicating drink called *soma* was offered as a sacrifice to the gods). Its base had to be constructed to precise dimensions if the sacrifice was to bear fruit. It had to be an isosceles trapezium ABCD where AD and BC were 24 and 30 *padas* (literally 'feet'). The altitude of the trapezium (i.e. the distance between the midpoints X and Y of AD and BC) had to be precisely 36 *padas*.

The instructions given for the construction of this altar in Apastamba's *Sulbasutra* are, in modern notation, as follows:

1 With the help of a rope mark out XY, which is precisely 36 *padas*.
2 Along this line, locate points P, R and Q such that XP, XR and XQ equal 5, 28 and 35 *padas* respectively.
3 Construct perpendiculars at X and Y.
4 Use the fact that the triangles APX, DPX, BRY and CRY are right-angled triangles with integral-valued sides to locate points A, B, C and D. In other words, make AXD 24 *padas* and BYC 30 *padas*. Join AB, BC, CD and DA.

Implied in these directions for construction are the following right-angled triangles with integral sides:

△APX and △DPX	with sides	5, 12, 13
△AOX and △DOX	with sides	12, 16, 20
△BRY and △CRY	with sides	8, 15, 17
△BOY and △COY	with sides	15, 20, 25
△AQX and △DQX	with sides	12, 35, 37
△BXY and △CXY	with sides	15, 36, 39

Figure 8.3 The layout of a *smasana* sacrificial altar.

AX = XD = 12 *padas*
BY = YC = 15 *padas*
XP = 5 *padas*
PR = 23 *padas*
RQ = 7 *padas*
QY = 1 *pada*
XY = 36 *padas*

Besides these integral 'Pythagorean triples', two involving fractions ($2\frac{1}{2}, 6, 6\frac{1}{2}$ and $7\frac{1}{2}, 10, 12\frac{1}{2}$) were also used to construct right-angled triangles. Furthermore, the construction of some altars required the use of triples such as $1, 1, \sqrt{2}$, $5\sqrt{3}, 12\sqrt{3}, 13\sqrt{3}$ and $15\sqrt{2}, 36\sqrt{2}, 39\sqrt{2}$. These numbers probably arose from ritual requirements which dictated constructions of altars whose areas were either integral multiples or fractions of the areas of other altars of the same shape. For example, the dimensions of a *soutramani* altar (one with a triangular base of sides $5\sqrt{3}, 12\sqrt{3}$ and $13\sqrt{3}$) were arrived at by starting with a 5, 12, 13 triangle, the unit of measurement being the *purusha* (nearly 2·5 metres, or the height of a man with his arms stretched above him).

The *Sulbasutras* were, however, primarily instruction manuals for geometric constructions: squares, rectangles, trapeziums and circles which had to conform to specified dimensions or areas. Any inaccuracy would make the consequent rituals and sacrifices ineffective. Here are three examples:

1 *To merge two equal or unequal squares to obtain a third square* The method is reported in all three *Sulbasutras*. Figure 8.4 shows the construction. In modern notation, let ABCD and PQRS be the two squares to be combined, and let DX be equal to SR. Join AX. The square on AX is equal to the sum of the squares ABCD and PQRS. The original explanation then points out that $DX^2 + AD^2 = AX^2 = SR^2 + AD^2$, which shows the use of the Pythagorean theorem.

2 *To transform a rectangle into a square of equal area* The result, from Baudhayana's *Sulbasutra*, is shown in Figure 8.5. ABCD is a rectangle of length AD and breadth AB. The procedure begins by completing the square ABKH. Let E and M be the midpoints of HD and KC respectively, so that EM bisects the rectangle HKCD. Move the rectangle EMCD so that its new position is KBJG. Complete the square KGFM. Draw an arc of radius JF to cut BC at W, and draw a line through W, parallel to MF, to cut JF at S. The required square is then the square on JS, JSTR. A demonstration that the square JSTR is equal in area to the rectangle ABCD follows easily from the Pythagorean theorem:

$$JS^2 = JW^2 - WS^2 = AJ^2 - BJ^2 = (AJ + BJ)(AJ - BJ)$$
$$= AD \cdot AB = \text{area of ABCD}$$

3 *Squaring a circle and circling a square* No geometric method can achieve this exactly; what the *Sulbasutras* provide are approximate constructions. In

Figure 8.4

Figure 8.6 ABCD is a given square of side a, centred at O. Join OD, and construct an arc from D to the point P, which lies on the line passing through O and the midpoint E of the side CD. Thus OD = OP. To construct a circle centred at O and of radius r equal in area to the square, we are advised to take the radius of the circle as the sum of half the length of the side of the square (i.e. $\frac{1}{2}a$) and one-third of the length of OP that remains outside the square.

In other words,

$$r = ON = OE + EN = OE + \tfrac{1}{3}EP = OE + \tfrac{1}{3}(OP - OE)$$

Now,

$$\sin\theta = \sin 45° = OE/OD = OE/OP = 1/\sqrt{2}$$

Therefore $OP = \sqrt{2}OE$ and so

$$r = OE + \tfrac{1}{3}(\sqrt{2}OE - OE)$$

Figure 8.5

Since $OE = \frac{1}{2}a$,

$$r = \frac{a}{2} + \frac{\frac{1}{2}a(\sqrt{2} - 1)}{3} = \frac{a}{6}(2 + \sqrt{2})$$

The area of the square is a^2 and that of the circle is πr^2, so

$$a^2 = \frac{\pi}{36}[a(2 + \sqrt{2})]^2$$

which implies a value of π of 3·088.

All three *Sulbasutras* give the following directions for converting a circle into a square: 'Divide the diameter into 15 parts and take 13 of these parts as the side of the square.' If d is the diameter of the circle and a the side of the required square, then

$$a = \frac{13}{15}d$$

which implies that $\pi = 3·004$.

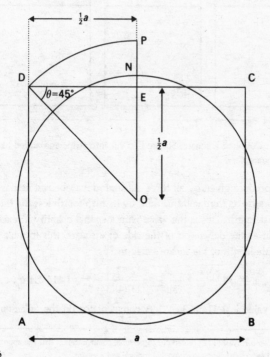

Figure 8.6

Irrational square roots: An approximation procedure

A remarkable achievement of Vedic mathematics was the discovery of a procedure for evaluating square roots to a high degree of approximation. The problem may have originally arisen from an attempt to construct a square altar twice the area of a given square altar.*

The problem, which the reader may wish to try, is one of constructing a square twice the area of a given square, A, of side 1 unit. It is clear that for the larger square, C, to have twice the area of square A, it should have side $\sqrt{2}$ units. Also, we are given a third square, B, of side 1, which needs to be dissected and reassembled so that by fitting cut-up sections of square C on square A, it is possible to make up a square close to the size of square C. Figure 8.7 shows diagrammatically what needs to be done.

Figure 8.7 Doubling a square: Square B is cut into strips and added to Square A, to give Square C.

The procedure given by all three *Sulbasutras* may be restated as: 'Increase the measure by its third and this third by its own fourth less the thirty-fourth part of that fourth. This is the value with a special quantity in excess.' If we take 1 unit as the dimension of the side of a square, this formula gives the approximate length of the square's diagonal as

$$\sqrt{2} = 1 + \frac{1}{3} + \frac{1}{(3)(4)} - \frac{1}{(3)(4)(34)} = 1\cdot414\,215\,6\ldots$$

The true value is $1\cdot414\,213\ldots$. A commentator on the *Sulbasutras*, Rama,

---

*Professor Takao Hayashi has pointed out in personal communication that the approximation for $\sqrt{2}$ was used for constructing a right-angled triangle and a square.

who lived in the middle of the fifteenth century AD, gave an improved approximation by adding two further terms to the equation:

$$-\frac{1}{(3)(4)(34)(33)} + \frac{1}{(3)(4)(34)(34)}$$

which gives a value correct to seven decimal places.

The *Sulbasutras* contain no clue as to how this remarkable approximation was arrived at. Many explanations have been proposed. A plausible one, put forward by Datta (1932), is as follows. Consider two squares, ABCD and PQRS, each of unit side (see Figure 8.8). PQRS is divided into three equal rectangular strips, of which the first two are marked 1 and 2. The third strip is subdivided into three squares, of which the first is marked 3. The remaining two squares are each divided into four equal strips marked 4 to 11. These eleven areas are added to the square ABCD as shown in Figure 8.8 to obtain a large square less a small square at the corner F. The side of the augmented square (AEFG) is

$$1 + \frac{1}{3} + \frac{1}{(3)(4)}$$

The area of the shaded square is $[1/(3)(4)]^2$, so that the area of the augmented square AEFG is greater than the sum of the areas of the original squares, ABCD and PQRS, by $[1/(3)(4)]^2$.

To make the area of the square AEFG approximately equal to the sum of the areas of the original squares ABCD and PQRS, imagine cutting off two very

Figure 8.8 How doubling the square may have led to the Indian approximation to $\sqrt{2}$ (after Datta, 1932).

narrow strips, of width x, from the square AEFG, one from the left side and one from the bottom. Then

$$2x \left[1 + \frac{1}{3} + \frac{1}{(3)(4)} \right] - x^2 = \left[\frac{1}{(3)(4)} \right]^2$$

Simplifying the above expression and ignoring x^2, an insignificantly small quantity, gives

$$x \simeq \frac{1}{(3)(4)(34)}$$

The diagonal of each of the original squares is $\sqrt{2}$, which can be approximated by the side of the new square as just calculated:

$$\sqrt{2} \simeq 1 + \frac{1}{3} + \frac{1}{(3)(4)} - \frac{1}{(3)(4)(34)}$$

What is particularly appealing about this line of reasoning is that there is other evidence from *Sulbasutra* geometry of the use of this 'concrete' mode of argument, which was described as the 'out–in complementarity' principle in our earlier discussion of Chinese geometry. This mode of demonstration requires neither a well-developed symbolic algebra nor a Greek-style procedure of deductive inference. In this instance, the Indian and Chinese geometric approaches exhibit similarities which would surely repay further study.

In Chapter 4, on Babylonian mathematics, we came across an approximation method for evaluating $\sqrt{2}$. The result, which in sexagesimal notation is 1;24,51,10, is more or less equivalent to the *Sulbasutra* value, and this prompts the question of whether the *Sulbasutra* procedure was derived from the Babylonian. However, we cannot provide an answer, for two reasons: there is little evidence of the sexagesimal base in the Indian mathematics of that period; and, unlike the isolated instance of the Babylonian evaluation of $\sqrt{2}$, operations with surds occur in the procedures in the *Sulbasutras*.

A curiosity: The mathematics of *sriyantra*

A hymn from *Atharavaveda* is dedicated to an object that closely resembles Figure 8.9. The *sriyantra* ('great object') belongs to a class of devices used in meditation, mainly by those belonging to the Hindu tantric tradition. The diagram consists of nine interwoven isosceles triangles: four point upwards, representing Sakti, the primordial female essence of dynamic energy, and five point downwards, representing Siva, the primordial male essence of static wisdom. The triangles are arranged in such a way that they produce 43 subsidiary triangles, at the centre of the smallest of which there is a big dot (known as the *bindu*). These smaller triangles are supposed to form the abodes of different gods, whose names are sometimes entered in their respective places. In common with many depictions of the *sriyantra*, the one shown here has outer rings consisting of an eight-petalled lotus, enclosed by a sixteen-petalled lotus, girdled in turn by three circles, all enclosed in a square with four doors, one on each side. The square represents the boundaries within which the deities reside, protected from the chaos and disorder of the outside world.

Tantric tradition suggests that there are two ways of using the *sriyantra* for meditation. In the 'outward approach', one begins by contemplating the *bindu* and proceeds outwards by stages to take in the smallest triangle in which it is enclosed, then the next two triangles, and so on, slowly expanding outwards through a sequence of shapes to the outer shapes in which the whole object is contained. This outward contemplation is associated with an evolutionary

Figure 8.9 The *sriyantra* (after Kulaichev, 1984, p. 280).

view of the development of the universe where, starting with primordial matter represented by the dot, the meditator concentrates on increasingly complex organisms, as indicated by increasingly complex shapes, until reaching the very boundaries of the universe from where escape is possible only through one of the four doors into chaos. The 'inward' approach to meditation, which starts from a circle and then moves inwards, is known in tantric literature as the 'process of destruction'.

The mathematical interest in the *sriyantra* lies in the construction of the central nine triangles, which is a more difficult problem than might first appear. A line here may have three, four, five or six intersections with other lines. The problem is to construct a *sriyantra* in which all the intersections are correct and the vertices of the largest triangles fall on the circumference of the enclosing circle. We shall not go into the details of how the Indians may have achieved accurate constructions of increasingly complex versions of the *sriyantra*, including spherical ones with spherical triangles. Bolton and Macleod (1977) offer a simple overview of the subject; Kulaichev (1984) goes into the 'higher' mathematics implicit in constructing different types of *sriyantra*.

There is, however, a curious fact about all the correctly constructed *sriyantras*, whether enclosed in circles (as in Figure 8.9) or in squares. In all such cases the base angle of the largest triangles (shown as θ in Figure 8.9) is about $51\frac{1}{2}°$. The monument that comes to mind when this angle is mentioned is the Great Pyramid at Gizeh in Egypt, built around 2600 BC. It is without doubt the most massive building ever to have been erected, having at least twice the volume and thirty times the mass of the Empire State Building in New York, and built from individual stones weighing up to 70 tonnes each. The slope of the face to the base (or the angle of inclination) of the Great Pyramid is $51° 50' 35''$.

It is possible from the dimensions of the Great Pyramid to derive probably the two most famous irrational numbers in mathematics. One is π, and the other is ϕ, the 'golden ratio' or 'divine proportion', given by $\frac{1}{2}(1 + \sqrt{5})$ (its value to five decimal places is $1.618\,03$). The golden ratio has figured prominently in the history of mathematics, both as a semi-mystical quantity (Kepler suggested that it should be named the 'divine proportion') and for its practical applications in art and architecture, including the Parthenon at Athens and a number of other buildings of Classical Greece. In the Great Pyramid, the golden ratio is represented by the ratio of the length of the face (the slope height), inclined at an angle θ to the ground, to half the length of the side of the square base, equivalent to the secant of the angle θ. The original dimensions of the Great Pyramid are not known exactly, because later generations removed the outer limestone casing for building material, but as far as we can tell the above two lengths were about 186·4 and 115·2 metres respectively. The ratio of these lengths is, to five decimal places, 1·618 06, in very close agreement with ϕ. The number ϕ has some remarkable mathematical properties. Its square is equal to itself plus one, while its reciprocal is itself minus one. But the most intriguing feature is its link with what are called the Fibonacci numbers.

The Fibonacci numbers are the sequence

$$0, 1, 1, 2, 3, 5, 8, 13, 21, 34, 55, 89, 144, 233, 377, 610, \ldots$$

where each number equals the sum of its two predecessors. This sequence crops up in a variety of natural phenomena — in patterns of plant growth and in the laws of Mendelian heredity, for example. It is easily shown that the ratio between successive Fibonacci numbers gets closer to ϕ the further up the sequence one goes. In the Fibonacci sequence given above, the ratio of 233 to 144 gives the value of ϕ calculated from the dimensions of the Great Pyramid.

The quantity π can also be found in the dimensions of the Great Pyramid. If its height (146·6 metres) is taken to be the radius of a circle, the perimeter of its base ($4 \times 230\cdot4 = 921\cdot6$ metres) is almost equal to the circumference of that circle ($2\pi r \simeq 921\cdot2$ metres). The product of π and the square root of ϕ is close to 4.

The largest isosceles triangle of the *sriyantra* design (Figure 8.9) is one of the face triangles of the Great Pyramid in miniature, showing almost exactly the same relationship between π and ϕ as in its larger counterpart. It would be idle to indulge in any further speculation.

Many of the accurate constructions of *sriyantras* in India are very old. Some are even more complicated than the one shown in Figure 8.9. There are those that consist of spherical triangles (Figure 8.10) for which the constructor, to achieve perfect intersections and vertices falling on the circumference of the circle enclosing the triangles, would require knowledge of 'higher mathematics [which] the medieval and ancient Indian mathematicians did not possess' (Kulaichev, 1984, p. 292). Kulaichev goes on to suggest that the achievement of such geometrical constructs in Indian mathematics may indicate 'the existence of unknown cultural and historical alternatives to mathematical knowledge, e.g. the highly developed tradition of special imagination'.

EARLY INDIAN NUMERALS AND THEIR DEVELOPMENT

Three early types of Indian numerals are shown in Table 8.2 in chronological order of appearance. The *Kharosthi* numerals are found in inscriptions dating to a period from the fourth century BC to the second century AD. Special symbols

Figure 8.10 A spherical form of the *sriyantra* (after Kulaichev, 1984, p. 291).

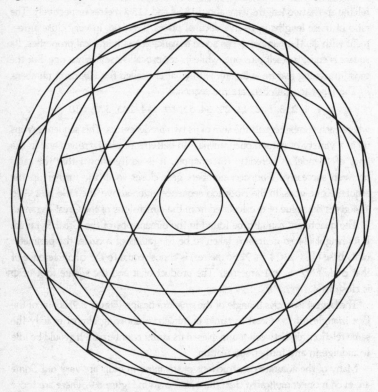

were used to show both 10 and 20. Numbers up to 100 were then built up additively; for larger numbers the multiplication principle came into operation, with special symbols for higher powers of 10. The *Brahmi* numerals were more highly developed. There were separate symbols for the digits 1, 4 to 9, and the number 10 and its higher powers. There were also symbols for multiples of 10 up to 90, and for multiples of 100 up to 900. The number 486, for example, would be written by using the symbols for 400, 80 and 6. It is possible that our symbols '2' and '3' are cursive versions of the *Brahmi* numerals (i.e. from ⟋ and ≡ may have evolved 2 and 3).

The earliest trace of *Brahmi* numerals is from the third century BC, on the Asoka pillars scattered around India, though more detailed pieces of evidence are found elsewhere later. At the top of Nana Ghat near Poona in Central India is a cave, which must once have been a resting-place for travellers; inscribed on the cave walls are *Brahmi* numerals which date back to 150 BC. Another version of the *Brahmi* numerals is found at Nasik, near present-day Bombay, from around AD 100. Both versions resemble each other, and it is thought that from them evolved first the *Bakhshali* number system and then the *Gwalior* system (*c.* AD 850), which is recognizably close to our present-day number system. In both the *Bakhshali* and *Gwalior* number systems, ten symbols were used to represent 1 to 9 and zero. With them it became possible to express any number, no matter how large, by a decimal place-value system.

The earliest appearance of the symbol that we associate with zero in India in a decimal place-value system is in an inscription from Gwalior dated 'Samvat 933' (AD 876) where the numbers 50 and 270 are given as $\mathcal{E}$o and $\mathcal{R}\mathcal{R}$o, respectively. Note the close similarity with our notation for 270. For earlier evidence, we have to turn to South East Asia when it was under the cultural influence of India. There, three inscriptions have been found bearing dates in the Saka era, which began in AD 78. A Malay inscription at Palembang in Sumatra from AD 684 shows 60 and 606 Saka as $\mathcal{O}$o and $\mathcal{O}$o $\mathcal{O}$, respectively, a Khmer inscription at Sambor in Cambodia from AD 683 gives 605 as $\mathcal{C} \bullet \mathcal{E}$, and an inscription at Ponagar, Champa (now southern Vietnam) from AD 813 represents 735 as $\mathcal{Z} \equiv \mathcal{E}$. If, however, the original version of the Bakhshali Manuscript dates from the third century AD, it would be the earliest evidence of a well-established number system with a place-value scale and zero which is also recognizably an ancestor of our present-day number system. In the Bakhshali Manuscript are found the following numbers:

330: $\mathbf{3}\mathbf{3}$o,   846 720: $\mathbf{3}\mathbf{4}\mathbf{3}\mathbf{7}\mathbf{3}$o,   947: $\mathbf{9}\mathbf{4}\mathbf{7}$

What we have here is a fully developed decimal place-value system incorporating zero.

Table 8.2 Three types of Indian numerals, in chronological order.

	1	2	3	4	5	6	7	8	9	10
Kharosthi	I	II	III	X	IX	IIX	IIIX	XX		γ
Brahmi	—	=	≡	⨯	$\hbar$	φ	γ	ζ	$\supset$	$\propto$
Gwalior	γ	$\mathcal{Z}$	$\mathcal{3}$	$\mathcal{8}$	$\mathcal{4}$	$\mathcal{C}$	$\mathcal{7}$	$\mathcal{C}$	$\mathcal{9}$	$\mathcal{7}$o

The Kharosthi numeral for nine is not known for certain.

The emergence of the place-value principle

Two important features of these early numerals may have been of significance in the subsequent development of Indian numerals. Ever since the Harappan period the number ten has formed the basis of numeration; there is no evidence of the use of any other base in the whole of Sanskrit literature. A long list of number-names for powers of ten are found in various early sources. For example, one of the four major *Vedas*, the *Yajurveda*, gives special names for powers of ten from 1, or 10^0 (*eka*), to 10^{12} (*parardha*). In the *Ramayana*, one of the most popular texts of Hinduism and roughly contemporaneous with the later *Vedas*, it is reported that Ravana, the chief villain of the piece, commanded an army whose total equalled

$$10^{12} + 10^5 + 36(10^4)$$

Facing them was the rival army of Rama, the hero of the Epic, which had

$$10^{10} + 10^{14} + 10^{20} + 10^{24} + 10^{30} + 10^{34} + 10^{40} + 10^{44}$$
$$+ 10^{52} + 10^{57} + 10^{62} + 5 \text{ men!}$$

Even though these numbers are fantastic, the very existence of names for powers of ten up to 62 indicates that the Vedic Indians were quite at home with very large numbers. This is to be compared with the ancient Greeks, who had no words for numbers above the myriad (10^4).

And these were by no means the largest numbers ever conceived in early India. The Jains, who came after the Vedic Indians, were particularly fascinated by even larger numbers which were intimately tied up with their philosophy of time and space. (We shall look at the Jaina contribution in detail in a later section.) For units of measuring time, the Jains suggested the following relationships:

$$1 \, purvis = 750 \times 10^{11} \text{ days}$$
$$1 \, shirsa \, prahelika = (8\,400\,000)^{28} \text{ purvis}$$

The last number contains 194 digits!

The early use of such large numbers eventually led to the adoption of a series of names for successive powers of ten. The importance of these number-names in the evolution of the decimal place-value notation cannot be exaggerated. The word-numeral system, later replaced by an alphabetic notation, was the logical outcome of proceeding by multiples of ten. Thus 60 799 is *sastim* (60) *sahsara* (thousand) *sapta* (seven) *satani* (hundred) *navatim* (nine ten times) *nava* (nine).

The decimal place-value system developed when a decimal scale came to be associated with the value of the places of the numbers arranged left to right or right to left. And this was precisely what happened in India, probably as early as the Vedic period. Word-numerals (*bhuta Samkhya*) continued to be used even after the introduction of special symbols for the numbers from one to nine, and zero, with different words standing for the same number to provide scope for the poetic expression of mathematical results. Thus zero was represented by *sunya* ('void') or *ambara akasa* ('heavenly space'), one by *rupa* ('Moon') or *bhumi* ('Earth'), two by *yama* ('twice') or *kara* ('hands'), and so on. With multiple word-numerals available, the choice of a particular word for a number would be dictated by literary considerations. The use of these alternatives continued for many years in both secular and religious writings, because it was aesthetically pleasing and offered an easier means of remembering rules.

There are traces of this system of numeration in Yavanajataka (AD 269), though the first clearest and detailed evidence of it is found in the works of the astronomer Varahamihira (d. AD 587). Thus, except for the actual symbols themselves, the present-day number system with distinct numerals for the numbers from zero to nine, the place-value principle and the use of the zero within the decimal base is essentially what we see in this early number system. In a sense, what is used as a symbol for a number, whether it be a letter, a word or a specially invented squiggle, is of little importance. Indeed, an unduly close association − or even identity − between a number and the symbol used to represent it may even be counter-productive, preventing the strength of the place-value principle from being fully exploited in elementary operations. Let us explore this idea more fully, looking at methods for long multiplication first expounded by Bharati Krishna Tirthaji.

Multiplication the Vedic way

Swami Bharati Krishna Tirthaji (1884–1960) was born in Tinnevely, a town near Madras. After his early education in his home town and in Madras, at the age of nineteen he sat and passed the MA examination as an external student of an American College based in New York, taking a variety of subjects including Sanskrit, philosophy, mathematics and history. In 1911 he left his job as a college lecturer, and was later initiated into a Hindu religious order of which he became the head in 1925. His fame rests on a book entitled *Vedic Mathematics* which was published in 1965, five years after his death. It is the only volume to have survived of the sixteen he was reputed to have written.

In it Tirthaji claims to have reconstructed sixteen *sutras* (or concise rules) and thirteen *'sub-sutras'* (or corollaries) from a *parisista* (appendix) of the *Atharvaveda*, one of the four *Vedas* mentioned earlier, which he then proceeded to apply to a wide range of problems. No one has been able to locate this *parisista*; it is certainly not among those that have been published. So the historical basis of *Vedic Mathematics* is as yet uncertain.

Irrespective of the difficulty of establishing its historical authenticity (the wisest course at present would be to keep an open mind), the methods used in the book provide an original and refreshing approach to subjects which are often dismissed as mechanical and tedious. But more important from our point of view are the new insights they provide into the scope and strength of place-value number systems such as ours. Some of the methods may even have been used by early Indian mathematicians before the full development of Indian numerals. To illustrate Tirthaji's approach, we consider two of the *sutras* applied to long multiplication.

The second of the sixteen *sutras*, which we shall call the *Nikhilam Sutra*, may be translated as:

All from 9 and the last from 10.

As it stands, this is cryptic to the point of incomprehension. Such extreme conciseness is characteristic of many of the mathematical texts of ancient and medieval India. In a sense, Tirthaji performed the same function as many commentators before him, to elaborate and illustrate the *sutras* with well-chosen examples. Let us take a few examples of multiplication where the application of this *sutra* would be particularly appropriate.

EXAMPLE 8.1 *Multiply 88 by 96.*
Solution Arrange the numbers in the following way:

NUMBER	DEFICIENCY
(100)	
88	12
96	4
84	48

So 88 × 96 = 8448

ALGEBRAIC EXPLANATION

Let x (or the base) be the power of 10 nearest the two numbers (100 in this case). Let a and b be the 'deficiencies' with respect to x of the two numbers to be multiplied. Then

$$(x - a)(x - b) = x(x - a - b) + ab$$

Applying this formula to 88×96 gives

$$(100 - 12)(100 - 4)$$
$$= 100(100 - 12 - 4) + 48$$
$$= 8400 + 48 = 8448$$

The power of 10 nearest the two numbers is 10^2, or 100, which we refer to as the 'base'. Subtract 88 and 96 from the base to get the 'deficiencies', 12 and 4 respectively. Lay out these calculations as given above. The product will consist of two components on either side of the vertical line.

The number 84 is obtained by cross-subtraction:

$$88 - 4 = 84 \quad \text{or} \quad 96 - 12 = 84$$

The number 48 is obtained by taking the product of the deficiencies:

$$12 \times 4 = 48$$

The final answer is obtained by merging the two parts to give 8448.

The reader is invited to calculate 79×85 by this procedure.

EXAMPLE 8.2 *Multiply 1038 by 1006.*
Solution   Arrange the numbers in the following way:

NUMBER		SURPLUS	ALGEBRAIC EXPLANATION
	(1000)		
1038		38	Let x be the base, and a and b the
1006		6	'surpluses'. Then
1044		228	$(x + a)(x + b) = x(x + a + b) + ab$

So $1038 \times 1006 = 1\,044\,228$

Applying this formula gives

$$(1000 + 38)(1000 + 6)$$
$$= 1000(1000 + 38 + 6) + 228$$
$$= 1\,044\,228$$

The method is the same as in Example 8.1, except that we deal here with 'surpluses'.

The reader may wish to multiply 105 by 112 by this procedure.

EXAMPLE 8.3   *Multiply 128 by 89.*
Solution   Arrange the numbers in the following way:

NUMBER	(100)	SURPLUS/(DEFICIENCY)
128		28
89		(11)
117		(308)
117 − 4		400 − 308
113		92

So 128 × 89 = 11 392

ALGEBRAIC EXPLANATION

Let x be the base, and a and b the 'surplus' and 'deficiency', respectively. Then

$$(x + a)(x - b)$$
$$= x(x + a - b) - ab$$

Applying this formula gives

$$(100 + 28)(100 - 11)$$
$$= 100 (100 + 28 - 11) - 308$$
$$= 11 392$$

The method is slightly more involved than in the first two examples. The numbers and their 'surplus' or 'deficiency' are arranged in the usual way. The left-hand side is obtained either by cross-subtraction (128 − 11 = 117) or by cross-addition (89 + 28 = 117). On the right-hand side, the product of the surplus and the deficiency gives a deficiency of 308, requiring 4 hundreds to be borrowed and subtracted from 117 on the left-hand side. The last step is to subtract the 308 that remains from the 400 borrowed from the left-hand side. Thus the net adjustment is to reduce the left-hand side by 4, leaving 113.

The reader is invited to try 112 × 94 by this procedure.

The second *sutra* we consider is the *Urdhva Tiryak Sutra*. It too is very concise, and may be rendered in English as:

Vertically and crosswise.

We illustrate the application of this *sutra* to multiplication by taking two examples.

EXAMPLE 8.4   *Multiply 36 by 53.*
Solution   Arrange the numbers in the following way:

TENS	UNITS

[15][9 + 30][18]

So 36 × 53 = 1908

ALGEBRAIC EXPLANATION
Use the algebraic identity

$$(ax + b)(cx + d) = acx^2 + (ad + bc)x + bd$$

to multiply a pair of two-digit numbers *ab* and *cd*. The algebraic validity of this *sutra* follows from putting $x = 10$.

The method is suggested by the words of the *sutra*, 'vertically and crosswise'. First, the numbers in the 'units' column are multiplied vertically (6 × 3 = 18). Next the 'units' and 'tens' digits are multiplied crosswise and added ((3)(3) + (6)(5) = 39). Then the numbers in the 'tens' column are multiplied vertically (3 × 5 = 15). Finally, the place-value adjustments are made by carrying over the relevant numbers.

The reader is invited to try 41 × 27 and 87 × 78 by this procedure.

This method can be applied however many digits there are in the two numbers. If the numbers of digits differ, then zeros are added to the left of the smaller number.

EXAMPLE 8.5   *Multiply 2341 by 683.*
Solution   Arrange the numbers in the following way:

THOUSANDS		HUNDREDS			TENS		UNITS
2		3			4		1
0		6			8		3
[2×0]	[12+0]	[16+0+18]	[6+0+24+24]	[9+6+32]	[12+8]	[1×3]
(0)	(12)	(34)	(54)	(47)	(20)	3
1	5	9	8	9	0	3

So 2341 × 683 = 1 598 903

The pattern of the calculation is shown below. The large dots represent digits and the lines joining them show the individual multiplications.

The reader may wish to try 257 × 128, 385 × 471 and 816 × 16, using the following order of calculations:

We have examined only one application of the two *sutras* discussed — to multiplication. The range of possible applications of the *sutras* is quite considerable, though, as is illustrated in Tirthaji's book. For example, the 'vertically and crosswise' *sutra* may be used to evaluate determinants, solve simultaneous linear equations, derive trigonometric functions, solve cubic and higher-order equations, and evaluate logarithms and exponentials. Of course, this is not to make the ludicrous claim that these subjects were known in ancient times, rather that the 'Vedic' approach not only provides a fresh insight into the versatility of our number system, but also that it can be extended to cover a fairly wide range of mathematical topics.

Even at the level of elementary arithmetical operations, this approach is seen to be useful. First, it provides fresh insight into elementary operations with our number system. There are benefits from looking at a number not just as itself, but also in relation to a suitable base. Thus, if asked to multiply 97 by 103, we can note that the first number is three less than 100, and the second is three in excess of 100. The operation is considerably simplified, as we saw in the examples discussed, if it takes place in relation to the defined 'surplus' and 'deficiency' relative to the base.

Second, these methods provide a clear and much-needed reminder that algebra is a generalization of arithmetic. For each of the above examples there is a clear algebraic rationale. But more importantly, they highlight the less obvious fact that any polynomial may be expressed in terms of a positional notation without the base being specified. Once this is accepted, we can see that the same algorithmic scheme may be applied to arithmetic and to algebraic problems. The artificial distinction that has grown up between the two subjects is a by-product of both increasing specialization and the traditional structuring of mathematical knowledge – it did not exist in any of the mathematical traditions examined in this book.

Finally, these methods encourage mental arithmetic, a subject neglected in the present-day study of elementary operations. Nobody would deny that the arithmetic of everyday life depends more on mental than on written computation. Research into the process of mental computation shows the importance of having a number of reference points and the ability to choose the right one for the right occasion. If one were asked to work out the product of 105 and 103, one could choose a reference number suitably close to the answer (i.e. 10 000), recognizing that the answer will be in excess of this reference number. The ability to make the final adjustment correctly is an important aspect of competence at mental arithmetic. The Vedic methods provide not only a whole range of reference points (identifying and using the base for a particular multiplication being one of the many skills), but also a coherent rationale for the method and a check on accuracy of the result obtained. With today's overdependence on calculators for simple computations, often at the expense of losing the skills of mental arithmetic, the 'Vedic' methods and similar approaches have considerable pedagogical value and may even reawaken the lost appetite for 'sums'.

JAINA MATHEMATICS

The rise of Buddhism and Jainism around the middle of the first millennium BC was in part a reaction to some of the excesses of Vedic religious and social practices. The resulting decline in offerings of Vedic sacrifices, which had played such a central role in Hindu ritual, meant that occasions for constructing altars requiring practical skills and geometric knowledge became few and far between. There was also a gradual change in the perception of the role of mathematics: from fulfilling the needs of sacrifical ritual, it became an abstract discipline to be cultivated for its own sake. The Jaina contribution to this change should be recognized. Unfortunately, sources of information on Jaina mathematics are scarce, though there are enough to show how original the work was.

A number of Jaina texts of mathematical importance have yet to be studied, and what we know of them is based almost entirely on later commentaries. Of particular relevance is the old canonical literature: *Surya Prajnapti, Jambu Dwipa Prajnapti, Sthananga Sutra, Uttaradhyayana Sutra, Bhagwati Sutra* and *Anuyoga Dwara Sutra*. The first two works are from the third or fourth century BC, and the others are from at least two centuries later. The *Sthananga Sutra* gives a list of mathematical topics which were studied at the time: the theory of numbers, arithmetical operations, geometry, operations with fractions, simple equations, cubic equations, biquadratic (quartic) equations, and permutations and combinations. This classification by the Jains was adopted by later mathematicians.

Given the paucity of existing evidence and the little scrutiny it has received, our survey of Jaina mathematics must be rather piecemeal. We shall examine four main areas in which the Jaina contribution was distinctive.

Theory of numbers

Like the Vedic mathematicians, the Jains had an interest in the enumeration of very large numbers which was intimately tied up with their philosophy of time and space. We mentioned earlier that they had devised a measure of time, called a *shirsa prahelika*, which equalled $756 \times 10^{11} \times (8\,400\,000)^{28}$ days! Other examples of the Jaina fascination with very large numbers are these two definitions: a *rajju* is the distance travelled by a god in six months if he covers a hundred thousand *yojanna* (approximately a million kilometres) in each blink of his eye; a *palya* is the time it will take to empty a cubic vessel of side one *yojanna* filled with the wool of new-born lambs if one strand is removed every century.

The contemplation of such large numbers led the Jains to an early concept of infinity which, if not mathematically precise, was by no means simple-minded. All numbers were classified into three groups, enumerable, innumerable and infinite, each of which was in turn subdivided into three orders:

1 *Enumerable*: lowest, intermediate and highest;
2 *Innumerable*: nearly innumerable, truly innumerable and innumerably innumerable; and
3 *Infinite*: nearly infinite, truly infinite and infinitely infinite.

The first group, the enumerable numbers, consisted of all the numbers from 2 (1 was ignored) to the highest. An idea of the 'highest' number is given by the following extract from the *Anuyoga Dwara Sutra*, from around the beginning of the Christian era:

> Consider a trough whose diameter is that of the Earth (100 000 *yojanna*) and whose circumference is 316 227 *yojanna*. Fill it up with white mustard seeds counting one after another. Similarly fill up with mustard seeds other troughs of the sizes of the various lands and seas. Still the highest enumerable number has not been attained.

(1 *yojanna* is about 10 kilometres.) But once this number, call it N, is attained, infinity is reached via the following sequence of operations:

$$N + 1, N + 2, \ldots, (N + 1)^2 - 1$$
$$(N + 1)^2, (N + 2)^2, \ldots, (N + 1)^4 - 1$$
$$(N + 1)^4, (N + 2)^4, \ldots, (N + 1)^8 - 1$$

and so on. Five different kinds of infinity are recognized: infinite in one direction, infinite in two directions, infinite in area, infinite everywhere and infinite perpetually. This was quite a revolutionary idea in more than one way:

1 The Jains were the first to discard the idea that all infinities were the same or equal, an idea still generally accepted in Europe until the work of Georg Cantor in the late nineteenth century.

2 The highest enumerable number (i.e. N) of the Jains corresponds to another concept developed by Cantor, aleph-null (the cardinal number of the infinite set of integers $1, 2, \ldots, N$), also called the first transfinite number. The Jains even attempted to define a whole system of transfinite numbers, of which aleph-null is the smallest. It was Cantor who defined the concept of a sequence of transfinite numbers and devised an arithmetic of such numbers. (It is not practicable to examine this fascinating area of mathematics here. Simple introductions to the concept of transfinite numbers and operations with such numbers are given by Stewart (1981, pp. 127–43) and by Sondheim and Rogerson (1981, pp. 148–59).)

3 In the Jaina work on the theory of sets (not discussed here, though references are given in the Bibliography), two basic types of transfinite number (i.e. the cardinal numbers of infinite sets) are distinguished. On both physical and ontological grounds, a distinction is made between *asmkhyata* and *ananta*, between rigidly bounded and loosely bounded infinities. With this distinction, the way was open for the Jains to develop a detailed classification of transfinite numbers and mathematical operations for handling transfinite numbers of different kinds. (For further details see Jain (1973, 1982) and Singh (1987).)

What is being suggested here is that further work on this aspect of Jaina mathematics may prove fruitful in future research on the foundations of mathematics. That Jaina literature from fifteen hundred to two thousand years ago may hold valuable clues to the very nature of mathematics is an exciting thought.

Indices and logarithms

Without a convenient notation for indices, the laws of indices cannot be formulated precisely. But there are some indications that the Jains were aware of the existence of these laws, and made use of related concepts.

The *Anuyoga Dwara Sutra* lists sequences of successive squares or square roots of numbers. Expressed in modern notation as operations performed on a certain number a, these sequences may be represented as:

$$(a)^2, (a^2)^2, [(a^2)^2]^2, \ldots$$

$$\sqrt{a}, \sqrt{\sqrt{a}}, \sqrt{\sqrt{\sqrt{a}}}, \ldots$$

In the same *sutra*, we come across the following statement on operations with power series or sequences: 'The first square root multiplied by the second square root, [is] the cube of the second square root; the second square root multiplied by the third square root, [is] the cube of the third square root.' Expressed in terms of a, this says that

$$a^{1/2} \times a^{1/4} = (a^{1/4})^3 \quad \text{and} \quad a^{1/4} \times a^{1/8} = (a^{1/8})^3$$

As a further illustration, the total population of the world is given as 'a number obtained by multiplying the sixth square by the fifth square, or a number which can be divided by two 96 times'. This gives a figure of $2^{64} \times 2^{32} = 2^{96}$, which in decimal form is a number of 29 digits!

Does this statement indicate that the laws of indices

$$a^m \times a^n = a^{m+n} \quad \text{and} \quad (a^m)^n = a^{mn}$$

were familar to the Jains? From the period around the eighth century AD, there is some interesting evidence in the *Dhavala* commentary by Virasenacharya to suggest that the Jains may have developed the idea of logarithms to base 2, 3 and 4 without using them for any practical purposes. The terms *ardhacheda*, *trakacheda* and *caturthacheda* of a quantity may be defined as the number of times the quantity can be divided by 2, 3 and 4, respectively, without leaving a remainder. For example, since $32 = 2^5$, the *ardhacheda* of 32 is 5. Or, in the language of modern mathematics, the *ardhacheda* of x is $\log_2 x$, the *trakacheda* of x is $\log_3 x$, and so on.

Permutations and combinations

A permutation is a particular way of ordering some or all of a given number of items. Therefore the number of permutations which can be formed from a group of unlike items is given by the number of ways of arranging them. As an example, take the letters a, b, and c, and find the number of permutations two letters at a time. Six arrangements are possible:

ab, ac, ba, ca, bc, cb

Instead of listing all possible arrangements, we can work out the number of permutations by arguing as follows: the first letter in an arrangement can be any of the three, while the second must be either of the other two letters. Consequently the number of permutations for two of a group of three letters is $3 \times 2 = 6$. The shorthand way of expressing this result is $_3P_2 = 6$.

A combination is a selection from some or all of a number of items; unlike permutations, order is not taken into account. Therefore, the number of combinations which can be formed from a group of unlike items is given by the number of ways of selecting them. To take the same illustration as above, the number of combinations of two letters at a time from a, b and c is three:

ab, ac, bc

Again, instead of listing all possible combinations, we can work out how many there are as follows: in each combination the first letter can be any of the three, but there is just one possibility for the second letter, so there are three possible combinations. (Thus although ab and ba are two different permutations, they amount to the same combination.) A shorthand way of expressing this result is $_3C_2 = 3$.

Permutations and combinations were favourite topics of study among the Jains. Statements of results, presumably arrived at by methods like the one just discussed, appear quite early in the Jaina literature. In the *Bhagabati Sutra* (*c.* 300 BC) are set forth simple problems such as finding the number of combinations that can be obtained from a given number of fundamental philosophical categories taken one at a time, two at a time, and three or more at a time. Others include calculation of the groups that can be formed out of the five senses, and selections that can be made from a given number of men, women and eunuchs. The *Bhagabati Sutra* gives the corresponding values correctly for selections of up to three at a time as

$$_nC_1 = n, \qquad _nC_2 = \frac{n(n-1)}{1 \cdot 2}, \qquad _nC_3 = \frac{n(n-1)(n-2)}{1 \cdot 2 \cdot 3}$$

$$_nP_1 = n, \qquad _nP_2 = n(n-1), \qquad _nP_3 = n(n-1)(n-2)$$

Values are given for $n = 2, 3, 4$, and there is then the following observation: 'In this way, 5, 6, 7, . . . , 10, etc., or an enumerable, unenumerable or infinite number of things may be specified. Taking one at a time, two at a time, . . . , ten at a time, as the number of combinations are formed, they must all be worked out.' Apart from the generalizations implied, the application of the principle to different kinds of infinities or different dimensions is noteworthy.

Even before the advent of Jainism there was some interest in the notion of permutations and combinations. Sushruta's great work on medicine, written around the second century AD, contained the statement that 63 combinations may be made out of 6 different tastes (*rasa*) – bitter, sour, saltish, astringent, sweet, hot – by taking the *rasa* one at a time, two at a time, three at a time, and so on. This solution of 63 combinations can easily be checked as follows:

$$_6C_1 + {_6}C_2 + {_6}C_3 + {_6}C_4 + {_6}C_5 + {_6}C_6 = 6 + 15 + 20 + 15 + 6 + 1 = 63$$

Another interesting example relates to the number of ways of combining different metres (*chandas*) in a poetic composition. In a book entitled *Chandasutra* from the third century AD, Pingala considered a method of estimating the number of combinations of short (*laghu*) and long (*guru*) sounds in a given poetical composition. During this period, the music of sound variations (*varnasangita*) was based mainly on these two sounds. Pingala considered a three-syllabic metre, for which the following different combinations of the sounds of *guru* and *laghu* could result: three *guru* sounds will occur once, two *guru* and one *laghu* three times, one *guru* and two *laghu* also three times, and three *laghu* sounds once.

If we represent *guru* by a and *laghu* by b, then the different combinations may be represented by the coefficients of the binominal expansion:

$$(a + b)^3 = a^3 + 3a^2b + 3ab^2 + b^3$$

For a four-syllabic metre, different combinations of the two sounds can be found by the same representation:

$$(a + b)^4 = a^4 + 4a^3b + 6a^2b^2 + 4ab^3 + b^4$$

This technique of finding the number of variations of sounds was useful as a means of testing the quality of different metres, and after Pingala it was commonly used for this purpose.

Around the end of the tenth century AD, Halayudha produced a commentary on Pingala's *Chandasutra* in which he introduced a pictorial representation of different combinations of sounds, enabling them to be read off directly. Figure 8.11 shows Halayudha's *meruprastara* (or pyramidal arrangement) for a binomial expansion of $(a + b)^n$, where $n = 0, 1, 2, 3, 4$. We came across the same triangular array of numbers, Pascal's triangle, in the last chapter on Chinese mathematics. However, there is no evidence that the triangle was used for any other purpose, such as numerical solutions to higher-order equations, as it was in China. Indeed, there is no evidence that the device was ever incorporated into Indian mathematics.

Sequences and progressions

Jaina interest in sequences and progressions developed out of their philosophical theory of cosmological structures. Schematic representations of the cosmos constructed according to this theory contained innumerable concentric rings of alternate continents and oceans, the diameter of each ring being twice that of the previous one, so that if the smallest ring had a diameter of 1 unit, the next largest would have a diameter of 2 units, the next 2^2 units, and so on to the nth ring of diameter 2^{n-1} units.

Arithmetic progressions were given the most detailed treatment. Separate formulae were worked out for finding the first term a, the common difference d, the number of elements n in the series and the sum S of the terms. This was well explored in a Jaina canonical text entitled *Trilokaprajnapti*. One of its problems is to find the sum of a complicated series consisting of 49 terms made up of 7 groups, each group itself forming a separate arithmetical progression, and the terms of each group forming another. (We shall not attempt a solution here; for details see Bag (1979).)

Figure 8.11 Halayudha's *meruprastara* (Pascal's triangle).

It was the elaborate treatment of mathematical series by Mahavira (*c.* AD 850) that paved the way for some notable work by medieval mathematicians in this area. We shall examine these developments briefly in the next chapter.

Geometry

The term *rajju* was used in two different senses by the Jaina theorists. In cosmology it was a frequently occurring measure of length, approximately 3.4×10^{21} kilometres according to the Digambara school. But in a more general sense it was the term the Jains used for geometry or mensuration, in which they followed closely the Vedic *Sulbasutras*. Their notable contribution was with measurements of the circle. In Jaina cosmography the Earth is a large circular island called the Jambu Island, with a diameter of 100 000 *yojanna*. While there are a number of estimates of the circumference of this island, including the rather crude 300 000 *yojanna*, an interesting estimate mentioned in both the *Anuyoga Dwara Sutra* and the *Triloko Sara*, from around the beginning the first millennium AD, is

$$316\,227 \text{ yojanna, } 3 \text{ krosa, } 128 \text{ danda and } 13\tfrac{1}{2} \text{ angula}$$

where 1 *yojanna* is about 10 kilometres, 4 *krosa* = 1 *yojanna*, 2000 *danda* = 1 *krosa*, and 96 *angula* (literally a 'finger's breadth') = 1 *danda*. This result is consistent with taking the circumference to be given by $\sqrt{10}\,d$, where $d = 100\,000$ *yojanna*. The choice of the square root of 10 for π was quite convenient, since in Jaina cosmography islands and oceans always had diameters measured in powers of 10.

MATHEMATICS ON THE EVE OF THE CLASSICAL PERIOD

In 1881, near a village called Bakhshali near the north-west border of India, a farmer digging in a ruined stone enclosure came across a manuscript written in an old form of Sanskrit, using Sarada characters, on seventy leaves of birch bark. The find was described as being as fragile as 'dry tinder', with a substantial part mutilated beyond repair. What remained was put in order and parts of it translated into English by Rudolph Hoernle; it now resides in the Bodleian Library at Oxford.

G. R. Kaye produced the first complete translation of this manuscript, published in 1933 together with a commentary by him. Unfortunately, there are serious errors in both his translation and interpretation – errors which have passed into histories of mathematics that cite his work. There has been much controversy over the manuscript's age, and here Kaye's pronouncement has been particularly unfortunate. On the basis of rather dubious literary evidence, Kaye argued that the Bakhshali Manuscript belonged to the twelfth century AD. The general consensus supports Hoernle's assessment that the manuscript was probably a later copy of a document composed at some time in the first few centuries of the Christian era. Hoernle's dating was based on a careful consideration of a number of aspects, including the mathematical content, the units of money given in some examples, the use of the symbol $+$ for the negative sign and the lack of reference to certain topics (especially the solution of indeterminate equations) which appeared in works known to have been written later. The manuscript may therefore be the next substantial piece of evidence, after Jaina mathematics, to bridge the long gap between the *Sulbasutras* of the Vedic period and the mathematics of the Classical period, which began around AD 500. It is also the earliest evidence we have of Indian mathematics free from any religious or metaphysical associations. Indeed, there is some resemblance between the manuscript and the Chinese *Chiu Chang* from a few centuries earlier, which we examined in Chapter 6, both in the topics discussed and in the style of presentation of results. It should, however, be added that the premier Chinese text is far more wide-ranging and 'advanced' than the Bakhshali work.

The Bakhshali Manuscript is a handbook of rules and illustrative examples together with their solutions. It is devoted mainly to arithmetic and algebra, with just a few problems on geometry and mensuration. Only parts of it have been restored, so we cannot be certain about the balance between different topics. The arithmetic examples cover fractions, square roots, profit and loss, interest and the rule of three, while the algebraic problems deal with simple and simultaneous equations, quadratic equations, and arithmetic and geometric progressions. There is no clue as to who was the author of the work.

The subject-matter is arranged in groups of *sutras*, and presented as follows. A rule is stated and then a relevant example is given, first in words and then in notational form. The solution follows, and finally we have the demonstration or 'proof'. This method of presentation is quite unusual in Indian mathematics. The few texts arranged in this way are invariably commentaries on earlier works. Since, in terms of its content, the Bakhshali Manuscript is but a prelude to more substantial work in the Classical period, we confine our discussion to a few novel features in this text. We begin with an examination of the system of notation used, as it is a recognizable precursor of later systems.

Notations and operations

The system of notation used in the Bakhshali Manuscript bears some resemblance to those used by later mathematicians such as Aryabhata I (*c.* AD 475–550), Brahmagupta (*c.* 628) and even Bhaskaracharya (*c.* 1114–85). But there is one important difference. In the Bakhshali text we find that the sign for a negative quantity looks exactly like the present 'plus' symbol used to denote addition or a positive quantity. This sign was placed after the number it qualifies. For example,

$$
\begin{array}{cc}
15 & 8+ \\
\\
4 & 3
\end{array}
$$

means 15/4 − 8/3. Later, the + sign was replaced by a dot over the number to which it referred. Incidentally, this is one of the clues which tell us that the Bakhshali Manuscript must have originated before the twelfth century. Another interesting aspect of the notation shown in the example above is the representation of fractions. It is similar to the present-day representation in that the denominator is placed below the numerator, but the line between the two numbers is missing.

This and other aspects of the notation and operations will be brought out if we take an example from the twenty-fifth *sutra*. There the following representation appears:

$$
\begin{array}{cccccccc}
\bullet & 1 & 1 & 1 \\
1 & 1 & 1 & 1 & bha & 32 & \qquad phala & 108 \\
& 3+ & 3+ & 3+
\end{array}
$$

Here the black dot is used very much in the same way as we use the letter x to denote the unknown quantity whose value we are seeking. A fraction is denoted by placing one number under another, without a line between them. A compound fraction is shown by placing three numbers under one another; thus the second column of the representation above denotes 1 minus 1/3, or 2/3. (Without the + sign, it would denote 1 plus 1/3.) Multiplication is usually indicated by placing the numbers side by side. Thus the representation above means (2/3) × (2/3) × (2/3), or 8/27. *Bha* is an abbreviation of *bhaga*, meaning 'part', and indicates that the number preceding it is to be treated as a denominator; *bha* is thus the symbol for division. The representation above therefore means

$$x = [(2/3)^3]^{-1} \times 32 = (27/8) \times 32 = 108$$

or, in words, 'a certain number (unknown) is found by taking the reciprocal of 8/27 and multiplying the result by 32. That number is 108.'

In the Bakhshali Manuscript the dot is also used to represent zero. The use of the same symbol to represent both an unknown quantity and a numeral is interesting. At the time the dot indicated an empty place, as its Sanskrit name shows: *sunya* means 'empty', or 'void'. From its Arabic translation to *sifr* comes the English word 'cipher'. It is this dual meaning that gives us a clue to the age of the text.

The symbol for addition is the abbreviation *yu* (for *yuta*). The whole operation is enclosed between lines, and the result is set down on the right of *pha*. Thus $3 + 6 = 9$ is represented as

$$\begin{array}{|ccc|cc}
\hline
3 & 6 & & \\
& & yu & & pha\ 9 \\
1 & 1 & & \\
\hline
\end{array}$$

The rule of three

One of the problems from the *Chiu Chang* (Example 6.2, discussed in Chapter 6) has a solution which suggests knowledge of the 'rule of three'. However, the first systematic treatment of this rule is found in the Bakhshali Manuscript. The problem to which the rule is applied is of a type familiar to schoolchildren today: for example, 'If 8 oranges cost 92 pence, what will 14 oranges cost?' The solution is

$$\frac{92 \times 14}{8} = \text{£1.61}$$

The method suggested in the Bakhshali Manuscript, which is also found in later works, may be stated in the following terms: If *p* (argument, or *pramana*) yields *f* (fruit, or *phala*), what will *i* (requisition, or *iccha*) yield? It is suggested that the three quantities be set down as follows:

where p and i are of the same denominations, and f is of a different denomination. For the required result the middle quantity is to be multiplied by the last quantity and divided by the first, to give the result as fi/p. The following example from the Bakhshali Manuscript illustrates the application of this rule.

EXAMPLE 8.6 *Two page-boys are attendants of a king. For their services one gets 13/6 dinaras a day and the other 3/2. The first owes the second 10 dinaras. Calculate and tell me when they have equal amounts.* Suggested solution Take the denominators 6 and 2, together with the number 10 that the first has to give. The lowest common multiple of 2, 6 and 10 is 30, so 30 is the *iccha* (requisition). Now apply the rule of three:

	p (day)	f (*dinaras*)	i (day)
1st page-boy	1	13/6	30
2nd page-boy	1	3/2	30

Required result $= fi/p = 65$ *dinaras*

Required result $= fi/p = 45$ *dinaras*

Note that if the first page-boy gives the second 10 *dinaras*, both will be left with 55 *dinaras*.

This rule, brought to Europe via the Arabs, came to be known as the Golden Rule. It was very popular in solving commercial problems during and after the Renaissance.

Extracting square roots

The Bakhshali Manuscript extended the work on square roots in the *Sulbasutras*, which we discussed earlier in this chapter, to give a more accurate formula for finding an approximate value of the square root of a non-square number. The relevant *sutra* reads:

> In the case of a non-square number, subtract the nearest square number; divide the remainder by twice [the nearest square]; half the square of this is divided by the sum of the approximate root and the fraction. This is subtracted, and will give the corrected root.

In symbolic form this rule is:

$$\sqrt{A} = \sqrt{a^2 + r} \simeq a + r/2a - \frac{(r/2a)^2}{2(a + r/2a)}$$

where a^2 is the perfect square nearest to A and $r = A - a^2$. For example:

$$\sqrt{41} \simeq 6 + 5/12 - \frac{(5/12)^2}{2(6 + 5/12)} \simeq 6{\cdot}4031 \quad \text{(to 4 decimal places)}$$

(The reader may wish to try using this rule to evaluate $\sqrt{3}$ and $\sqrt{5}$.) The formula was also applied in the manuscript to calculate the values of $\sqrt{105}$, $\sqrt{487}$, $\sqrt{889}$ and $\sqrt{339009}$. The formula may be compared with the approximation procedure for finding the square root of a non-square integer usually attributed to the Hellenistic mathematician Heron (*c.* AD 200), although we came across a close antecedent of the procedure in Babylonian mathematics in Chapter 5. Heron's formula for finding the square root of A is

$$\sqrt{A} \simeq \tfrac{1}{2}(a^* + A/a^*)$$

where a^* is a first approximation to the square root of A, and can be a non-integer (this is not possible with the Bakhshali method). Like the Babylonian procedure, Heron's formula permits successive approximation (also not possible with the Bakhshali method). It is easily seen that the Bakhshali formula and Heron's formula produce the identical result for the square root of 3, for $a = 1$ and $a^* = 1{\cdot}5$.

Indeterminate equations: Their first appearance

The following is one of a number of similar problems found in the manuscript:

EXAMPLE 8.7 *Three persons possess 7 asavas, 9 hayas and 10 camels, respectively [asavas and hayas are two breeds of horse]. Each gives two animals, one to each of the others. They are then equally well off. Find the price of each kind of animal and the total value of the livestock possessed by each person.*

Solution (in symbolic terms) Let x_1, x_2 and x_3 be the prices of an *asava*, a *haya* and a camel respectively. Then, from the information given in the question,

$$5x_1 + x_2 + x_3 = x_1 + 7x_2 + x_3 = x_1 + x_2 + 8x_3 = k$$

or

$$4x_1 = 6x_2 = 7x_3 = k$$

and we seek values of x_1, x_2, x_3 and k which are positive integers.

To get integer solutions, we take k to be any multiple of the lowest common multiple of 4, 6 and 7. In the Bakhshali Manuscript k is taken as $4 \times 6 \times 7 = 168$. Then the price of an *asava* is 42, the price of a *haya* is 28 and the price of a camel is 24. The total value of livestock in the possession of each person is 262.

From these humble beginnings, over the next thousand years there was to be a systematic development of indeterminate analysis, which will be examined in the next chapter.

An unusual series

Besides the many arithmetic series found in the Bākhshali Manuscript, there are some more unusual series. The next problem has been reconstructed from a much mutilated birch-bark strip.

EXAMPLE 8.8 *A certain king distributes 57 dinaras among five wise men. He gives a certain amount to the first man and subsequently each time doubles the money he gives to the previous one. On finding that he still has some money left, he gives to the first, whatever he gave to the first four on the previous occasion, to the second, whatever he gave to the first three, to the third, whatever he gave to the first two, and to the fourth whatever he gave to the first one previously. The fifth wise man receives no money in this round since the king has none left. Find how much each of the five wise men receive.*

Solution (in symbolic terms) Let u represent the amount given to the first wise man in the first round. Subsequent disbursement may be represented in the form of the following series:

$$u + 2u + 2^2u + 2^3u + 2^4u + S_4 + S_3 + S_2 + S_1 = 57$$

where

$$S_r = u + 2u + 2^2u + 2^3u + \ldots \quad \text{(to } r \text{ terms, with } r = 1, 2, 3, 4)$$

The solution in the text implies that $u = 1$, so the five men receive gifts to the value of 16, 9, 7, 9 and 16 *dinaras* respectively.

The state of Indian mathematics at the middle of the first millennium AD, as represented by the Bakhshali Manuscript, may be summarized as follows.

1 While the mathematics of the Vedic age and of the Jains were in part inspired by religion, the mathematics of this period became more practical and

secular, being applied to everyday problems. Examples of profit and loss, computation of the average impurities of gold, wages or gifts to be paid to subordinates, and speeds and distances to be covered form the subject-matter of the Bakhshali Manuscript.

2 Whereas the writers of the *Sulbasutras* had already devised rules to find approximate values of $\sqrt{2}$, these rules were now more elaborate and were being used to obtain the square root of any number to a greater degree of accuracy.

3 There is some evidence that work on series begun during the Jaina period was continued.

4 This period marks the beginning of the great interest in indeterminate analysis. Such an interest did not arise solely from the demands made by astronomical calculations. Other problems, some of them of a recreational nature, were also catered for. The examples discussed in the Bakhshali Manuscript and the solutions offered are not difficult, but they mark the beginning of a study which was to reach an advanced level during the so-called Classical period of Indian mathematics.

5 There is evidence of a well-developed place-value number system which included zero (represented by a dot). The ease with which the system is used in the manuscript suggests that the system predates the document by a few hundred years.

6 In contrast to the vast majority of Indian mathematical works composed before and after this manuscript, the method of exposition follows a systematic order: (i) statement of the rule (*sutra*), (ii) statement of the example(s) (*udaharana*), and (iii) demonstration of the operation (*karana*) of the rule. Most of the other sources of Indian mathematics we know of contain concise statements of the rules, usually without any attempt at deriving or demonstrating them. These were left to subsequent commentators or teachers to explain.

9 Indian Mathematics:
The Classical Period and After

◇◇

From the last chapter it is clear that our evidence of mathematical activities after the Vedic period, as represented by Jaina canonical literature and the Bakhshali Manuscript, is imperfect and incomplete. Our knowledge of the development of mathematics and astronomy between the *Sulbasutras* and the period of Aryabhata I (*c.* AD 500) is therefore fairly limited. Yet this hiatus in our knowledge is particularly puzzling given the wealth of evidence we have for the same period in other fields, notably in medicine and chemistry, and in philosophy where outstanding work was produced by the Nyaya and Mimamasa schools.

Various explanations have been offered for this apparent discontinuity. The virtual disappearance of Vedic sacrifices removed, as it were, the *raison d'être* for continued interest in geometry. The sheer size of the Indian subcontinent would have restricted communication between different parts of the country, with an adverse effect on the transmission of mathematical ideas, which were widely scattered and normally restricted to certain families. If a particular generation showed little interest or aptitude the family's mathematical knowledge might be lost for ever. Mathematical ideas were transmitted orally in a verse form which could easily be memorized. This *sutra* form was specially suited for this purpose, but to the uninitiated it required elaborate explanations – without commentaries, the *sutras* often made little sense. This form of transmitting knowledge had the result of confining mathematical pursuits to a tiny elite. This elitism, born of the caste system, is probably one of the reasons why Indian mathematics floundered for a few centuries after its impressive beginnings. (With other disciplines such as medicine and chemistry, knowledge was concentrated in schools and did not suffer in this way.)

The revival, which came in the middle of the first millennium AD, also established channels of communication both within India, where mathematical work was concentrated in three centres of learning – Kusum Pura and Ujjain in the north and Mysore in the south – and within other cultures, first Persia, and later the Arab world and China. The scene was set for the transmission of Indian mathematical ideas to the West and the incorporation of important Babylonian and Hellenistic ideas, mainly from Alexandria, into Indian astronomy.

The few centuries preceding and following the beginning of the Christian era saw the emergence of a class of texts called the *Siddhantas*. Contained in them were important changes in astronomical methods and practices. The traditional system, based on tracking the movement of the planets in relation to 27 (or 28) stars chosen as reference points, was dispensed with and replaced by the twelve signs of the zodiac. More sophisticated mathematical methods were used to determine the periods of planetary revolutions. The mean longitudes were calculated from the number of days that had elapsed since the beginning of the present Kaliyuga era (Friday 18 February 3102 BC, on the Gregorian calendar). Different measures of the duration of the day and year were correctly determined. Planetary positions were computed using Ptolemaic epicycles and deferents; eclipses were calculated and the results corrected for parallax (the apparent displacement of celestial objects resulting from the changing location of the observer as the Earth moves in its orbit). These computations required a wide range of mathematical techniques, including certain innovative methods of plane and spherical trigonometry and applications of indeterminate equations. Some of these methods will be examined in this chapter.

MAJOR INDIAN MATHEMATICIAN-ASTRONOMERS

Aryabhata I (b. AD 476)

In his best known work *Aryabhatiya*, Aryabhata I states that he composed it during the 3600th year of the Kaliyuga, when he was twenty-three years old. So he must have been born in AD 476 and completed his work in 499 (i.e. $3601 - 3102 = 499$). There are a number of conjectures about his birthplace, ranging from the south (Kerala, Tamil Nadu, Andhra Pradesh) to the north-east (Bihar, Bengal). But all we know is that he wrote his great work in Kusum Pura, near modern Patna in Bihar. It was then the imperial capital of the Gupta empire, and had been an important centre of learning since the Jaina period; the great Jaina metaphysician-scientist Umasvati (c. AD 200) recorded that a famous school of mathematics and astronomy had stood there before his time.

The *Aryabhatiya* is short and concise, and is essentially a systematization of the results contained in the older *Siddhantas*. The mathematical section consists of just thirty-three verses. It is, however, of particular importance not only because of the picture it gives of the state of mathematical knowledge of the period, but also for the impetus it gave to the future study of the subject. It contains details of an alphabet-numeral system of notation,* rules for arithmetical operations, and methods of solving simple and quadratic equations, and indeterminate equations of the first degree. The book pays some attention to trigonometry and introduces the sine and versine (i.e. $1 - \text{cosine}$) functions – a notable innovation on earlier work both in and outside India. The Indian contribution to trigonometry will be discussed later in this chapter.

Aryabhata hit upon $3 \cdot 1416$ as a close approximation to the ratio of the circumference of a circle to its diameter, a fact mentioned earlier in our discussion of the history of π in Chapter 7. He also gave correct general rules for computing the sum of natural numbers, and of their squares and cubes. There was, on the evidence of a later commentator, a non-extant work under the title *Arya Siddhanta*, a more detailed examination of astronomy (and possibly trigonometry), but there are serious doubts about its authorship.

Aryabhata's place as the premier and pioneering mathematician of India will have to be reassessed in the light of recent discoveries about the scope and quality of Jaina mathematics. As the mathematical activities of the post-Vedic and pre-Classical period become better known, Aryabhata will come to be seen mainly as an astronomer who had a great influence on those who came after him. It is therefore appropriate that India's first artificial satellite, designed and built in India, and launched in the USSR on 19 April 1975, was named 'Aryabhata'.

Two astronomers and commentators who followed Aryabhata extended his work. As an astronomer, Varahamihira (*c.* 505–587) is remembered for a revised version of the Indian calendar which he corrected for the amount of precession that had accumulated since the preparation of *Surya Siddhanta* (one of the

---

*Such a system involves allocating specific numbers to specific letters of the alphabet. In Aryabhata's scheme, the twenty-five 'classified' consonants of the Devanagari (a Sanskrit script) were allocated the numbers 1 to 25. The numbers 30, 40, . . . , 90 were denoted by the seven 'unclassified' consonants, while the ten vowels were used to denote powers of 10 from 100 onwards. Here are two examples from Aryabhata's best-known work, *Aryabhatiya*:

$$\textit{Khya } \text{रूय} = \text{ख} + \text{य} : 2 + 30 = 32$$
$$\textit{Khyu } \text{रूयउ} = \text{ख} + \text{य} \times \text{उ} : (2 + 30) \times 10^3 = 32\,000$$

This interesting method of numeration was adopted mainly because it lent itself to the expression of mathematical information in a poetic form. This was to remain a characteristic of Indian mathematical writing until the advent of modern mathematics.

widely known earlier *Siddhantas*). Mathematically, his work is interesting for its detailed exposition of trigonometry. It gives a number of relations between three functions, *jya* (Indian sine, or half-chord), *kojya* (Indian cosine, or cosine-chord of an arc) and *utkramajya* (Indian versine), which we shall examine in the section on trigonometry later in this chapter. Also, extending the work of Aryabhata I, he gave values of different *jyas* in a quadrant drawn at a fixed interval (i.e. an Indian sine table).

Bhaskara I (*c.* 600) was one of the most competent exponents of Aryabhata's astronomy. His three major works consisted of two treatises on astronomy and a commentary on the *Aryabhatiya*; his notable contribution to mathematics was his solution of indeterminate equations of the first degree, which was to have a significant influence on a later mathematician-astronomer, Brahmagupta.

Brahmagupta (b. 598)

After Varahamihira, the best known mathematician-astronomer of the Ujjain school is Brahmagupta. He wrote two works, the first and more important, *Brahma Sputa Siddhanta*, when he was thirty. The book is a comprehensive astronomy text, several chapters of which deal with mathematics. Brahmagupta called the twelfth chapter *Ganita* ('arithmetical calculation'), although it includes a discussion of mathematical series and a few geometric topics. The eighteenth chapter, *Kuttaka* (literally 'pulverizer', but broadly translated as 'algebra'), contains solutions of indeterminate equations of the first and second degree. Scattered through his second book, an astronomical treatise entitled *Khanda Khadyaka*, are further developments in trigonometry, including a method of obtaining the sines of intermediate angles from a given table of sines. The method employed is equivalent to the Newton–Stirling interpolation formula up to second-order differences. We shall return to this later.

Brahmagupta holds a special place in the history of mathematics. As we shall see in the next chapter, it was partly through a translation of his *Brahma Sputa Siddhanta* that the Arabs, and then the West, became aware of Indian astronomy and mathematics. This was to have momentous consequences for the development of the two subjects.

Mahavira (*c.* 850)

Mahavira was the best-known Indian mathematician of the ninth century. A Jain by religion, he was familiar with Jaina mathematics, which he included and refined in his book *Ganita Sara Samgraha*. Mahavira also knew the works of

Aryabhata and his commentators, and Brahmagupta. Unlike his predecessors, Mahavira was not an astronomer – his work was confined to mathematics. He was a member of the mathematical school at Mysore in southern India.

In *Ganita Sara Samgraha* he gives a lucid classification of arithmetical operations and a number of examples to illustrate the rules. His contributions include:

1 a detailed examination of operations with fractions, with some ingenious methods for decomposing integers and fractions into unit fractions (a subject of practical utility for the ancient Egyptians, as we saw in Chapter 3);

2 an extension and systematization of the Jaina work on permutations and combinations, for both of which he provides the well-known general formulae;

3 solutions of different types of quadratic equation, as well as an extension of his predecessors' work on indeterminate equations; and

4 geometric work on right-angled triangles whose sides are rational and, something unusual in Indian mathematics, attempts (albeit unsuccessful) to derive the formulae for the area and perimeter of an ellipse.

The book was widely used in southern India and translated into Telegu, a regional language, during the eleventh century.

Mahavira's contribution may be looked at in two ways. *Ganita Sara Samgraha* may be seen as the culmination of Jaina work on mathematics (indeed it is the only substantial treatise on Jaina mathematics that we have). Alternatively, Mahavira can be seen as summarizing and extending the mathematical content of the works of his predecessors such as Aryabhata, Bhaskara I and Brahmagupta. He was very conscious of the debt he owed those who came before him. In the introductory chapter of his book, he wrote:

> With the help of the accomplished holy sages, who are worthy to be worshipped by the lords of the world . . . I glean from the great ocean of the knowledge of numbers a little of its essence, in the manner in which gems are [picked] from the sea, gold from the stony rock and the pearl from the oyster shell; and I give out according to the power of my intelligence, the *Sara Samgraha*, a small work on arithmetic, which is [however] not small in importance.

Sridhara (c. 900)

There remains some controversy over Sridhara's time and place of birth; some scholars suggest he came from Bengal, others from southern India. What is definitely known is that he wrote *Patiganita*, a work on arithmetic and mensuration. He deals with elementary operations, including extracting square and cube roots, and fractions. Eight rules are given for operations involving zero (but not division). His methods of summation of different arithmetic and geometric series were to become standard references in later works.

Probably a contemporary of Sridhara was the astronomer Aryabhata II, who lived around the middle of the tenth century. In his major astronomical treatise of eighteen chapters, *Maha Siddhanta*, there is a clear treatment of *kuttaka*, which had by then come to mean the solution of indeterminate equations.

Bhaskara II (b. 1114)

Bhaskara II, or Bhaskaracharya ('Bhaskara the Teacher') as he is still popularly known in India, was the most distinguished mathematician-astronomer produced by the school at Ujjain. His fame rests on three works: *Lilavati*, *Bijaganita* and *Siddhanta Siromani*. The last, a highly influential astronomical work, was written in 1150 when he was thirty-six years old.

Lilavati, which is based on the works of Brahmagupta, Sridhara and Aryabhata II, shows a profound understanding of arithmetic. Bhaskaracharya's work on fundamental operations, his rules of three, five, seven, nine and eleven, his work on permutations and combinations, and his rules of operations with zero together speak of a maturity, a culmination of five hundred years of mathematical progress.

In 1587, on the instructions of the Mughal emperor Akbar, the court poet Fyzi translated *Lilavati* into Persian. Fyzi tells a charming story of the book's origin. Lilavati was the name of Bhaskaracharya's daughter. From casting her horoscope, he discovered that the auspicious time for her wedding would be a particular hour on a certain day. He placed a cup with a small hole at the bottom in a vessel filled with water, arranged so that the cup would sink at the beginning of the propitious hour. When everything was ready and the cup was placed in the vessel, Lilavati suddenly out of curiosity bent over the vessel and a pearl from her dress fell into the cup and blocked the hole in it. The lucky hour passed without the cup sinking. Bhaskaracharya believed that the way to console his dejected daughter, who now would never get married, was to write her a manual of mathematics!

Bhaskaracharya's *Bijaganita* contains problems on determining unknown quantities, evaluating surds and solving simple and quadratic equations, and certain general rules which went beyond Sridhara in dealing with the solution of indeterminate equations of the second degree, and even equations of the third and fourth degree. Bhaskaracharya's 'cyclic' method for solving indeterminate equations of the form $ax^2 + bx + c = y$ was rediscovered in the West by William Brouncker in 1657.

In *Siddhanta Siromani*, Bhaskaracharya demonstrates his knowledge of trigonometry, including the sine table and relationships between different trigonometric functions. In his work can be traced certain preliminary concepts of the infinitesimal calculus and analysis, concepts which would be taken up by the Kerala school of mathematics in their work on infinite series some two hundred years later.

He won such a great reputation that his manuscripts were still being copied and commented upon as late as the beginning of the nineteenth century. A medieval temple inscription refers to him in the following terms (and here there reappears the imagery of the peacock which provides the inspiration for the title of the present book):

> Triumphant is the illustrious Bhaskaracharya whose feats are revered by both the wise and the learned. A poet endowed with fame and religious merit, he is like the crest on a peacock.

It was generally believed until recently that mathematical developments in India came to a virtual halt after Bhaskaracharya. This opinion has had to be revised in the light of recent research on what one could describe as medieval Indian mathematics. A number of the manuscripts of this period are yet to be published, or even subjected to critical scrutiny. But we are able to identify the following notable contributors to mathematics during the medieval period.

Narayana Pandit (*c.* 1370)

Narayana lived during the reign of Firoz Shah (1355–88) and composed *Ganita Kaumadi*, a treatise on arithmetic, and *Bijaganita Vatamsa*, a work on algebra, both of which were heavily influenced by Bhaskaracharya. There were a number of later commentaries on both texts. The topics contained in Narayana's books include laws of signs, mathematical operations with zero, approximation methods for finding the square root of a non-square number and a diagrammatic method of representing different mathematical series, to be discussed in the last section of this chapter.

Madhava of Sangamagramma (*c.* 1340–1425)

Madhava was probably the greatest of the Indian medieval astronomer-mathematicians, but he has come to the fore only in recent years as a result of research into Kerala mathematics. It was Madhava who 'took the decisive step onwards from the finite procedures of ancient mathematics to treat their limit-passage to infinity, which is the kernel of modern classical analysis' (Rajagopal and Rangachari, 1978, p. 101).

Sangamagramma was a village with a temple dedicated to a deity of the same name, and situated near Cochin in Kerala. This place-name is often given when referring to Madhava so as to distinguish him from others like the astrologer Vidya Madhava. Later astronomers called him Golavid (or 'Master of Spherics'). Of his works that have survived, all are astronomical treatises; for his mathematical contributions we rely on reports by his contemporaries and successors. These contributions, which include infinite-series expansions of circular and trigonometric functions and finite-series approximations, are discussed in a later section.

Nilakantha Somayaji (1445–1545)

Nilakantha, who lived to the ripe old age of 100, was a student of the eminent astronomer Paramesvara of Vatasreni. He came from a Nambuthri Brahmin family in South Malabar, Kerala. He was a versatile scholar, but, like Madhava, all his surviving works are on astronomy. The mathematical sections of his most notable work, *Tantra Samgraha*, elaborate and extend the contributions that are attributed to Madhava.

The Kerala school of mathematics and astronomy continued for another two centuries, producing detailed commentaries on the works of classical mathematicians such as Aryabhata and Bhaskaracharya as well as continuing the work on trigonometry and infinite series begun by Madhava. In a notable work in Malayalam (the regional language of Kerala) entitled *Yuktibhasa*, Jyesthadeva (*c.* 1550) provides a detailed summary of the mathematical contributions made by the Kerala school. It is unusual in Indian mathematics since it contains derivations of most of the theorems and formulae stated in the text.

In the rest of this chapter we survey the major contributions of the Indian mathematicians of the Classical (from Aryabhata I to Bhaskaracharya) and medieval (from after Bhaskaracharya until about 1600) periods. Our approach will be a thematic one, with the exception of the contribution of the Kerala school, considered in a separate section. It may help whenever necessary to refer both to the map in Figure 8.1 and to the Indian chronology in Table 8.1.

INDIAN ALGEBRA

It was briefly indicated in the last chapter that the *Sulbasutras* and the later Bakhshali Manuscript contain some early algebra, including the solution of linear, simultaneous and even indeterminate equations. But it is only from the time of Aryabhata I that algebra grew into a distinct branch of mathematics. Different names were used for this area of mathematics. Brahmagupta called it *kuttaka ganita*, or simply *kuttaka*, which later came to refer to a particular branch of algebra dealing with the techniques of indeterminate analysis. And this subject was considered so important that for a period it became a blanket term to describe solutions of all types of equation. Only with Pruthudakaswami (*c.* 850) did algebra acquire a new name, *bijaganita* (literally 'the science of calculation with unknown elements').

A significant feature of early Indian algebra which distinguishes it from other mathematical traditions was the use of symbols, such as a dot (in the Bakhshali Manuscript) or the letters of the alphabet, to denote unknown quantities. In fact it is this very feature of algebra that one immediately associates with the subject today. The Indians were probably the first to make systematic use of this method of representing unknown quantities. A general term for any unknown was *yavat tavat*, which was shortened to the algebraic symbol *ya*. In Brahmagupta's work appear Sanskrit letters, the abbreviations of names of different colours, which he used to represent several unknown quantities. Thus the letter *ka* stood for *kalaka*, meaning 'black', and the letter *ni* for *nilaka*, meaning 'blue'.

Simple operations were also indicated by abbreviations or symbols. We saw in Chapter 8 that addition was represented in the Bakhshali Manuscript by placing *yu* (which stood for *yuta*, 'addition') between the terms to be added, and subtraction by placing the sign $+$ after the term to be subtracted. Multiplication was indicated by placing *gu* (for *ganita*), and division by putting *bha* (for *bhaga*), between the two terms. A square root was indicated by *mu* (for *mula*) after the term. There were several other similar abbreviations for other operations.

In Pruthudakaswami's commentary on Brahmagupta's *Brahma Sputa Siddhanta* appears the following representation:

$$yava\ 0 \quad ya\ 10 \quad ru\ 8$$
$$yava\ 1 \quad ya\ 0 \quad ru\ 1$$

Here *ya* is an abbreviation for *yavat tavat* (the unknown quantity, or *x*) and *yava* is an abbreviation for *yavat avad varga* (the square of the unknown quantity, or x^2); *ru* stands for *rupa* (the constant term). In other words, this is what we would now write as

$$10x + 8 = x^2 + 1$$

Solutions of determinate equations

A geometric solution to a linear equation in one unknown (an equation like $3x + 8 = 23$) is found in Baudhayana's *Sulbasutra*, while an algebraic solution appears for the first time in the Bakhshali Manuscript. The method used was an inversion method where one works backwards from a given piece of information – an approach particularly favoured by Arab mathematicians five hundred years later, which may have reached them from India. (An illustration of the procedure is given later in this section.) What is interesting is that, at least from the time of Aryabhata I, the method of false position, which we came across first in Egyptian mathematics and later in Chinese mathematics, is not used in Indian algebra. Could it be that, with the early development of algebraic symbolism, there was simply no need for it in Indian mathematics?

Quadratic equations make their first appearance in the *Sulbasutras* in the forms $ax^2 = c$ and $ax^2 + bx = c$. No solution is given. For an equation of the form $ax^2 + bx - c = 0$, the Bakhshali Manuscript offers the solution (in modern notation)

$$x = \frac{\sqrt{b^2 - 4ac} - b}{2a} \tag{9.1}$$

The first explicit statement of a general rule appears in a work by Sridhara which is unfortunately lost, though the rule is preserved in quotations by Bhaskaracharya and others. It is:

> Multiply both sides [of the equation] by a known quantity equal to four times the coefficient of the square of the unknown; add to both sides a known quantity equal to the square of the coefficient of the unknown; then [extract] the square root.

This solution, obtained by transforming the left-hand side of the quadratic equation

$$ax^2 + bx = c$$

by multiplying both sides by $4a$, adding b^2 and finally taking the square root, is a variant of (9.1). There is no evidence that Sridhara used both signs of the radical, but Mahavira was certainly aware of both possibilities. He gave the solution (in modern notation) as

$$x = \frac{-b/a \pm \sqrt{(b/a - 4c/b)b/a}}{2}$$

In the works of Mahavira, Bhaskaracharya and others are found a number of fascinating problems, clearly devised to stimulate the interest of the reader. Let us consider a few of these, about linear and quadratic equations, beginning with one from Aryabhata whose solution used the method of 'algebraic inversion'.

EXAMPLE 9.1   *O beautiful maiden with beaming eyes, tell me, since you understand the method of inversion, what number multiplied by 3, then increased by three-quarters of the product, then divided by 7, then diminished by one-third of the result, then multiplied by itself, then diminished by 52, whose square root is then extracted before 8 is added and then divided by 10, gives the final result of 2?*

Solution   We start with the answer 2 and work backwards. When the problem says divide by 10, we multiply by that number; when told to add 8, we subtract 8; when told to extract the square root, we take the square; and so on. It is precisely the replacement of the original operation by the inverse that gives the method its name of 'inversion'.

Therefore the original number is obtained thus:

$$[(2)(10) - 8]^2 + 52 = 196$$
$$\sqrt{196} = 14$$
$$\frac{(14)(3/2)(7)(4/7)}{3} = 28$$

EXAMPLE 9.2   *From a swarm of bees, a number equal to the square root of half the total number of bees flew out to the lotus flowers. Soon after, 8/9 of the total swarm went to the same place. A male bee enticed by the fragrance of the lotus flew into it. But when it was inside the night fell, the lotus closed and the bee was caught inside. To its buzz, its consort responded anxiously from outside. O my beloved! How many bees are there in the swarm?* (Bhaskaracharya's *Lilavati*)

Solution   Bhaskaracharya's approach is equivalent to solving the following equation:

$$\sqrt{(1/2)x} + (8/9)x + 2 = x$$

where x is the total number of bees in the swarm. It is to be assumed from the question that the male bee and his consort were late arrivals from the same swarm – hence the 2 in the above equation. Only $x = 72$ bees is given as an admissible solution. However, in the following problem from *Bijaganita*, Bhaskaracharya admits that more than one solution is valid:

EXAMPLE 9.3 *Inside a forest, a number of apes equal to the square of one-eighth of the total apes in a pack are playing noisy games. The remaining 12 apes, who are of a more serious disposition, are on a nearby hill and irritated by the shrieks coming from the forest. What is the total number of apes in the pack?*

Solution The solutions $x = 16$ and $x = 48$ are equally admissible, according to Bhaskaracharya.

EXAMPLE 9.4 *Three merchants find a purse lying on a road. One of them says, 'If I keep this purse, I shall be twice as rich as both of you together.' 'Give me the purse and I will be thrice as rich,' says the second, while the third exclaims, 'I shall be much better off than either of you if I keep the purse. I shall become five times as rich!' How much money is there in the purse? How much money has each merchant?* (Mahavira's *Ganita Sara Samgraha*)

Solution The solution starts by setting up the following relationships:

$$m + x = 2(y + z)$$
$$m + y = 3(x + z)$$
$$m + z = 5(x + y)$$

where m is the amount of money in the purse, and x, y and z are the amounts of money in the possession of the three merchants.

The final solution is given in the form of ratios since there is no unique solution set:

$$m:x:y:z = 15:1:3:5$$

Interest in such indeterminate problems, with no unique solution, had been a characteristic of Indian mathematics ever since the Vedic period.

Indeterminate equations

It is in Aryabhata I's work, *Aryabhatiya*, that we come across the first unequivocal statement of a problem of indeterminate analysis and its solution. The problem arose, just as it did in China, in the field of astronomy, where there was a need to determine the orbits of planets. Expressed in simple mathematical terms, the problem is one of finding a general positive integral solution to an indeterminate equation of the form

$$by = ax \pm c \qquad (9.2)$$

where a, b and c are integral values, and x and y are the unknowns.

It was solved by the method of *kuttaka*. This word is derived from *kutt*, meaning 'to crush, grind or pulverize', and may describe a method which is a successive process of breaking down into smaller and smaller pieces: making the values of the coefficients (*a* and *b* in equation (9.2)) smaller and smaller. All the great mathematicians of the period dealt with the *kuttaka*, and it is one of the very few topics in Indian mathematics to be made the subject of a special monograph, entitled *Kuttakara Siromani*, written by a commentator on Aryabhata I named Devaraja.

As we have already examined problems about indeterminate equations of the first degree (in Chapter 7), we concentrate here on the climax of Indian work in this area – the solution of indeterminate equations of the second degree. Brahmagupta considered the following two equations, the second of which is a special case of the first:

$$ax^2 \pm c = y^2 \qquad (9.3)$$

$$ax^2 + 1 = y^2 \qquad (9.4)$$

where *a* and *c* are known as the multiplier and augment, and *x* and *y* as the smaller and larger roots. Equation (9.4) is a form of Pell's equation, wrongly named by Euler after an English mathematician, John Pell (1610–85). Brahmagupta was probably the first mathematician to give solutions to both equations (9.3) and (9.4) in rational integers. His approach is ingenious and general; here, though, we give a simple example to illustrate his approach. The algebraic intricacies are gone into by Bag (1979, pp. 193–228).

EXAMPLE 9.5   *Solve the equation $8x^2 + 1 = y^2$.*
Solution   Brahmagupta's method may be expressed in the following way: From inspection, it is obvious that the smallest integral solution (root) for *x* is 1, and the *y* that corresponds to this solution is 3, the minimum solution (root) for *y*.

Now arrange this information as follows:

SMALLER ROOT	LARGER ROOT	AUGMENT
1	3	1
1	3	1

Multiply crosswise as indicated by the arrows and add the products. (Note the resemblance between this procedure and the 'vertically and crosswise' multiplication *sutra* we looked at in Chapter 8.) Thus

$$3 + 3 = 6 = x \quad \text{so} \quad y = 17$$

Arrange the old and new sets of values of x and y together with the augment in the following way:

SMALLER ROOT   LARGER ROOT   AUGMENT

1   3   1

6   17   1

Multiply crosswise the first two columns, and add to obtain $x = 35$. The corresponding value for y is found by substituting into the original equation, which gives $y = 99$.

Proceeding along these lines, we can construct the following sequence of diagrams to obtain larger and larger solution sets:

SMALLER ROOT   LARGER ROOT   AUGMENT

1   3   1

35   99   1

SMALLER ROOT   LARGER ROOT   AUGMENT

1   3   1

204   577   1

Thus the solution sets for (x, y) are

$$(1, 3), (6, 17), (35, 99), (204, 577), (1189, 3363), \ldots$$

The last column of the above diagrams comes into play when, in the process of calculation, we no longer obtain perfect squares, as in the next example.

EXAMPLE 9.6   Solve the equation $11x^2 + 1 = y^2$.

Solution   Follow the same procedure as before. Take the smaller root $x = 1$.

The left-hand side of the equation is not a perfect square. If, however, the augment is -2 rather than 1, the left-hand side becomes 9, a perfect square. The diagrammatic representation is then as follows:

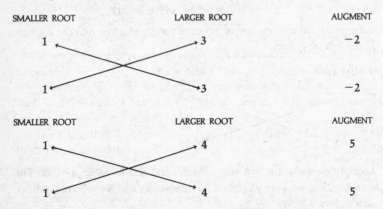

SMALLER ROOT	LARGER ROOT	AUGMENT
1	3	-2
1	3	-2

SMALLER ROOT	LARGER ROOT	AUGMENT
1	4	5
1	4	5

Take the upper diagram, multiply crosswise and add to get $x^* = 6$. Find the product of the smaller roots (i.e. 1×1), multiply the product by 11 (1×11) and add to it the product of the larger root (3×3) to give

$$11 + 9 = 20 = y^*$$

But $x^* = 6$ and $y^* = 20$ satisfy the equation

$$11x^2 + 4 = y^2$$

which has an augment of 4 and is not the same as the equation we started with. Now, the product of the assumed augment is $-2 \times -2 = 4$. Dividing 4 by 4 gives an augment of 1. Thus to obtain the values of x and y that correspond to the original augment 1, divide x^* and y^* by 2 to give one solution set for (x, y) as $(3, 10)$.

If we begin with the smaller root $x = 1$ and the larger root $y = 4$, the augment is 5. Following exactly the same procedure as before, and operating with the lower diagram, the resulting solution set for (x, y) is $(8/5, 27/5)$.

To generate another solution set, proceed as before but use the following diagram:

SMALLER ROOT	LARGER ROOT	AUGMENT
3	10	1
8/5	27/5	1

The new solution set for (x, y) is $(161/5, 534/5)$.

It is worth noting that this method was first used by Brahmagupta as early as AD 600, though it is usually attributed to Euler, who named it *theorem elegantissimum*. The sheer ingenuity and versatility of the approach is also highlighted by the fact that it was not until 1767 that Lagrange gave a complete solution to Pell's equation, using continued fractions.

Jayadeva (c. 1000) was one of the first to point out that, while Brahmagupta's approach would easily produce an infinite number of solutions with an augment of ± 1, ± 2 or ± 4, with all other augments a trial-and-error process was necessary. In his book *Sundari*, Jayadeva states a general method for solving indeterminate equations of the kind just discussed. The method was refined and provided with a mathematical rationale by Bhaskaracharya about a hundred years later. The Jayadeva–Bhaskaracharya method was known as the *chakravala*, or 'cyclic', method because the same set of operations is repeated over and over again. It bears a close resemblance to the so-called 'inverse cyclic method' based on continued-fraction expansions that attracted the attention of European mathematicians of the calibre of Pierre de Fermat (1601–65), Leonhard Euler (1707–83), Joseph Lagrange (1736–1813) and Évariste Galois (1811–32). The method must be regarded as a purely Indian creation, for there is no record of it at all in Chinese mathematics. (For details of the Indian cyclic method, see Bag (1979, pp. 217–23) and Selenius (1975).)

There is a problem of considerable historical interest, for which Bhaskaracharya offers the first complete solution. The problem is to solve

$$61x^2 + 1 = y^2 \quad \text{for minimum } x \text{ and } y$$

He gives the solution $x = 226\,153\,980$ and $y = 1\,766\,319\,049$. It was precisely this problem that Fermat set as a challenge to his friend Frénicle de Bessy in 1657. We do not know whether Frénicle de Bessy took up the challenge; the problem was finally solved by Lagrange about a hundred years later. A comparison between Lagrange's and Bhaskaracharya's methods is quite illuminating. Lagrange's method requires the calculation of twenty-one successive convergents of the continued fraction for the square root of 61, while the Jayadeva–Bhaskaracharya approach gives the solution in a few easy steps. Selenius's (1975, p. 180) assessment of the method is interesting:

> The method represents a best approximation algorithm of minimal length that, owing to several minimization properties, with minimal effort and avoiding large numbers always automatically produces the [best] solutions

to the equation . . . The *chakravala* method . . . anticipated the European methods by more than a thousand years. But no European performances in the whole field of algebra at a time much later than Bhaskara's, nay nearly up to our times, equalled the marvellous complexity and ingenuity of *chakravala*.

INDIAN TRIGONOMETRY

The origins of trigonometry are obscure. There are certain problems in the Ahmes Papyrus (*c.* 1650 BC) about measuring the steepness of the face of a pyramid by the ratio of the 'run' to the 'rise' (the horizontal departure of the oblique face from the vertical per unit height). This ratio, known as the *seqt* of the pyramid, is equivalent to the cotangent of the angle made by the face of the pyramid and its base. As discussed in the last chapter, this angle was kept constant at around 52° in the Great Pyramid at Gizeh and many other Egyptian pyramids. There is also the conjecture, discussed in Chapter 4, that a column of numbers contained in a Babylonian cuneiform tablet, Plimpton 322, is a table of secants, but this must be considered a far-fetched idea, especially since more plausible explanations exist. Also, there is no evidence that the Babylonians of that period were familiar with the concept of an angle. However, we cannot be so dismissive about the possibility that the Babylonians of the 'New' period may have constructed a form of proto-trigonometry for astronomical purposes. The Babylonian astronomers of the first millennium BC were known to have accumulated a large number of observations which survived to provide the Greeks and then the Alexandrians with the impetus for early work on trigonometry.

The beginnings of a systematic study of the relationships between the angles (or arcs) of a circle and the lengths of chords subtending them are usually attributed to the Alexandrian Hipparchus (*c.* 150 BC), who was also credited with a twelve-part treatise dealing with the construction of a table of chords of arcs of a circle. Ptolemy (*c.* AD 100) is believed to have used this table to construct his own, which gave the lengths of the chords of all central angles of a given circle in half-degree intervals from $\frac{1}{2}°$ to 180°. The radius of a circle was divided into 60 equal parts, and the chord lengths were then expressed sexagesimally in terms of one of these parts as a unit. Ptolemy's table has entries like crd 36° = 0;37,4,55, which means that the length of the chord of a central angle of 36° (see Figure 9.1) is equal to 37 small parts of the radius (37/60), plus 4/60 of one of these small parts and 55/3600 more of one of the small parts. The division of a circumference into 360° goes back to the period

Figure 9.1

← crd 36° →

of the Babylonians. Again, it was the sexagesimal system that led Ptolemy to subdivide the diameter of his trigonometric circle into 120 parts, each of these in turn being split into 60 minutes and each minute into 60 seconds.

The sources of Indian trigonometry

While the work of the Alexandrians Hipparchus (*c.* 150 BC), Menelaus (*c.* AD 100) and Ptolemy (*c.* AD 150) in astronomy laid the foundations of trigonometry, further progress was piecemeal and spasmodic. From about the time of Aryabhata I (*c.* AD 500), the character of the subject changed, and it began more to resemble its modern form. Subsequently it was transmitted to the Arabs, who introduced further refinements. From the Arabs, the knowledge spread to Europe, where a detailed account of existing trigonometric knowledge first appeared under the title *De triangulis omni modis*, written in 1464 by Regiomontanus.

In early Indian mathematics, trigonometry formed an integral part of astronomy. References to trigonometric concepts are found in the *Surya Siddhanta* (*c.* AD 400), Varahamihira's *Pancha Siddhanta* (*c.* AD 500) and Brahmagupta's *Brahma Sputa Siddhanta* (AD 628). A detailed and systematic study of the subject was made by Bhaskaracharya in his *Siddhanta Siromani*. He felt that the title *acharya* (i.e. master or teacher) in astronomy could be given only to those who possessed sufficient knowledge of trigonometry. Infinite expansions of trigonometric functions, building on Bhaskaracharya's work, are found in the work of Madhava and Nilakantha, discussed in the next section.

The development of trigonometric functions

On account of their shapes, the arc of a circle (e.g. the arc ACB in Figure 9.2) was known as the 'bow' (*capa*) and its full-chord (e.g. the line segment AMB in Figure 9.2) as the 'bow-string' (*samasta-jya*). In their study of trigonometric functions, Indian mathematicians more often used the half-chord (e.g. the line segment AM or MB). The half-chord was known as *jya-ardha*, later abbreviated to *jya* to become the Indian sine. Three functions were developed, whose modern equivalents are defined here with reference to Figure 9.2:

$$\text{jya}\,\alpha = \text{AM} = r\sin\alpha$$
$$\text{kojya}\,\alpha = \text{OM} = r\cos\alpha$$
$$\text{ukramajya}\,\alpha = \text{MC} = \text{OC} - \text{OM} = r - r\cos\alpha$$
$$= r(1 - \cos\alpha) = r\,\text{vers}\,\alpha$$

To calculate jya α ($r\sin\alpha$) for angles $\alpha \leqslant 90°$, Varahamihira suggested the following formulae:

$$\text{jya}\,30° = \tfrac{1}{2}r, \qquad \text{jya}\,60° = \tfrac{1}{2}\sqrt{3}r, \qquad \text{jya}\,90° = r$$

With the help of these formulae he calculated the values of $r\sin\alpha$ ranging from 3°45' in twenty-four multiples to 90°.

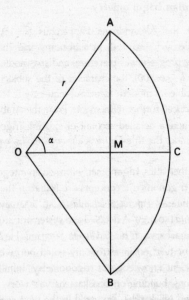

Figure 9.2

Bhaskara I (*c.* 600), in his *Maha Bhaskariya*, gave the following approximate formula for calculating the Indian sine of an acute angle without the use of a table:

$$r \sin \alpha \simeq \frac{r(180° - \alpha)\alpha}{(r/4)[40\,500 - (180 - \alpha)\alpha]}$$

which is equivalent to

$$\sin \beta \simeq \frac{16\beta(\pi - \beta)}{5\pi^2 - 4\beta(\pi - \beta)}$$

where β radians correspond to α degrees. If now $\beta = \pi/3$ and then $\pi/7$,

$$\sin(\pi/3) \simeq 0.8643\ldots \quad \text{and} \quad \sin(\pi/7) \simeq 0.4314\ldots$$

which are both correct to the second decimal place.

The same degree of accuracy is achieved when the values of $\sin \pi$, $\sin(\pi/2)$ and $\sin(\pi/4)$ are obtained from the above approximation formula. Bhaskara I ascribed this formula to Aryabhata I. It occurs in Brahmagupta's *Brahma Sputa Siddhanta* and in several later works.

Some other trigonometric relations found in the astronomical texts of the Classical period are shown below, together with the names of the author in whose work they first appear. (For ease of expression we take $r = 1$, so that the Indian sine becomes equal to the modern sine.)

$$\sin(n + 1)\alpha - \sin n\alpha = \sin n\alpha - \sin(n - 1)\alpha - (1/225)\sin n\alpha$$

Aryabhata I

$$\cos \alpha = \sin(\tfrac{1}{2}\pi - \alpha) \qquad \text{Varahamihira}$$

$$\sin^2 \alpha + \cos^2 \alpha = 1 \qquad \text{Varahamihira}$$

$$\sin^2 \alpha = \tfrac{1}{4}(\sin^2 2\alpha + \text{versin}^2 2\alpha) = \tfrac{1}{2}(1 - \cos 2\alpha) \qquad \text{Varahamihira}$$

$$1 - \sin^2 \alpha = \cos^2 \alpha = \sin^2(\tfrac{1}{2}\pi - \alpha) \qquad \text{Brahmagupta}$$

$$\sin(\tfrac{1}{4}\pi + \tfrac{1}{2}\alpha) = \sqrt{\tfrac{1}{2}(1 \pm \sin \alpha)} \qquad \text{Aryabhata II}$$

$$\sin(A \pm B) = \sin A \cos B \pm \cos A \sin B \qquad \text{Bhaskaracharya}$$

The construction of sine tables

Various relationships between the sine of an arc and its integral and fractional multiples were used to construct sine tables for different arcs lying between 0 and 90°. These tables were used for astronomical calculations, for example to compute exact locations of planets. The general formula by the name of Aryabhata I above was used to compute tables of half-chords in a quadrant divided into 24 equal parts, so that the smallest arc is 3°45' or (2.25'). Since the Indian sines are not the ratios of the corresponding half-chords and the radius, but represent the half-chords themselves, their values obviously depend on the length of the radius chosen. The values for the radius adopted by Varahamihira, Aryabhata I and Brahmagupta were 120′, 3438′ and 3270′ (or 900′), respectively. The values of the twenty-four sines given in the *Pancha Siddhanta* and *Aryabhatiya* are given together with equivalent modern values in Table 9.1. The accuracy of Aryabhata's sines is quite remarkable.

Table 9.1 Indian sines: values given by Varahamihira and Aryabhata I, and their modern equivalents.

Angle, θ	Varahamihira		Aryabhata		Modern value of $\sin\theta$
	$r\sin\theta$ $(r = 120′)$	Computed $\sin\theta$	$r\sin\theta$ $(r = 3438′)$	Computed $\sin\theta$	
3°45′	7′51″	0·06542	225′	0·06545	0·06540
7°30′	15′40″	0·13056	449′	0·13060	0·13053
11°15′	23′25″	0·19514	671′	0·19517	0·19509
15°	31′ 4″	0·25889	890′	0·25962	0·25882
18°45′	38′34″	0·32139	1105′	0·32141	0·32143
22°30′	45′56″	0·38278	1315′	0·38249	0·38268
26°15′	53′ 5″	0·44236	1520′	0·44212	0·44229
30°	60′	0·50000	1719′	0·50000	0·50000
33°45′	66′40″	0·55556	1910′	0·55556	0·55556
37°30′	73′ 3″	0·60875	2093′	0·60878	0·60876
41°15′	79′ 7″	0·65931	2267′	0·65910	0·65935
45°	84′51″	0·70708	2431′	0·70710	0·70711
48°45′	90′13″	0·75181	2585′	0·75189	0·75184
52°30′	95′13″	0·79347	2728′	0·79348	0·79335
56°15′	99′46″	0·83139	2859′	0·83159	0·83147
60°	103′56″	0·86611	2978′	0·86620	0·86602
63°45′	107′38″	0·89694	3084′	0·89703	0·89687
67°30′	110′53″	0·92402	3177′	0·92408	0·92388
71°15′	113′38″	0·94694	3256′	0·94706	0·94693
75°	115′56″	0·96611	3321′	0·96597	0·96593
78°45′	117′43″	0·98097	3372′	0·98080	0·98079
82°30′	119′	0·99167	3409′	0·99156	0·99144
86°15′	119′45″	0·99792	3431′	0·99796	0·99786
90°	120′	1·00000	3438′	1·00000	1·00000

Adapted from Table 3.4 in Bose *et al.* (1971, p. 200).

The first known variant of a sine table was Ptolemy's. He gave a table of chords within a circle of diameter 60 units and expressed it in sexagesimal units. The arc ranges from $\frac{1}{2}°$ to 180° at half-degree intervals. With the help of this table, the corresponding length of the chord can be calculated when the length of the arc is known, and vice versa. The chord lengths were given by Ptolemy in terms of the circle's diameter, whereas in India they were given in terms of its radius. There is some controversy as to whether the Indian sine table was derived from Ptolemy's table. What seems more likely is that both Ptolemy and the Indian astronomers were indebted to a common source – Babylonian astronomy.

In the year 665, when Brahmagupta was 67 years old, he wrote an astronomical treatise entitled *Khanda Khadyaka*. In its ninth chapter he shows how to interpolate the sines of intermediate angles from a sine table. Table 9.2 is a partial table of sines, where *r* is taken to be 150 units. (Brahmagupta takes the radius of the circle as 900'.) We can use Brahmagupta's interpolation formula to find the sine of 67°:

the interval is: $\qquad\qquad h = 15°$ or 900'

the residual angle is: $\qquad\quad \Delta\theta = (67 - 60) = 7°$ or 420'

the relevant tabulated values are: $D_p = 24$, $D_{p+1} = 15$

where D_p and D_{p+1} are the corresponding functional differences taken from Table 9.2.

Table 9.2 Calculation of sines using Brahmagupta's interpolation formula.

Angle (degrees)	Indian sine (jya)	First difference, D_p	Second difference, $D_{p+1} - D_p$
0	0		
		39	
15	39		−3
		36	
30	75		−5
		31	
45	106		−7
		24	
60	130		−9
		15	
75	145		−10
		5	
90	150		

Adapted from Bag (1979, p. 257).

The value given by the interpolation formula is then

$$\frac{\Delta\theta}{h}\left(\frac{D_{p+1} + D_p}{2} + \frac{\Delta\theta}{h}\frac{D_{p+1} - D_p}{2}\right)$$

Applying the Brahmagupta interpolation formula gives

$$\frac{7}{15}\left[\frac{15 + 24}{2} + \frac{7}{15}\left(\frac{15 - 24}{2}\right)\right] = 8.12$$

Hence

$$\text{jya}\,67° = 130 + 8.12 = 138.12$$

which is close to $150 \sin 67° = 138.08$, the modern value. The reader is invited to estimate $150 \sin 78°$, using the Brahmagupta interpolation formula, and check how good the approximation is to the modern value of 146.72.

The rule is equivalent to the Newton–Stirling formula to second-order differences, often expressed as

$$f(a + xh) = f(a) + x\frac{[\Delta f(a) + \Delta f(a - h)]}{2} + \frac{x^2 \Delta^2 f(a - h)}{2!}$$

where Δ is the first-order forward-difference operator (i.e. the column of D_p values given in Table 9.2), Δ^2 is the second-order difference operator (i.e. the column of $D_{p+1} - D_p$ values in Table 9.2) and $x = \Delta\theta/h$.

Two centuries after Brahmagupta, the astronomer Govindaswami (*c.* 800–850), one of the earliest Kerala commentators on the works of Aryabhata and Bhaskara I, produced a rule for second-order interpolation to compute intermediate functional values. This proved to be a particular case (up to second order) of the general Newton–Gauss interpolation formula:

$$f(a + xh) = f(a) + x\Delta f(a) + \tfrac{1}{2}x(x - 1)[\Delta f(a) - \Delta f(a - h)]$$

Later work on computing sine and cosine functions, mainly in the Kerala school, produced expressions which are equivalent to the modern Taylor series approximations to second order, and predate Taylor by more than 300 years. We shall now turn to the work of the Kerala mathematicians from about the fourteenth to the seventeenth century.

KERALA MATHEMATICS

Along the south-west coast near the tip of the Indian peninsula lies a narrow strip of land known as Kerala. It has figured prominently in history, not only as a stop-over for travellers and explorers such as ibn Battuta and Vasco da Gama arriving from across the Arabian Sea, but as a centre of maritime trade, with its rich variety of spices greatly in demand, even as early as the time of the Babylonians. While most of India was in political upheaval during the first part of the second millennium AD, Kerala was a place of relative tranquillity, sheltered by the high mountains of the Western Ghats to the east, and the Arabian sea to the west. In recent years, Kerala has played a central role in the reconstruction of medieval Indian mathematics.

As we have seen, it was generally believed until recently that mathematics in India made no progress after Bhaskaracharya, that later scholars seemed content to chew the cud, writing endless commentaries on the works of the venerated mathematicians who preceded them, until they were introduced to modern mathematics by the British. However true this may have been for the rest of India, in Kerala the period between the fourteenth and seventeenth centuries marked a high point in the indigenous development of astronomy and mathematics. The quality of the mathematics available from the texts that have been studied is of such a high level compared with what was produced in the Classical period that it seems impossible for the one to have sprung from the other – there must be 'missing links' to bridge the gap between the two periods. There is no 'convenient' external agency, a Greece or Babylonia, that we can invoke to explain the Kerala phenomenon. Indeed, the only point of comparison is with later discoveries in European mathematics which were anticipated by Kerala astronomer-mathematicians two hundred to three hundred years earlier. And this leads us to ask whether the developments in Kerala had any influence on European mathematics. To answer this question, there is a need for a careful examination of the nature of the contacts between this most accessible of areas and the Europeans who came here in the wake of Vasco da Gama. There is some evidence, mentioned by Lach (1965), of a transfer of technology and products from Kerala to Europe. A lot more research on archival material from maritime, commercial and religious sources is required before the matter can be satisfactorily resolved.

The story of the discovery of Kerala mathematics sheds some fascinating light on the character of historical scholarship of the period. In 1835, Charles Whish published an article in which he referred to four works – Nilakantha's *Tantra Samgraha*, Jyesthadeva's *Yuktibhasa*, Putumana Somayaji's *Karana Paddhati* and Sankara Varman's *Sadratnamala* – as being among the main astronomical and mathematical texts of the Kerala school. While there were some doubts about Whish's views on the dating and authorship of these works, his main conclusions are still broadly valid. Writing about *Tantra Samgraha*, he

claimed that this work 'laid the foundation for a complete system of fluxions'.*
The *Sadratnamala*, a summary of a number of earlier works, he says 'abounds
with fluxional forms and series to be found in no work of foreign countries'.
The Kerala discoveries include the Gregory and Leibniz series for the inverse
tangent, the Leibniz power series for π, and the Newton power series for the
sine and cosine, as well as certain remarkable rational approximations of
trigonometric functions, including the well-known Taylor series approximations for the sine and cosine functions. And these results had apparently
been obtained without the use of infinitesimal calculus.

Whish's findings remained unnoticed for over a century until a series of
articles in the 1940s by Rajagopal and his collaborators highlighted the
contributions of Kerala mathematics. None of their results has as yet percolated
into the standard Western histories of mathematics. Boyer (1968, p. 244) writes
that 'Bhaskara was the last significant medieval mathematician from India, and
his work represents the culmination of earlier Hindu contributions.' And
according to Eves (1983, p. 164), 'Hindu mathematics after Bhaskara made only
spotty progress until modern times.'

Our present information on the mathematicians of the Kerala school is
fragmentary. We have already met Madhava. His work on power series for π
and for sine and cosine functions is referred to reverentially by a number of the
later writers, although the original sources remain undiscovered or unstudied.
Nilakantha (1445–1555) was mainly an astronomer, but his *Aryabhatiya Bhasya*
and *Tantra Samgraha* contain work on infinite-series expansions, problems of
algebra and spherical geometry. Jyesthadeva (*c.* 1550) wrote *Yuktibhasa*, one
of those rare texts in Indian mathematics or astronomy that gives detailed
derivations of many theorems and formulae in use at the time; it is also unusual
in being written in a regional language rather than in Sanskrit. This work is
mainly based on the *Tantra Samgraha* of Nilakantha. A joint commentary on
Bhaskaracharya's *Lilavati* by Narayana (*c.* 1500–75) and Sankara Variar (*c.*
1500–60), entitled *Kriyakramakari*, also contains a discussion of Madhava's
work. The *Karana Paddhati* by Putumana Somayaji (*c.* 1660–1740) provides a
detailed discussion of the various trigonometric series. Finally there is Sankara
Varman, the author of *Sadratnamala*, who lived at the beginning of the
nineteenth century and may be said to have been the last of the notable names
in Kerala mathematics. His work in five chapters contains, appropriately, a

---

*'Fluxion' was the term used by Isaac Newton for the rate of change (derivative) of a continuously
varying quantity, or function, which he called a 'fluent'. What Whish was saying was that here
clearly is the essence of a form of calculus.

summary of most of the results of the Kerala school, without any proofs though with some attempts at providing a rationale.

Astronomy provided the main motive for the study of infinite-series expansions of π and rational approximations for different trigonometric functions. For astronomical work, it was necessary to have both an accurate value for π and highly detailed trigonometric tables. In this area Kerala mathematicians made the following discoveries:

1 the power series for the inverse tangent, usually attributed to Gregory and Leibniz;

2 the power series for π, usually attributed to Leibniz, and a number of rational approximations to π; and

3 the power series for sine and cosine, usually attributed to Newton, and approximations for sine and cosine functions (to the second order of small quantities), usually attributed to Taylor; this work was extended to a third-order series approximation of the sine function, usually attributed to Gregory.

Apart from the work on infinite series, there were extensions of earlier work, notably of Bhaskaracharya:

1 the discovery of the formula for the circum-radius of a cyclic quadrilateral which goes under the name of l'Huilier's formula (and will be examined in a later section);

2 the use of the Newton–Gauss interpolation formula (to the second order) by Govindaswami, which was briefly discussed in the section on Indian trigonometry; and

3 the statement of the mean value theorem of differential calculus, first recorded by Paramesvara (1360–1455) in his commentary on Bhaskaracharya's *Lilavati*, to be discussed in the section on Indian calculus.

The Madhava–Gregory series for the inverse tangent

The power series for $\tan^{-1} x$ is

$$\tan^{-1} x = x - \frac{x^3}{3} + \frac{x^5}{5} - \ldots, \quad \text{for } x \leqslant 1 \qquad (9.5)$$

and is generally known as the Gregory series for the inverse tangent after the Scottish mathematician, James Gregory, who derived it in 1667. Madhava is credited with the following rule by various texts including Jyesthadeva's *Yuktibhasa* (c. 1550) and the *Kriyakramakari* of Narayana (c. 1500–75) and Sankara Variar (c. 1500–60). The first of these two sources gives the rule as follows.

The first term is the product of the given Sine and radius of the desired arc divided by the Cosine of the arc. The succeeding terms are obtained by a process of iteration when the first term is repeatedly multiplied by the square of the Sine and divided by the square of the Cosine. All the terms are then divided by the odd numbers $1, 3, 5, \ldots$. The arc is obtained by adding and subtracting [respectively] the terms of odd rank and those of even rank. It is laid down that the [Sine of the] arc or that of its complement whichever is smaller should be taken here [as the given Sine]. Otherwise, the terms obtained by this above iteration will not tend to the vanishing magnitude.

The use of capital letters in Sine and Cosine in this extract indicates that we are dealing with the Indian sine and cosine, where $\mathrm{Sin}\,\theta = r \sin \theta$ and $\mathrm{Cos}\,\theta = r \cos \theta$, r being the radius. The condition given at the end of this rule may be interpreted as ensuring that $r \sin \theta$ is less than $r \cos \theta$, or that $\tan \theta$ (i.e. x) should be less than 1 to ensure absolute convergence of the series. Thus Madhava's rule given above may be written as

$$r\theta = \frac{r(r \sin \theta)}{1(r \cos \theta)} - \frac{r(r \sin \theta)^3}{3(r \cos \theta)^3} + \frac{r(r \sin \theta)^5}{5(r \cos \theta)^5} - \cdots$$

or

$$\theta = \tan \theta - \frac{\tan^3 \theta}{3} + \frac{\tan^5 \theta}{5} - \cdots \qquad (9.6)$$

This is equivalent to the Gregory series (9.5) for the inverse tangent.

It would be far beyond the scope of this book to give the geometric derivation of this series found in the two texts cited above. The method used corresponds to what is known today as the method of expansion and term-by-term integration. (Details of the derivation of this and other series discussed below are given by Sarasvati (1963, 1979).)

When $x = 1$ in equation (9.5) or $\theta = 45° = \pi/4$ in equation (9.6), it is easily shown that the Madhava–Gregory series reduces to Euler's series:

$$\frac{\pi}{4} = 1 - \frac{1}{3} + \frac{1}{5} - \frac{1}{7} + \cdots$$

This series occurs in both Nilakantha's *Tantra Samgraha* and Putumana's *Karana Paddhati*. For $x = 1/\sqrt{3}$ or $\theta = 30°$, i.e. for an arc which is 1/12th of the circumference, we get an approximation for π, attributed to Madhava and found in both the *Yuktibhasa* and the *Kriyakramakari*:

$$\pi = \sqrt{12}\left(1 - \frac{1}{3 \times 3} + \frac{1}{5 \times 3^2} - \frac{1}{7 \times 3^3} + \cdots\right)$$

It was the incommensurability of π that gave the impetus to further developments of finite-series approximations to the circumference of a circle.

In the *Tantra Samgraha*, Nilakantha gives the following rule, again attributed to Madhava:

$$\pi d = 4d - \frac{4d}{3} + \frac{4d}{5} - \ldots \mp \frac{4d}{2n - 1} \pm \frac{4dn}{(2n)^2 + 1} \qquad (9.7)$$

where d is the diameter of the circle and n the number of terms. The same text suggests replacing the last term of the series by the following term to improve accuracy:

$$\frac{4d(n^2 + 1)}{n[(n + 1)^4 + 1]}$$

It is likely that Madhava used the approximation formula (9.7) with this new term when he estimated the value of π correct to 11 places of decimals, for a circle of diameter $d = 9 \times 10^{11}$ and $n = 75$. Madhava's calculated value of the circumference is 2 827 433 388 233 units (implying a value for π of 3·141 592 653 59), as reported in the *Kriyakramakari* (Sarma, 1972, p. 26).

A translation of Madhava's verse statement of the circumference gives a flavour of how numbers were recorded in verse. In the notational system, known as *bhuta samkhya*, certain objects were traditionally used to represent numerals, either singly or in pairs, reading from right to left. Thus the circumference 2 827 433 388 233 was recorded as

> Gods (33), eyes (2), elephants (8), snakes (8), fires (3), three (3), qualities (3), *vedas* (4), *naksatras* (27), elephants (8), and arms (2) – the wise say that this is the measure of the circumference when the diameter of a circle is 900 000 000 000.

These approximations are not to be found in any other mathematical literature: they are unique to Kerala.

The Madhava–Newton power series for the sine and cosine

Madhava may also have discovered the sine and cosine series, about three hundred years before Isaac Newton. These series made their first appearance in Europe in 1676 in a letter written by Newton to the Secretary of the Royal Society, Henry Oldenburg. The series are:

$$\sin x = x - \frac{x^3}{3!} + \frac{x^5}{5!} - \frac{x^7}{7!} + \ldots$$

$$\cos x = 1 - \frac{x^2}{2!} + \frac{x^4}{4!} - \frac{x^6}{6!} + \ldots$$

(where ! indicates factorial, e.g. $3! = 3 \times 2 \times 1$). The rules given for generating the two power series may be explained in terms of Figure 9.3 as

$$\text{Indian sine} = r \sin \theta = AC$$
$$\text{Indian cosine} = r \cos \theta = OC$$

Madhava's approach is to express $r \sin \theta$ and $r \cos \theta$ in terms of r and a. (We shall not attempt to follow the long and complex steps in the derivation of the sine and cosine series in the *Yuktibhasa*. Details of the original explanation, involving the use of the method of 'expansion and term-by-term integration', are given by Rajagopal and Venkataraman (1949), and Sarasvati (1963).)

The final results, in the notation of Figure 9.3, are

$$r \sin \theta = a - \frac{a^3}{3!r^2} + \frac{a^5}{5!r^4} - \frac{a^7}{7!r^6} + \frac{a^9}{9!r^8} - \ldots$$

For $r = 1$ and $a = \theta$, we can express the series in its modern form:

$$\sin \theta = \theta - \frac{\theta^3}{3!} + \frac{\theta^5}{5!} - \frac{\theta^7}{7!} + \frac{\theta^9}{9!} - \ldots$$

Similarly, the cosine power series is

$$r \cos \theta = r - \frac{a^2}{2!r} + \frac{a^4}{4!r^3} - \frac{a^6}{6!r^5} + \frac{a^8}{8!r^7} - \ldots$$

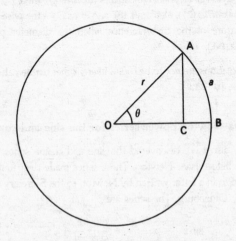

Figure 9.3

and for $r = 1$ and $a = \theta$,

$$\cos\theta = 1 - \frac{\theta^2}{2!} + \frac{\theta^4}{4!} - \frac{\theta^6}{6!} + \frac{\theta^8}{8!} - \dots$$

These power series were probably used to construct accurate sine and cosine tables for astronomical calculations. In a table of values of half-sine chords reportedly calculated by Madhava for twenty-four arcs drawn at equal intervals for a quadrant of a given circle, the values are correct in almost all cases to the eighth or ninth decimal place. Such an accuracy was not to be achieved in Europe for another two hundred years.

Approximations for sine and cosine functions to the second power of small quantities are also attributed to Madhava. In modern notation, Madhava's results can be written as

$$\sin(x + h) \simeq \sin x + (h/r)\cos x - (h^2/2r^2)\sin x$$
$$\cos(x + h) \simeq \cos x - (h/r)\sin x - (h^2/2r^2)\cos x$$

where h is the small quantity and r the radius. These results are but special cases of one of the most familiar expansions in mathematics, the Taylor series, named after Brook Taylor (1685–1731).

In our rather brief look at Kerala mathematics, the name that keeps recurring is that of Madhava of Sangamagramma. His brilliance is generously acknowledged by those who came after him, and the effects of his teaching on the works of Paramesvara, Nilakantha, Jyesthadeva and others are there to see. It would be quite in keeping with Indian tradition if, in holding him in such awe, his successors were to have credited him with more than his share of discoveries. Of his teachers we know nothing. Madhava's outstanding contributions, in the area of infinite-series expansions of circular and trigonometric functions, and finite-series approximations to them, predate European work on the subject by two to three hundred years.

We may consider Madhava to have been the founder of mathematical analysis. Some of his discoveries in this field show him to have possessed extraordinary intuition, making him almost the equal of a more recent intuitive genius, Srinivasa Ramanujan (1887–1920), who spent his childhood and youth at Kumbakonam, not very far from Madhava's birthplace. Ramanujan also showed a considerable intuitive grasp of infinite-series expansions, particularly of trigonometric and circular functions, as his *Notebooks* (1985) now testify. Do we see in these notes the vestiges of a 'hidden' indigenous mathematical tradition which was not submerged by the influx of modern mathematics from the West? Before this question can be answered much needs to be done, beginning with the retrieval and study of some of Madhava's mathematical works; to date only a few of his minor astronomical works have been studied.

There is an interesting Chinese connection which would merit further investigation. In the middle of the eighteenth century Ming Antu wrote a book on geometry which contained power series expansions for trigonometric functions and π. The derivations of the Gregory and Newton formulae contained in his book bear an uncanny resemblance to the work of the Kerala school. The inference is usually that the Chinese were introduced to these results by the Jesuit missionary Pierre Jartoux at the beginning of the eighteenth century, and that the proofs were subsequently arrived at independently. However, the Kerala–Chinese link should also be examined. We mentioned in the historical introduction to Chapter 6 that the early Ming period saw considerable maritime contact between China and parts of Asia and Africa. Kerala has a long history of trade contacts with China. One piece of tangible evidence of technology having been transferred from China to Kerala is the Chinese fishing net, which is still in use today. Would it be far-fetched to suggest that the contacts between the two areas also had a less tangible dimension?

OTHER NOTABLE CONTRIBUTIONS

It is clearly impossible, given the scope of this book, to examine the whole range of subjects covered by Indian mathematicians over a period of two thousand five hundred years. In this section we consider three areas where contributions were notable or unusual, though they may have had only a limited impact outside India. We begin with medieval approaches to mathematical series, and then discuss briefly some special topics in Indian geometry. We conclude by assessing the preliminary notions of the infinitesimal calculus to be found in the work of Bhaskaracharya.

Geometric representation of arithmetic series

Interest in number sequences that follow particular laws has been shown by several mathematical cultures, beginning – as we saw in Chapter 3 – with the Egyptians. In India, it was not until Mahavira's time (c. AD 850) that a systematic examination of the properties of different series was first attempted. We concentrate here on one memorable aspect of Indian work in this area: the study of the properties of different arithmetic series through diagrams (or *sredhiksetras*), which aroused interest well into the fifteenth century.

In Nilakantha's commentary on the *Aryabhatiya*, mentioned earlier, an arithmetic series (see Figure 9.4(a)) is represented by piling rectangular strips, of unit width and lengths equal to the number of units in each term of the series, on top of each other, with the shortest strip at the top and the longest at the bottom. If two of these *sredhiksetras* are joined together, one inverted to fit with the other, as shown in Figure 9.4(a), the resulting figure will be a rectangle with height n units (the number of terms, or rectangular strips) and length $a + f$ units, where a and f are the first and last terms of the series, respectively. Since the area of this rectangle is $n(a + f)$ square units, the area of one of the *sredhiksetras* will be the sum of the series, $\frac{1}{2}n(a + f)$.

Nilakantha proceeds to demonstrate the relationship

$$\sum \frac{n(n + 1)}{2} = \frac{n(n + 1)(n + 2)}{6}$$

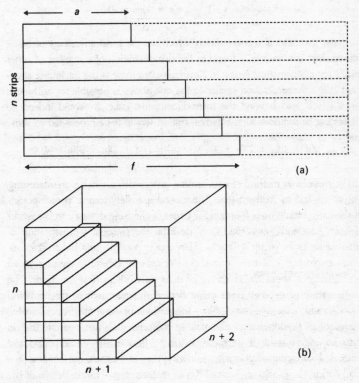

Figure 9.4

by taking six *sredhiksetras*, representing the sum of n natural numbers, and combining pairs of them to form three rectangular strips with sides of n, $n + 1$ or $n + 2$, and thickness 1 unit. If one of these strips is placed flat on the ground, as in Figure 9.4(b), and the other two are held vertically touching the edge of the first, so that the sides of the section along the top of the strip are $n + 1$ and $n + 2$, then the three strips define a rectangular block of sides n, $n + 1$ and $n + 2$. The inside of the block is filled with the set of rectangles formed by joining three pairs each of the *sredhiksetras* representing the sums of $n - 1$, $n - 2$, $n - 3$, . . . , 2, 1 natural numbers. A solid cuboid measuring n by $n + 1$ by $n + 2$ is the final result. The volume of this cuboid is equal to $n(n + 1)(n + 2)$. This may be equated with the 'contents' of the cuboid, known to be $6 \sum \frac{1}{2} n(n + 1)$, so that each of the *sredhiksetras* occupies one-sixth of this:

$$6 \sum \tfrac{1}{2} n(n + 1) = n(n + 1)(n + 2)$$

so

$$\sum \frac{n(n + 1)}{2} = \frac{n(n + 1)(n + 2)}{6}$$

A similar geometrical representation, which has the great advantage of being immediately convincing, is found in the demonstration of a number of other results for mathematical series. In *Kriyakramakari* there is the intriguing statement that a demonstration similar to the one above is possible for arithmetic series which lead beyond the three-dimensional cube. It would indeed be interesting to see how such a demonstration would have proceeded to show that

$$\sum \frac{n(n + 1)(n + 2)(n + 3)}{4!} = \frac{n(n + 1)(n + 2)(n + 3)(n + 4)}{5!}$$

The *sredhiksetra* method of representing mathematical series is an interesting feature of Indian mathematics. The closest parallel from a contemporary mathematical tradition is found in the China of the Sung dynasty, in the works of Shen Kao and Yang Hui. They describe the pictorial representation of different series as 'piling up stacks'. Here again, possible Sino-Indian links are worth exploring. However, one should always be careful in making broad generalizations about the character of different mathematical traditions. The subtle nature of geometric reasoning behind this approach to mathematical series should make one wary of any suggestions of a hypothetical 'Oriental' mathematics, predominantly algebraic in character. It is just possible that in India an undercurrent of geometry began to flow in the Vedic period and continued, surfacing occasionally, as during the Jaina epoch, but coming to a head during the medieval phase of Kerala mathematics. (Further details of the *sredhiksetra* geometry are given by Sarasvati (1963, 1979).)

Special topics in Indian geometry

After the impressive start in mensuration in the *Sulbasutras*, subsequent geometrical developments were on the whole patchy and at times pedestrian. In both Jaina mathematics and the works of the Classical mathematicians there was considerable emphasis on simple rules of mensuration, but little sign of the level of sophistication found in Chinese geometry (see Chapters 6 and 7). But in one area of geometry, the Indian contribution was notable: the study of the properties of a cyclic quadrilateral (i.e. a quadrilateral inscribed in a circle). In the *Brahma Sputa Siddhanta* Brahmagupta gives the following two results:

1 The area of a cyclic quadrilateral is given by the product of half the sums of the opposite sides, or by the square root of the product of four sets of half the sum of the sides (respectively) diminished by the sides.

2 The sums of the products of the sides about the diagonal should be divided by each other and multiplied by the sum of the opposite sides. The square roots of the quotients give the diagonals of a cyclical quadrilateral.

In modern notation, and with reference to Figure 9.5, these rules may be expressed as follows. Let a, b, c and d be the sides of a cyclic quadrilateral of area A; let $s = \frac{1}{2}(a + b + c + d)$ be the semi-perimeter, and x and y the diagonals. Then

$$A = \sqrt{(s - a)(s - b)(s - c)(s - d)}$$

$$x = \sqrt{\frac{(ab + cd)(ac + bd)}{ad + bc}}, \qquad y = \sqrt{\frac{(ad + bc)(ac + bd)}{ab + cd}}$$

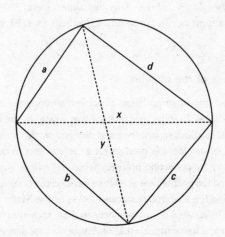

Figure 9.5

The first statement of the expressions for the diagonals in Western mathematics is found in 1619 in the work of Willebrord Snell, some thousand years later.

The derivations of these results are first referred to in a tenth-century commentary on Brahmagupta's work, but find their full expression in the sixteenth-century Kerala text *Yuktibhasa*. This makes use of Ptolemy's theorem, which states that the product xy of the diagonals of a cyclic quadrilateral is equal to the sum of the products of the two pairs of opposite sides, $ac + bd$. Bag (1979) and Sarasvati (1979) give details of the proofs.

Notable extensions in this area are contained in Narayana's *Ganita Kaumadi* in the fourteenth century and Paramesvara's *Lilavati Bhasya*, a detailed fifteenth-century commentary on Bhaskaracharya's *Lilavati*. The cyclic quadrilateral was an important device used by the Kerala school for deriving a number of important trigonometric results, including

$$\sin^2 A - \sin^2 B = \sin(A + B) \cdot \sin(A - B)$$

and

$$\sin A \cdot \sin B = \sin^2 \tfrac{1}{2}(A + B) - \sin^2 \tfrac{1}{2}(A - B)$$

A new rule is found in the work of Paramesvara for obtaining the radius r of the circle in which a cyclic quadrilateral of sides a, b, c and d is inscribed:

$$r = \sqrt{\frac{(ab + cd)(ac + bd)(ad + bc)}{(a + b + c - d)(b + c + d - a)(c + d + a - b)(d + a + b - c)}}$$

A detailed demonstration of this result is found in a later commentary on Bhaskaracharya's *Lilavati*, entitled *Kriyakramakari* (Gupta, 1977). This result makes its first appearance in European mathematics in 1782 in the work of l'Huilier.

The beginnings of the calculus

One of the most important problems of ancient astronomy was the accurate prediction of eclipses. In India, as in many other countries, the occasion of an eclipse had great religious significance, and rites and sacrifices were performed. It was a matter of considerable prestige for an astronomer to demonstrate his skills dramatically by predicting precisely when the eclipse would occur.

In order to find the precise time at which a lunar eclipse occurs, it is necessary first to determine the true instantaneous motion of the Moon at a particular point in time. The concept of instantaneous motion, known as *tatkalika-gati* in Indian astronomy, is found in the works of Aryabhata I and Brahmagupta. They calculated this quantity from the formula (in modern notation)

$$u' - u = v' - v \pm e(\sin w' - \sin w) \qquad (9.8)$$

where u, v and w denote the Moon's true longitude, mean longitude and mean anomaly at a particular time; u', v' and w' are these same quantities after a specific interval of time; and e is the eccentricity, or sine of the greatest equation of the orbit. The use of sine tables and interpolation formulae would then yield values of the sines of angles over very short intervals.

Manjula (*c.* AD 930) was the first Indian astronomer to recognize that equation (9.8) could also be expressed as

$$u' - u = v' - v \pm e(w' - w)\cos w \qquad (9.9)$$

since

$$(\sin w' - \sin w) = (w' - w)\cos w$$

In modern notation, we would write equation (9.9) as

$$\delta u = \delta v \pm e \cos w \, \delta w$$

Bhaskaracharya extended this result to obtain the differential of $\sin w$ as

$$d(\sin w) = \cos w \, dw$$

Bhaskaracharya proceeded to use this equation to work out the position angle of the ecliptic, the other quantity required for predicting the time of an eclipse.

This result in itself was a notable technical achievement in the astronomy of the period, but it may well be much more than this. It may seem far-fetched to claim, on this evidence alone, that Bhaskaracharya was one of the first mathematicians to conceive of the differential calculus, but there is further evidence to be found in his *Siddhanta Siromani*:

1 In computing the instantaneous motion of a planet, the time interval between successive positions of the planet was no greater than a *truti*, or 1/33 750 of a second, and his measure of velocity was expressed in this 'infinitesimal' unit of time.

2 Bhaskaracharya was aware that when a variable attains the maximum value, its differential vanishes.

3 He also showed that when a planet is either at its furthest from the Earth or at its closest, the equation of the centre* vanishes. He therefore concluded that for some intermediate position the differential of the equation of the centre is equal to zero.

---

*This is a measure of how far a planet is from the position it is predicted to be in by assuming it to move uniformly. The predicted and actual positions differ because planetary orbits are elliptical, whereas uniform motion implies a circular orbit.

In (3) there are traces of the 'mean value theorem', which today is usually derived from Rolle's theorem (1691).

Later mathematicians, particularly the Kerala school, continued the work of Bhaskaracharya. Nilakantha (1443–1543) derived an expression for the differential of an inverse sine function, and Acyuta Pisarati (*c.* 1550–1621) gave the rule for finding the differential of the ratio of two cosine functions. The idea of using the integral calculus to find the value of π, and the areas of curved surfaces and the volumes enclosed by them, is implicit in the method of exhaustion which we have examined in earlier chapters. Such ideas and their development are also found in the works of Bhaskaracharya, Narayana Pandit (*c.* 1350) and Jyesthadeva (*c.* 1550), but there are few novel features in the Indian treatment of the subject.

Outside Europe, there has been only one other country apart from India in which some form of calculus developed. In Japan, during the seventeenth century, Seki Kowa developed a form of calculus called *yenri* which was primarily used in circle measurements (Mikami, 1913; Smith and Mikami, 1914). In India itself, where the concept of differentiation was understood from the time of Manjula, differential calculus was applied only to astronomy and certain problems in mensuration and did not spread across the broad spectrum of mathematics. This spread has been an important factor in promoting the phenomenal development of modern mathematics during the last few hundred years. The crucial concept of the 'limit' of a function or a sum is essentially a modern idea, not to be found in Indian or any pre-modern mathematics. But its absence should not make one ignore the advances made during the Classical and medieval phases of Indian mathematics.

10 Prelude to Modern Mathematics: The Arab Contribution

◇◇◇

HISTORICAL BACKGROUND

The year AD 622 is a momentous one in world history. It was then that the Prophet Muhammad fled from Mecca and took refuge in Medina, about 350 kilometres away. He had incurred the wrath of pilgrims who had come to worship at a shrine called the Kabbah – a shrine then dedicated to many gods. Muhammad's preaching of a monotheistic faith, which he claimed had been directly revealed to him by the Archangel Gabriel, had aroused considerable hostility, contributing to his decision to flee his birthplace. Eight years later he returned at the head of an army, and two years after that he died. But he had already created a whirlwind that would eventually lead to the establishment of Islamic rule over areas stretching from North Africa in the south to the borders of France in the west, right across Persia and the Central Asian plains to the borders of China in the east, and down to Sind in northern India. Figure 10.1 shows the extent of the empire at its height, and the location of places referred to in the course of this chapter. Much of this vast territory was brought under Islamic rule in less than a hundred years.

Such a rapid expansion was possible for two reasons. First, there was something quite irresistible about the passion and egalitarian character of early Islam which fired the imagination and won the devotion of many who came across it for the first time. And second, in a number of areas through which the Islamic forces passed, local rulers were so unpopular with their subjects that the conquering armies were welcomed as liberators. With the physical conquest completed, the government of this vast territory passed into the hands of the caliphs, men who were generally regarded as the political and military successors of the Prophet. They belonged to one of two dynasties: the Umayyad and the Abbasid. The Umayyads were the early rulers of the Eastern empire, with their capital at Damascus, but in the year 750 they were overthrown and power passed to the Abbasids. This was not the end of the Umayyad dynasty, though.

Figure 10.1 Map of the 'Arab' world.

Among the few who escaped was a young man of twenty named Abdel Rahman, who reached Spain and re-established Umayyad power there. For the next three centuries Spain was to be the centre of Muslim power in the West, with its political and intellectual capital at Córdoba.

The Abbasids differed from the Umayyads in one important respect. While the latter came from the Arabian Peninsula, the former were more cosmopolitan, welcoming new converts from many different ethnic groups. In 762 the second of the Abbasid caliphs, al-Mansur, moved his capital to Baghdad and began the process of building it into a new Alexandria. This ambitious programme of construction was carried out during the caliphates of Harun al-Rashid (786–809), and his son al-Mamun (809–33), and included an observatory, a library, and an institute for translation and research named *Bait al-Hikma* ('House of Wisdom') that was to be the intellectual centre of the Arab world for the next two hundred years. Within its walls lived some of the greatest scientists of the period. It housed translators, busy rendering into Arabic scientific classics written in Sanskrit, Pahlavi (the classical language of Persia), Syriac and Greek. These translations were often carried out under the patronage of the caliph himself, or notable families in Baghdad, and the classic texts were acquired through missions personally sponsored by the caliphs and sent to places near and far. Early collections included Greek manuscripts from the Byzantine empire, translations of Babylonian astronomy by the Syriac schools of Antioch and Damascus, and the remains of the Alexandrian library in the hands of the Nestorian Christians at Edessa.

It is a mistake, however, to overemphasize the role of Baghdad at the expense of earlier pre-Islamic centres of scientific learning. By doing so, two questions are left unresolved: why was there a greater willingness to accept Indian rather than Hellenistic astronomy and mathematics during the earlier period of Arab rule? And why were so many of the early scholars in Baghdad from Central Asia, especially from regions such as Khwarizm and Khurasan, in present-day Soviet Central Asia? Long before the Arab conquest, there were scientific and translation centres in Syria as well as in Sassanid Persia. In astronomy, for example, the first translation of Ptolemy's *Almagest* and its important commentary by Theon of Alexandria (*c.* AD 300) was from Greek into Syriac. Three important components of Persian astronomy were carried into the Islamic period:

1 Syrian astronomy, inspired mainly by Hellenistic influences and notably those of Ptolemy;

2 much pre-Sassanid Persian astronomy, dating back to the Babylonian astronomy of the Seleucid period and earlier; and

3 Indian astronomy transmitted to Central Asia, probably during the first and second centuries AD, when such regions as Khwarizm, Parthia, Sogdiana and Bactria, as well as much of northern India, were part of the great Kushan empire. Against this background was composed the first *zij* of Sassanid Persia. The word *zij*, probably a distortion of the Pahlavi word *zeh* ('bow-string'), came later to be applied to Arabic works on calendar construction, and tables of movement of the Sun, Moon and planets, as well as trigonometric and geographical tables.

Thus the scientific culture that developed in Baghdad arose from an interaction of these different traditions. By far the greatest contribution of the Arabs, as we shall see, was to continue this creative synthesis. This they pursued with an openness of mind and a clearer understanding than had been shown by any of the earlier scientific cultures of the need in mathematics and other sciences to balance empiricism and theory. The process of synthesis was aided by the creative tension between two main traditions of astronomy and mathematics represented in Baghdad, even from the early years of Islamic rule. One tradition derived directly from Indian and Persian sources and is best exemplified in the astronomical tables and the algebraic approach to mathematics. One of the greatest exponents of this tradition, who left an indelible mark on the subsequent development of Arab mathematics, was al-Khwarizmi. To him mathematics had to be useful and help with practical concerns such as determining inheritances, constructing calendars or religious observances. The other tradition looked to Hellenistic mathematics, with its strong emphasis on geometry and deductive methods. One of the best-known exponents of this school was Thabit ibn Qurra, who was both an outstanding translator of Greek texts and an original contributor to geometry and algebra. That the two traditions eventually merged is evident in the work of later Arab mathematicians such as Omar Khayyam and al-Kashi.

We shall examine the contributions of these and other Arab mathematicians in three areas in which the results of Arab creativity and synthesis are most apparent: the introduction and popularization of our present-day numerals, the bringing together of the geometric and algebraic approaches to the solution of equations, and the first systematic treatment of trigonometry. First, though, let us look at the lives and achievements of the mathematicians themselves.

MAJOR ARAB MATHEMATICIANS
Muhammad ibn Musa al-Khwarizmi

Abu Jafar Muhammad ibn Musa al-Khwarizmi (to give him his full name, which means Muhammad, the father of Jafar and the son of Musa, from Khwarizm) was born in about 780. The name 'al-Khwarizmi' suggests that either he or his family came from Khwarizm, east of the Caspian Sea in what is today Soviet Central Asia. Little is known of his early life. There is a reference to al-Khwarizmi as 'al-Majusi' in a book entitled *History of Kings and Envoys* by al-Tabari, an Arab historian of the ninth century. In the Pahlavi language, a Zoroastrian was sometimes referred to as 'magus', from which come our words 'magi' and 'magician'. There is, therefore, the view that al-Khwarizmi may have been of Zoroastrian descent and had acquired his early knowledge of Indian mathematics and astronomy from Zoroastrian clergy, some of whom were reputed to be well acquainted with these subjects.

In about 820, after establishing a reputation as a talented scientist in Merv, the capital city of the Eastern provinces of the Abbasid caliphate, he was invited by Caliph al-Mamun to move to Baghdad, where he was appointed first astronomer and then head of the library at the House of Wisdom. He continued to serve other caliphs, including al-Wathiq during his short rule from 842 to 847. There is a story, told by the historian al-Tabari, that when al-Wathiq lay seriously ill he asked al-Khwarizmi to cast his horoscope and find out whether he would live. Al-Khwarizmi assured the caliph that he would live another fifty years, but al-Wathiq died within ten days. Whether this story illustrates al-Khwarizmi's highly developed sense of survival or his ineptness as a fortune-teller, it is difficult to say. We know little else of his later life except that he died in about 850.

However, we do have information on his scientific work, most of which was in algebra, arithmetic, astronomy and geography. He wrote a number of books, of which the two most influential – to be referred to often in this chapter – are *Hisab al-jabr w'al-muqabala* (Calculation by Restoration and Reduction), and *Algorithmi de numero indorum* (Calculation with Indian Numerals), whose original Arabic version no longer exists – we know it only in Latin translation. The first book, which we shall call the *Algebra*, was the starting-point for Arab work in algebra, and indeed gave the subject its name. It is a subtle blend of a variety of mathematical traditions including the Babylonian, Indian and Greek. The second book, which we shall call the *Arithmetic*, served to introduce the decimal positional number system which had developed in India a few hundred years earlier. It was also the first book on Arab arithmetic to be translated into Latin, and gave currency to the word 'algorithm', derived from the name of the author, and frequently used today to describe any systematic procedure for calculation.

Al-Khwarizmi also constructed a *zij* (a set of astronomical tables) that was to remain important in astronomy for the next five centuries. The origin of this *zij* is interesting, for it shows how Indian mathematics and astronomy entered the Arab world directly for the first time. An Arab historian, al-Qifti (*c.* 1270), reported that in the year AH 156 (or AD 773)* a man well versed in astronomy, by the name of Kanaka, came to Baghdad as a member of a diplomatic mission from Sind, in northern India. He brought with him Indian astronomical texts, including *Surya Siddhanta* and the works of Brahmagupta. Caliph al-Mansur ordered that some of these texts should be translated into Arabic and, according to the principles given in them, that a handbook be constructed for use by Arab astronomers. The task was delegated to al-Fazari, who produced a text which came to be known by later astronomers as the *Great Sindhind*. The word *sindhind* is derived from the Sanskrit word *siddhanta*, meaning an astronomical text. It was mainly on the basis of this text, as well as some other elements from Babylonian and Ptolemaic astronomy, that al-Khwarizmi constructed his *zij*. Unfortunately, the original Arabic text is no longer extant. But a Latin translation, made in 1126 from an edited version produced by Maslama al-Majriti (a Spanish astronomer who lived in Córdoba in about the year 1000), became one of the most influential astronomical texts in medieval Europe.

Finally there is al-Khwarizmi's geographical work, in particular his contribution to cartography. He was a member of a team set the following tasks by Caliph al-Mamun:

1 to measure the length of one degree of longitude at the latitude of Baghdad (the result obtained was quite accurate, at 91 kilometres);
2 to use astronomical observations to find the latitude and longitude of 1200 important places on the Earth's surface, including cities, lakes and rivers; and
3 to collate the personal observations of travellers on the physical features of different areas of the caliphate, and travelling times between them.

Al-Khwarizmi incorporated these and additional findings in his book *The Image of the Earth*, considered to be the most important work on the subject after Ptolemy's *Geography*. He corrected Ptolemy's overestimate of the length of the Mediterranean Sea, and provided more detailed and accurate descriptions of the geography of Asia and Africa.

In other fields, al-Khwarizmi was the first scholar to work on a history of the Arab caliphates. His mathematical texts are still recommended reading in some

---

*The Muslim era begins in the year that Muhammad fled from Mecca (AD 622, the year of the *Hejira*). The Muslim year is a lunar year and therefore about eleven days shorter than the Western calendar year, so an AH date ('after Hejira') cannot be converted to an AD date simply by adding 622.

Arab countries, not for their mathematical content but for their legal acumen. His book on algebra contained an analysis of property relations and the distribution of inheritance according to Islamic law, and rules for drawing up wills. His practical bent of mind was also reflected in the three expeditions he led to India and Byzantium to make scientific observations and collect primary research material. He was but the first of a succession of remarkable scientists who were largely responsible for laying the foundations of modern science. These included Thabit ibn Qurra (836–901), al-Razi (*c.* 865–901), al-Haytham (*c.* 965–1039), al-Biruni (973–1051), ibn Sina (980–1037), Umar al-Khayyami or Omar Khayyam (*c.* 1048–1126), Sharaf al-Din al-Tusi (d. 1213), Nasir al-Din al Tusi (1201–74) and al-Kashi (d. 1429), some of whom are better known in the West by their Latin names. Let us look very briefly at two of these who were not primarily mathematicians, but were nevertheless highly influential in other scientific areas.

Al-Razi, known in Latin as Rhazes, was the greatest clinical physician of the Arab world. Ibn Sina, better known in Europe as Avicenna, is generally regarded as the most able of the Arab philosopher-scientists; he was reputed to have mastered all the sciences by the age of eighteen. He was a man of enormous energy who, in spite of preoccupations with state affairs, managed to produce about two hundred books, some dictated on horseback as he rode with his ruler to battle! His best-known work is the *Canon of Medicine*, translated into Latin and then other European languages, and used for centuries as a primary text in Arab and European universities. His scientific, philosophical and theological views had a significant impact on medieval European thinkers such as Albertus Magnus, St Thomas Aquinas and Roger Bacon.

Al-Haytham, known in Latin as Alhazen, wrote more than two hundred books on mathematics, physics, astronomy and medicine, and commentaries on Aristotle and Galen. But his major work was in optics. It included an early account of refraction, a mathematical approach to finding the focal point of a concave mirror, and a refutation of the theory put forward by both Euclid and Ptolemy that human vision works by the eye sending out rays to the object being observed. Al-Haytham's work had a significant impact on Roger Bacon and Johann Kepler.

Thabit ibn Qurra

Abul Hassan Thabit ibn Qurra Marwan al-Harrani was born in Harran in northern Mesopotamia, probably in 836, and died in 901. Little is known of his early life except that when he reached adulthood he became a money-changer.

Thabit belonged to a religious sect who were descendants of Babylonian star-worshippers and produced eminent scholars in both astronomy and mathematics. The sect called themselves Sabeans (after a Chaldean sect whose possession of sacred books was recognized by the Quran) so as to avoid being persecuted as polytheists. Either his unorthodox religious beliefs or a quarrel with his community led him to leave Harran and head for Baghdad. He was soon befriended by a member of a wealthy and influential Baghdad family, under whose patronage he joined a circle of scholars and translators. Thabit's command of languages, including Arabic, Greek and Syriac, soon established him as one of the foremost translators in Baghdad. His notable translations of Greek mathematical texts included Euclid's *Elements*, several works by Archimedes, parts of Apollonius' *Conics* and Ptolemy's *Almagest*; many were in turn rendered into Latin by Gherardo of Cremona in the twelfth century, in which form they were to have a momentous impact on medieval Europe.

Thabit's passion for translation led him to set up a school for translators in Baghdad, the members of which included his son and some of the greatest translators of the day. Two of his grandsons followed the family tradition and became outstanding translators and mathematicians in their own right. Ibrahim ibn Sina ibn Thabit ibn Qurra's commentary extending Archimedes' work on the quadrature of the parabola has been described as one of the most innovative approaches known before the emergence of the integral calculus. Whether it was the range of translations that he undertook or the unusual breadth of mind possessed by the scholars of the period, Thabit himself became highly competent in a number of subjects which included – apart from mathematics – medicine, astronomy, philosophy, physics, geography, botany, agriculture and meteorology. However, it is for his work in mathematics that he is best remembered.

His notable contributions in mathematics include a rule for discovering pairs of 'amicable numbers' (to be discussed in a later section); a 'dissection' proof of the Pythagorean theorem, strangely reminiscent of the Chinese approach which we discussed earlier; work on spherical trigonometry; an attempt to prove Euclid's parallel postulate; and his contributions to the mensuration of parabolas and paraboloids, which some look upon as providing the essential link between Archimedes and later European mathematicians such as Cavalieri, Kepler and Wallis. As a geometer he had no equal in the Arab world, and it is clear from the above list that his algebraic strength was also considerable. Clearly we shall not be able to explore all the contributions of this remarkable polymath, but we shall consider briefly some of his geometric work and his rule for generating amicable numbers.

Omar Khayyam

The Rubaiyat of Omar Khayyam, a number of quatrains freely translated into English by Edward Fitzgerald in the middle of the last century, is one of the best-known and most translated books in world literature. But what is not widely known outside the Islamic world is that the poet was also a distinguished mathematician, astronomer and philosopher.

Abul-Fath Umar ibn Ibrahim al-Khayyami was born in about 1040 at Nishapur in Khurasan, now part of Iran. This region had already produced two distinguished figures – Firdausi (*c.* 940–1020), a poet, and ibn Sina. It is quite possible that these two men in different ways had a considerable influence on the young Omar. The name al-Khayyami would indicate that either Omar or his family were tent-makers. Little else is known about his childhood or his youth.

In 1070 he wrote his great work on algebra. In it he classified equations according to their degree, and gave rules for solving quadratic equations, which are very similar to the ones we use today, and a geometric method for solving cubic equations with real roots. (We shall be looking at Omar's solution of cubic equations in a later section.) He also wrote on the triangular array of binomial coefficients known as Pascal's triangle.

In 1074, Omar was appointed by Sultan Malik Shah as one of the eight learned men charged with the task of revising astronomical tables and reforming the calendar. The committee of eight produced a new calendar, according to which eight out of every thirty-three years were made into leap years. This adjustment produces a more accurate measure of a solar year than does our present Gregorian calendar year.

Three years later, Omar wrote *Sharh ma ashkala min musadarat kitab Uqlidis* (Explanations of the Difficulties in the Postulates of Euclid). An important part of the book is concerned with Euclid's famous parallel postulate, which had also attracted the interest of Thabit ibn Qurra. Al-Haytham too had attempted a demonstration of the postulate; Omar's attempt was a distinct advance. (The Arab work on the parallel postulate is outside the scope of this book; it is discussed by Al-Daffa and Stroyls (1984) and Bonala (1955).)

Omar's other notable work in geometry was on the theory of proportions. It provides a good example of how he and other Arab mathematicians absorbed and then reinterpreted Greek geometry. Euclid's theory of proportion, as set forth in the *Elements*, had both a geometric and an arithmetic dimension. Euclid's arithmetic definition of the equality of ratios is that, given any two numbers m and n, two ratios a/b and c/d are said to be equal if the following conditions are satisfied:

1 if $ma > nb$ then $mc > nd$

2 if $ma = nb$ then $mc = nd$

3 if $ma < nb$ then $mc < nd$

Omar had reservations about this definition, partly because of its inductive character (i.e. each and every combination of m and n must be tried before the definition can be declared 'foolproof') and partly because it implied a restrictive view of what constitutes 'number'. Instead, Omar suggested that two or more ratios may be defined as equal if they can be reduced to a ratio of integer numbers to a very high level of accuracy. Thus, the ratio of the diagonal of a square to its length (which is the square root of 2) or the ratio of the circumference of a circle to its diameter (which is π) can never be equated to any other ratio. In Omar's view, such numbers are not excluded under Euclid's definition of equal ratios, and so the definition is flawed. In fact, though, Euclid considered only rational numbers in his arithmetical interpretation of ratios; Omar's contribution was to extend the concept of a number to include positive irrational numbers. It was the transmission of Omar's ideas via Nasir al-Din al-Tusi (1201–74) to European mathematics that provided the impetus for a more rigorous examination of the concept of real numbers. René Descartes (1596–1650) established the correspondence between the geometric and arithmetic concepts of real numbers, but it was Georg Cantor (1845–1918) who would eventually provide a rigorous definition of a real number.

Omar Khayyam died in Nishapur in 1123. Unlike the image of him we may get from the *Rubaiyat*, of a hedonist who lived only for the present, he was a retiring scholar, a poet, a Sufi and a gnostic. He was that rare combination – an outstanding poet and a mathematician.

Jamshid al-Kashi

Ghiyath al-Din Jamshid al-Kashi was born at Kashan, a town in Persia not far from Isfahan, in the latter half of the fourteenth century. Two centuries of turmoil had followed the Mongol invasion under Hulegu Khan, grandson of Genghis Khan. The caliphate at Baghdad was destroyed in 1256 and a century later a new empire was created under Timur (also known as Tamerlane). But mathematical activities continued throughout this period, probably because of the patronage offered by Hulegu Khan to astronomers of the calibre of al-Tusi, and by Ulugh Beg, grandson of Timur, to a later group which included al-Kashi.

Little is known of al-Kashi's life until 1406, when he began a series of observations of lunar eclipses from his birthplace, Kashan. At Samarkand, Ulugh Beg had established an observatory and a *madrassa* (a school of advanced study in science or theology), and it is probable that al-Kashi was invited to join a group of scientists there. We know that in 1414 he revised a set of astronomical tables produced by Nasir al-Din al-Tusi and dedicated it to Ulugh Beg, who was no mean astronomer himself. Samarkand under the enlightened rule of Ulugh Beg had become an intellectual centre where, as al-Kashi observed in a letter to his father, 'the learned are gathered together, and teachers who hold classes in all the sciences are at hand, and the students are all at work on the art of mathematics'.

Al-Kashi's strength lay in prodigious calculations. His approximation for π, correct to 16 places of decimals, was obtained by circumscribing a circle by a polygon having 3×2^{28} (805 306 368) sides! His best known work, *Miftah al-hisab* (The Calculator's Key), completed in 1427, provides us with a compendium of the best of Arab arithmetic and algebra. Its contents include the first systematic exposition of decimal fractions; a method of extracting the nth root of a number, similar to the so-called Horner's method and thus probably derived from the Chinese; and the solution of a cubic equation to obtain a value for the sine of one degree. On al-Kashi's death in 1429, Ulugh Beg praised the mathematician's achievements, in the preface to his own *zij*, calling him 'the admirable mullah known among the famous of the world, who had mastered and completed the science of the ancients, and who could solve the most difficult questions'.

THE ARAB ROLE IN THE RISE AND SPREAD OF INDIAN NUMERALS

In Chapter 8 we saw how the Indian numerals evolved and spread to South East Asia; we now take up the story of their spread westwards. The Arabs were the leading actors in this drama. The first evidence of the westward migration of Indian numerals is found in the following (rather aggrieved) passage from the fragments of a book in Syriac. It was written in 662 by a Nestorian bishop, Severus Sebokht, who came from Keneshra in the upper reaches of the Euphrates. He had written previously on both geography and astronomy, and was hurt by the arrogance of some Greek (or Byzantine) scholars who looked down on his people. He wrote:

> I will omit all discussion of the science of the Indians, a people not the same as the Syrians; of their subtle discoveries in astronomy, discoveries that are more ingenious than those of the Greeks and the Babylonians; and of their

valuable methods of calculation which surpass description. I wish only to say that this computation is done by means of nine signs. If those who believe, because they speak Greek, that they have arrived at the limits of science, [would read the earlier texts], they would perhaps be convinced, even if a little late in the day, that there are others also who know something of value.

This supports the view that, even before the beginning of Arab rule, knowledge of Indian numerals had spread westwards, probably as a result of widespread interest in Indian astronomy. Christian sects, particularly the Nestorian and Syrian Orthodox denominations, needed to calculate an accurate date for Easter, and various astronomical texts were examined with this problem in mind. (It was a problem that continued to occupy mathematicians, including Gauss, down to the nineteenth century.) There is also the possibility, given the thriving commercial relations between Alexandria and India, that the Indian numeral system had reached the shores of Egypt as early as the fifth century AD. It would have been regarded as a useful commercial device rather than a system that might become more widely used or accepted; it would not have been adopted for scientific and astronomical calculations by Alexandrian scientists, who used the Babylonian sexagesimal system.

After the Arab conquest, Indian numerals probably arrived at Baghdad in 773 with the diplomatic mission from Sind to the court of al-Mansur. In about 820 al-Khwarizmi wrote his famous *Arithmetic*, the first Arab text to deal with the new numerals. As mentioned earlier, the original Arabic version of the text is now lost, and we know it only through several partial Latin translations undertaken five hundred years later by John of Seville and Robert of Chester. The text contained a detailed exposition of both the representation of numbers and operations using Indian numerals. Al-Khwarizmi was at pains to point out the usefulness of a place-value system incorporating zero, particularly for writing large numbers.

The earliest extant Arab texts to examine arithmetical operations with Indian numerals are Abul Hassan al-Uqlidisi's *Kitab al-fusul fil-hisab al-Hindi* (The Book of Chapters on Indian Arithmetic, 952) and Kushyar ibn Labban's *Usul hisab al-Hind* (Principles of Indian Reckoning, c. 1010). Operations are given in both the Indian decimal place-value system and the Babylonian sexagesimal system. Al-Uqlidisi's book is particularly notable for the first use of decimal fractions in computing with the new numerals; ibn Labban introduces 'Indian' reckoning as part of a general discussion of Indian methods used in astronomical calculations. Some aspects of both works will be highlighted in the next section.

Other references to Indian numerals are found in works by later Arab writers. The opinions of the tenth-century polymath Abu Rayhan al-Biruni are particularly valuable since he was one of the few Arab scientists who had lived in India and who knew Sanskrit. Two of his books, *Risalah* (Book of Numbers) and *Rasum al-Hind* (Indian Arithmetic), contain an authoritative assessment of Indian numeration as well as some corrections to earlier Arab works on the subject.

A minor incident from the autobiography of ibn Sina (*c.* 987–1037) shows us how the use of the Indian numerals was spreading. When he was about ten years old a group of missionaries belonging to a small Islamic sect came to Bukhara from Egypt, and it was from these people that ibn Sina learnt 'Indian arithmetic'. There is also a story of the young ibn Sina being taught 'Indian calculation' and algebra by a vegetable vendor. What these stories illustrate is that by the beginning of the eleventh century Indian numeration was being used from the borders of Central Asia to the southern reaches of the Arab empire in North Africa and Egypt – and not just by scholars.

In the transmission of Indian numerals to Europe, as with almost all knowledge obtained from the Arabs, Spain and (to a lesser extent) Sicily played the role of intermediaries, being the two areas in Europe which had been under Arab rule for many years. (This was one of the important aspects emphasized in our examination of the spread of mathematical knowledge in Chapter 1.) So it is not surprising to find that the oldest record of Indian numeration in Europe is found in a monastery in northern Spain dating from the year 976. This manuscript, known as the *Codex Vigilanus*, is now kept in a museum in Madrid. The relevant passage reads:

> So with computing symbols. We must realize that the Indians had the most penetrating intellect, and other nations were way behind them in the art of computing, in geometry, and in other free [?] sciences. And this is evident from the nine symbols with which they represented every rank of number at every level.

There follows a set of symbols, now known as the West Arabic or Gobar numerals, from which our present numerals derive. The shapes of these numerals are shown in Figure 10.2, which outlines the evolution of our numerals from some of the earlier forms. The Indian numerals from which the two main forms of Arabic numerals (East and West) were derived quite likely resembled those found in the Gwalior inscription of 876 which we discussed in Chapter 9. The Western version of the Arabic numerals that stemmed from the Indian figures were called Gobar numerals – presumably because, as the

Figure 10.2 The evolution of present-day numerals (after Open University, 1976. p. 53).

meaning of 'Gobar' suggests, these symbols were written on a board containing sand or dust, a practice that was popular in India. The Gobar numerals were widely used in the western part of the Arab empire, including Spain and Sicily; indeed, they are still found in parts of North Africa. The eastern Arabic numerals may have come to the Arabs by a more indirect route which included Persia. In the early years the differences between the two types of Arabic numerals were slight, but they grew with the passage of time. A striking feature of the evolution of Indian numerals from the Gwalior script to the present form, as shown in Figure 10.2, is how little they have changed on passing through the hands of one culture after another. In several instances what changed was the orientation of a symbol, not its form.

The oldest date to appear in the new numerals in Europe is on a Sicilian coin from the reign of the Norman king, Roger II. On it the year is expressed as AH 533 (AD 1138). The use of the Muslim date is not surprising, since Roger II encouraged the pursuit of Arab learning in his kingdom.

We now come to a landmark in the spread of Indian numerals: the appearance of one of the most influential mathematical texts in medieval Europe, the *Liber abaci* (Book of the Abacus) by Fibonacci (1170–1250). The young Fibonacci grew up in North Africa, where his father was in charge of a customs-house. He was first introduced to Indian numerals by his Arab teachers there. As a young man, he travelled extensively around the Mediterranean, visiting Egypt, Syria, Greece, Sicily and southern France, observing the various computational systems used by the merchants of different lands. He quickly recognized the enormous advantage of the Indian system, and introduced the new numerals with the following words:

> The nine Indian numerals are 9, 8, 7, 6, 5, 4, 3, 2, 1. With these nine and with the sign 0, which in Arabic is called *sifr*, any desired number can be written . . .

It was chiefly through this work that the Indian numerals came to be widely known in Christian Europe. For a long time they were used alongside the Roman numerals. The change from the latter to the former was a slow process with a number of false starts, primarily because the abacus remained popular for carrying out calculations, and traders and others engaged in commercial activities were reluctant to adopt a new system which was difficult to comprehend. At times there were diktats from above to discourage the use of the new numerals. In 1299, for example, the city of Florence passed an ordinance prohibiting the use of the new numerals since they were more easily altered (e.g. by changing 0 to 6 or 9) than Roman numerals or numbers written out in words. As late as the end of the fifteenth century, the Mayor of Frankfurt ordered his officials to refrain from calculating with Indian numerals. And even after the decimal numeral system was well established, Charles XII of Sweden (1682–1718) tried in vain to ban the decimal system and replace it with a base 64 system for which he devised sixty-four symbols! But these were all temporary setbacks. Once the contest between the abacists (those in favour of the use of the abacus or some mechanical device for calculation) and the algorists (those who favoured the use of the new numerals) had been won by the latter, it was only a matter of time before the final triumph of the new numerals, with bankers, traders and merchants adopting the system for their daily calculations.

ARAB ARITHMETIC

Arithmetical operations

The first systematic treatment of arithmetical operations is found in al-Khwarizmi's *Arithmetic*, in which he discusses the place-value system and rules for performing the four arithmetical operations. In later works, notably those of ibn Labban and al-Uqlidisi referred to in the last section, there are computational schemes for carrying out operations, such as long multiplication, which were reproduced in medieval Latin texts. Al-Kashi's *The Calculator's Key* presents a comprehensive treatment of arithmetical methods, including operations with decimal fractions (to be discussed later).

We have not the space here to examine in any detail how the Arabs carried out the basic operations; in any case, it was not very different from ours. However, we shall look at a method they popularized for long multiplication known as the 'sieve' or 'lattice' method, which is of historical significance. It is probably of Indian origin; and traces of it are found in the 'grating' methods explained in the *Treviso Arithmetic* of 1478, in the mechanical device known as 'Napier's rods' or 'Napier's bones' after its inventor, John Napier (1550–1617) of Scotland, and the nineteenth-century Japanese Sokuchi method. (These later multiplication methods are discussed by Smith (1923/5).) Even today the 'lattice' provides a useful diversion in learning long multiplication.

Figure 10.3 shows how the lattice method is used to multiply 1958 by 546. The numbers to be multiplied are entered as shown. The number of paths (a) to (g) is 7, the sum of the digits in the two numbers to be multiplied. The product to be entered in each cell is obtained by multiplying the numbers of the row and column in which it lies and arranging the result with the 'units' digit below and the 'tens' digit above the cell's diagonal. For example,

Figure 10.3 Multiplication by the 'lattice' method.

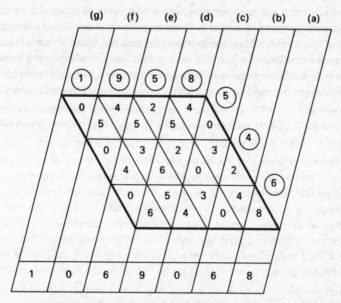

in the top right-hand cell of the square is entered 40, the product of 8 and 5, arranged so that 4 is above the diagonal and 0 below it. After all the cells have been filled, additions are made along the diagonal paths (a) to (g), starting at (a) and continuing to (g), carrying over if necessary, and the results written at the bottom. For example, summing the numbers in path (c) gives $0 + 3 + 0 + 3 + 4 = 10$, so 0 is written at the bottom and 1 carried over to (d). The answer appears at the bottom of Figure 10.3 as 1 069 068.

The reader is invited to multiply 1990 by 365 using this method.

Decimal fractions

In previous chapters we have seen the versatility of the Babylonian sexagesimal system for representing fractions, and the Chinese facility in manipulating fractions using their rod numerals. However, neither the Babylonians nor the Chinese had a device or symbol for separating integers from fractions. The credit for such a symbol must go to the Arabs.

Decimal fractions make their first appearance in Arab mathematics in *The Book of Chapters on Indian Arithmetic*, written in Damascus in the year 952 or 953 by Abul Hassan al-Uqlidisi. We know little about the author, except that he made a living by copying the works of Euclid (hence the 'Uqlidisi'). It is not clear whether it was he or some previous scholar who was responsible for the discovery, since he does state at the beginning of his text that he had attempted to incorporate in it the best methods from the past. Al-Uqlidisi's book is in four parts, the first two of which deal with computational methods using Indian numerals. It is in the second part that decimal fractions appear for the first time. He considers the problem of successively halving 19 five times, and gives the answer as $\overset{1}{0}59375$, the vertical mark on 0 indicating that the decimal fraction part of the number starts with the digit to the right. This notation is perfectly general, so that our 0·059 375 would be denoted by $\overset{1}{0}059375$.

There is evidence that al-Uqlidisi was aware of the method of multiplying decimal fractions by whole numbers. However, it is not until two centuries later that we find al-Samawal (1172) at ease with decimal fractions in problems of division and root extraction. In the fifteenth century, al-Kashi provided a comprehensive and systematic treatment of arithmetic operations with decimal fractions. Devices for separating integers from fractions now included not only the vertical mark placed on the last digit of the integer part, but also the use of different colours, or a numerical superscript giving the number of decimal places specified: thus $\overset{2}{36}$ would indicate the decimal fraction 0·36. Al-Kashi's multiplication procedure is identical to the one we use today.

During the fifteenth century this method of representing decimal fractions came to be known outside the Islamic world as the Turkish method, probably after a Turkish colleague of al-Kashi known as Ali Qushji who provided an explanation. Knowledge of the method then spread to Vienna, where in 1562 it appeared in a collection of Byzantine problems. It is quite likely that the Dutch mathematician Simon Stevin (1548–1620), who is often credited with the first systematic exposition of decimal fractions, may have learnt of the so-called Turkish method from this Byzantine text or a similar source. In place of the short vertical line over the last digit of the integer part of the number that was the original notation of al-Uqlidisi, Stevin used a cipher, so that the number 6·8145 would be represented as $\overset{0}{68145}$. To his contemporary John Napier we owe the present convention of using a decimal point to separate the integer and fractional parts.

Another contribution of al-Uqlidisi's was to adapt the Indian dustboard techniques of computation to methods suitable for pen and paper. It was a common practice, both in India and in the Arab world, to do arithmetical calculations by writing number symbols in sand or dust, rubbing out intermediate steps as one proceeded. Al-Uqlidisi had strong reservations about this procedure, not for any shortcomings in the method itself, but because dust-board calculations were carried out by street astrologers and other 'good-for-nothings' to earn their livelihood! He suggested the use of pen and paper for those who did not want to be associated with such company. Indeed, the 'pen and paper' methods for carrying out multiplication and division which found their way into medieval Latin works owe more to al-Uqlidisi than to al-Khwarizmi.

Mathematics in the service of Islamic law: Problems of inheritance

The second half of al-Khwarizmi's *Algebra* contains a series of problems about the Islamic law of inheritance. These laws are fairly straightforward in that when a woman dies her husband receives one-quarter of her estate, and the rest is divided among the children such that a son receives twice as much as a daughter. However, if a legacy is left to a stranger, the division gets more complicated. The law on legacies states that a stranger cannot receive more than one-third of the estate without the permission of the natural heirs. If some of the natural heirs endorse such a legacy but others do not, those who do must between them pay, *pro rata*, out of their own shares, the amount by which the stranger's legacy exceeds one-third of the estate. In any case, the legacy to the stranger has to be paid before the rest is shared out among the natural heirs. Clearly, it is possible to construct problems of varying degrees of complexity which illustrate different aspects of the law. Here is a simple example considered by al-Khwarizmi:

EXAMPLE 10.1 *A woman dies leaving a husband, a son and three daughters. She also leaves a bequest consisting of 1/8 + 1/7 of her estate to a stranger. Calculate the shares of her estate that go to each of her beneficiaries.*

Solution   The stranger receives $1/8 + 1/7 = 15/56$ of the estate, leaving $41/56$ to be shared out among the family. The husband receives one-quarter of what remains, i.e. $1/4$ of $41/56 = 41/224$. The son and the three daughters receive their shares in the ratio $2 : 1 : 1 : 1$, i.e. the son's share is two-fifths of the estate after the stranger and husband have been given their bequests. So, if the estate is divided into $5 \times 224 = 1120$ equal parts, the shares received by each beneficiary will be:

Stranger:	15/56 of 1120 or 300 parts
Husband:	41/224 of 1120 or 205 parts

Son: 2/5 of (1120 − 505) or 246 parts

Each daughter: 1/5 of (1120 − 505) or 123 parts

It is important, however, not to exaggerate the role of mathematics in the service of Islam. Undoubtedly, religious requirements such as implementing the law of inheritance, determining the direction of Mecca for daily prayer, or identifying the beginning and end of the period of fasting (Ramadan) gave a special impetus to the development of certain areas of mathematics. But there is a limit to the usefulness of mathematics for such purposes, and to the stimulus that such activities provided for the further development of mathematics.

The theory of numbers

In the theory of numbers, as in other fields, the Arabs managed to produce a creative synthesis of the ideas they obtained from different mathematical traditions — notably India and the Hellenistic world. This is best exemplified in the work of ibn Sina (or Avicenna, as he came to be known in Europe). Although he is widely known for his work on medicine, his mathematical work is little appreciated outside the Islamic world. But his other important work, entitled *Alai* in Persian and *Kitab al-shifa* in Arabic (Book of Physics), contains sections on arithmetic. They begin with a discussion, based on Greek and Indian sources, of different types of number (e.g. odd, even, deficient, perfect and abundant numbers*) and an explanation of different arithmetical operations, including the rule for 'casting out nines'.† He states two rules which are

---

* 6 is perfect since the sum of its proper divisors is $1 + 2 + 3 = 6$.

8 is deficient since the sum of its proper divisors is $1 + 2 + 4 < 8$.

12 is abundant since the sum of its proper divisors is $1 + 2 + 3 + 4 + 6 > 12$.

†An originally Indian method of checking addition and multiplication. It uses the well-known property that the sum of the digits of any natural number when divided by 9 produces the same remainder as when the number itself is divided by 9. For example, in checking that the product of 436 and 659 is 287 324,

1 Add the digits of 436 to get 13, whose digits are then added to get 4.

2 Add the digits of 659 to get 20, whose digits are then added to get 2.

3 Add the digits of the product 287 324 to get 26, whose digits add to 8.

So 'casting out nines' leaves remainders of 4, 2 and 8 respectively, and since $4 \times 2 = 8$, the multiplication is probably correct.

Figure 10.4

9	7	5	3	1
19	17	15	13	11
29	27	25	23	21
39	37	35	33	31
49	47	45	43	41

not found in earlier texts. The first is a rule for summing a square array of odd numbers:

> If successive odd numbers are placed in a square table, the sum of the numbers lying on the diagonal will be equal to the cube of the side; the sum of the numbers filling the square will be the fourth power of the side. (Al-Daffa and Stroyls, 1984, pp. 77–8)

Ibn Sina illustrates this rule by the square shown in Figure 10.4. The diagonals of the square add up to

$$9 + 17 + 25 + 33 + 41 = 125 = 1 + 13 + 25 + 37 + 49$$

which is equal to the cube of the 'side', 5^3. The total sum of the numbers of the square is $625 = 5^4$, the fourth power of the side. There is the clear implication here that ibn Sina knew that the sum of successive odd numbers starting with 1 is equal to the square of the number of odd numbers being added. For example,

$$1 + 3 + 5 + 7 + 9 + 11 = 36$$

which is the square of 6, the number of odd numbers added.

The second rule is for summing a triangular array of odd numbers:

> If successive odd numbers are placed in a triangle, the sum of the numbers taken from one row equals the cube of the [row] number. (Al-Daffa and Stroyls, 1984, p. 78)

The triangular array of the odd numbers from 1 to 30 is shown in Figure 10.5. It is easily seen that the sum of the numbers in, say, the third row is 27, the cube of the row number.

Figure 10.5

Figurate numbers were studied by several Arab mathematicians, particularly those who favoured a geometric rather than an algorithmic approach to numbers; al-Khwarizmi represented the latter, Thabit ibn Qurra the former approach. Thabit was one of the first Arab mathematicians to recognize the value of a geometric interpretation of an algebraic problem. However, his most notable contribution to the theory of numbers apparently had no geometric motivation – this was his derivation of a formula for generating pairs of amicable numbers.

A pair of natural numbers, M and N, are defined as amicable if each is equal to the sum of the proper divisors (i.e. all the divisors of a number, including 1, but not itself) of the other. The smallest pair of amicable numbers is 220 and 284. The proper divisors of 220 are

$$1, 2, 4, 5, 10, 11, 20, 22, 44, 55, 110$$

the sum of which is 284. Similarly, the proper divisors of 284 are

$$1, 2, 4, 71, 142$$

the sum of which is 220.

In his *Maqala fi istikhraj al-adad al-mutahabba bi-suhulat al-maslak ila dhalika* (Book on the Determination of Amicable Numbers), Thabit provides the following formula for deriving pairs of amicable numbers. Let p, q and r be distinct prime numbers given by

$$p = 3 \times 2^{n-1} - 1, \qquad q = 3 \times 2^n - 1, \qquad r = 9 \times 2^{2n-1} - 1$$

where n is greater than 1. Then M and N will be a pair of amicable numbers, given by

$$M = 2^n pq, \qquad N = 2^n r$$

For $n = 2$,

$$p = 3 \times 2 - 1 = 5, \qquad q = 3 \times 2^2 - 1 = 11, \qquad r = 9 \times 2^3 - 1 = 71$$

As p, q and r are all prime, they may be used to yield M and N as

$$M = 2^2 \times 5 \times 11 = 220, \qquad N = 2^2 \times 71 = 284$$

which is the smallest pair of amicable numbers. For $n = 3$, $q = 287$, which is not prime (it is divisible by 7), so the formula cannot be applied. For $n = 4$ it is found that $p = 23$, $q = 47$ and $r = 1151$. These are all primes and so the next pair of amicable numbers generated from Thabit's formula is

$$M = 2^4 \times 23 \times 47 = 17\,296, \qquad N = 2^4 \times 1151 = 18\,416$$

Thabit only obtained the first pair of amicable numbers from his rule. It was a later Arab mathematician, ibn al-Banna (1256–1321), who found the above pair corresponding to $n = 4$. Some six hundred years after Thabit, Fermat rediscovered this rule and the pair for $n = 4$, and shortly after this Descartes too rediscovered the rule and set $n = 7$ to yield $9\,363\,584$ and $9\,437\,056$. It turns out that Thabit's rule generates pairs of amicable numbers for $n = 2$, 4 and 7, but for no other value of n below 20 000. Fortunately there are other lines of attack. Euler found more than 60 pairs, using methods he developed himself – methods that still form the basis for present-day search techniques. Over a thousand pairs are now known. Curiously, the second smallest pair, 1184 and 1210, was overlooked by all the famous 'amicable number chasers' – it was discovered in 1866 by an Italian schoolboy!

Extraction of roots

The Arabic word for 'root', *jadhir*, was the term introduced by al-Khwarizmi to denote the unknown in an equation. With the help of the terms *mal* (second power, literally 'wealth') and *kab* (cube), he was able to describe equations of different degree: for example,

$$mal\ mal = x^4, \qquad mal\ kab = x^5, \qquad kab\ kab = x^6$$

and so on. There are close parallels between the way the Arabs extracted square and cube roots (by seeking numerical solutions to quadratic and cubic equations) and the methods used in other mathematical traditions. Both the Babylonian (*c.* 1800 BC) numerical method of extracting square roots, which we examined in Chapter 4, and the closely related method found in the *Sulbasutras* (*c.* 500 BC), discussed in Chapter 8, were known to the Arabs, probably through the work of the Alexandrian mathematician Diophantus (*c.* AD 250). We also saw, in Chapters 6 and 7, that the Chinese had developed quite sophisticated procedures for obtaining approximate solutions of $x^n = A$, for any integral value of n and to any degree of accuracy, based on variants of Horner's method and using the binomial coefficients of Pascal's triangle. Later Arab mathematicians worked with methods very similar to those used by the Chinese to extract roots of the second and higher order.

Perhaps Arab mathematics should be best remembered for its synthesis of geometry (mainly from Euclid) and algebra from the East, which began with al-Khwarizmi's geometric solution of certain quadratic equations and culminated in Omar Khayyam's geometric solution of cubic equations.

ARAB ALGEBRA

The word *al-jabr* appears frequently in Arab mathematical texts that followed al-Khwarizmi's influential *Hisab al-jabr w'al-muqabala*, written in the first half of the ninth century. There were two meanings associated with *al-jabr*. The more common was 'restoration', as applied to the operation of adding equal terms to both sides of an equation so as to remove negative quantities, or to 'restore' a quantity which is subtracted from one side by adding it to the other. Thus an operation on the equation $2x + 5 = 8 - 3x$ which led to $5x + 5 = 8$ would be an illustration of *al-jabr*. There was also another, less common meaning: multiplying both sides of an equation by a certain number to eliminate fractions. Thus if both sides of the equation $(9/4)x + 1/8 = 3 + (5/8)x$ were multiplied by 8 to give the new equation $18x + 1 = 24 + 15x$, this too would be an instance of *al-jabr*. The common meaning of *al-muqabala* is the 'reduction' of positive quantities in an equation by subtracting equal quantities from both sides. So for the two equations above, applying *al-muqabala* would give

$$5x + 5 = 8$$
$$5x + 5 - 5 = 8 - 5$$
$$5x = 3$$

and

$$18x + 1 = 24 + 15x$$
$$18x - 15x + 1 - 1 = 24 - 1 + 15x - 15x$$
$$3x = 23$$

The words *al-jabr* and *al-muqabala*, linked by *wa*, meaning 'and', came to be used for any algebraic operation, and eventually for the subject itself. Since the algebra of the time was almost wholly confined to the solution of equations, the phrase meant exactly that.

A geometric approach to the solution of equations

Al-Khwarizmi distinguished six different types of equation by using certain word conventions mentioned earlier. The unknown quantity (which we now denote by x) was referred to as 'root' or 'thing', and the constant was known as 'number'. So the six different types of equation were:

1 roots equal squares: $bx = ax^2$
2 roots equal numbers: $bx = c$
3 squares equal numbers: $ax^2 = c$
4 squares and roots equal numbers: $ax^2 + bx = c$
5 roots and numbers equal squares: $bx + c = ax^2$
6 squares and numbers equal roots: $ax^2 + c = bx$

where a, b and c are positive integers. Al-Khwarizmi provided rules for solving these equations and in a number of cases the geometric rationale for these solutions. Let us take one example of a type (4) equation (squares (*mal*) and roots equal numbers) which is interesting historically since it recurs in later Arab and medieval European texts.

EXAMPLE 10.2   *Solve mal and 10 root equals 39.*
(Or, in modern notation, solve $x^2 + 10x = 39$.)
Suggested solution

AL-KHWARIZMI'S EXPLANATION	EXPLANATION IN MODERN NOTATION
1 You halve the 'number' of roots: Result 5.	$x^2 + 10x = 39$
2 This you multiply by itself: Result 25.	
3 Add this to 39: Result 64.	$(x + 5)^2 = 39 + 25 = 64$
4 Take the square root of this: Result 8.	$x + 5 = 8$
5 Subtract from 8 the result given in Step 1: Result 3.	$x = 3$
This is the root of the square (the square itself is 9).	

(The negative root, $x = -13$, is ignored.)

Variants of this rule are found in both Babylonian and Indian mathematics, and there is every likelihood that this algorithm came from either or both of these sources.

The Crest of the Peacock

The real novelty of the Arab approach is contained in the following statement of al-Khwarizmi's. After giving numerical solutions for all six types of equation, he goes on:

> We have said enough, so far as numbers are concerned, about the six types of equation. Now it is necessary to demonstrate *geometrically* the truth of the same problems which we have explained in numbers.

We can illustrate al-Khwarizmi's geometric approach by returning to the above exercise. In Figure 10.6, ABCD is a square of side x. AD and AB are extended to E and F such that $DE = BF = 5$. The square AFKE is completed, and DC is extended to G and BC to H. From the diagram it is clear that the area of AFKE is equal to

$$x^2 + 10x + 25, \quad \text{or} \quad (x + 5)^2$$

Adding 25 to both sides of the equation $x^2 + 10x = 39$ gives

$$x^2 + 10x + 25 = 39 + 25 = 64$$

from which one of the sides of AFKE, say EK, is found to be $x + 5 = 8$, and so EH $= x = 3$.

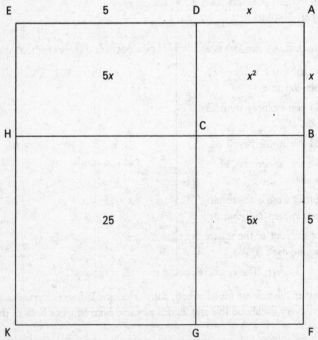

Figure 10.6 Al-Khwarizmi's geometric solution of a quadratic equation.

Al-Khwarizmi's geometric demonstration for each type of equation is based on a specific example. Thabit ibn Qurra presented the first general demonstration, using two of Euclid's theorems, in a short work entitled *Kitab filtatli listikhrag amal al-masail handasiya* (On the Correct Solution of Algebraic Problems by Geometric Methods), of which only a single manuscript has survived. We can illustrate this demonstration for the type (4) equation discussed above – squares and roots equal numbers, or $x^2 + bx = c$.

In Figure 10.7,

$$\text{ABCD} + \text{BHFC} = \text{AHFD} = x^2 + bx$$

The application of a result from Euclid's *Elements* (Book II, Proposition 5) on the equivalence of areas gives

$$\text{AHFD} + \text{BG}^2 = \text{AG}^2 \quad \text{if G is the midpoint of BH}$$

Now,

$$\text{AHFD} = x^2 + bx = c, \quad \text{BG}^2 = (\tfrac{1}{2}b)^2, \quad \text{AG}^2 = c + (\tfrac{1}{2}b)^2$$

and therefore

$$x^2 + bx + (\tfrac{1}{2}b)^2 = c + (\tfrac{1}{2}b)^2 \tag{10.1}$$

Applying another of Euclid's propositions (*Elements*, Book II, Proposition 6) to the left-hand side of equation (10.1) gives

$$x^2 + bx + (\tfrac{1}{2}b)^2 = (x + \tfrac{1}{2}b)^2 \tag{10.2}$$

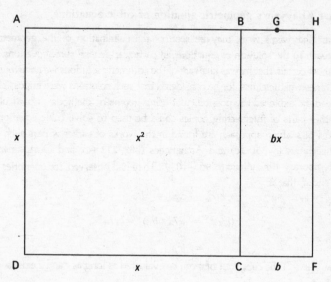

Figure 10.7

Substituting (10.2) in (10.1) gives

$$(x + \tfrac{1}{2}b)^2 = c + (\tfrac{1}{2}b)^2$$

Therefore

$$x = \sqrt{c + (\tfrac{1}{2}b)^2} - \tfrac{1}{2}b$$

The validity of this result can be checked by applying it to Example 10.2 to solve for x. Thabit also provided a similar geometric proof for type (6) equations, squares and numbers equal to roots, or $x^2 + c = bx$.

To sum up, the Arab work on solving quadratic equations is yet another illustration of their ability to bring together two strands of mathematical thinking – the geometric approach which had been carefully cultivated by the Greeks, and the algebraic/algorithmic methods which had been used to such effect by the Babylonians, Indians and Chinese. The Arabs went far beyond the ingenuity and calculating skills of the Babylonians, of which we have seen ample evidence in Chapter 4. By devising an efficient system of classifying equations, the Arab mathematicians, starting with al-Khwarizmi, reduced all equations to three main types. For each type they offered solutions as well as a geometric rationale, thereby laying the foundation of modern algebra. Thus were the *ahl al-jabr* (the 'algebra people'), in Thabit's words, and the 'geometry people' brought together.

Omar Khayyam's geometric solution of cubic equations

Omar Khayyam's work may be seen as the culmination of the geometric approach to the solution of equations, in particular general cubic equations. It is almost certain that he was unaware of the arithmetic solutions we have found in Chinese mathematics, for he considered that such solutions were impossible. Instead, he explored the possibility of using geometric methods, in particular whether parts of intersecting conics could be used to solve cubic equations. Traces of such an approach are found in the works of earlier writers such as Menaechmus (*c.* 350 BC) and Archimedes (287–212 BC), and Omar's near-contemporary al-Haytham (*c.* 965–1039). They had observed, for quantities a, b, c and d, that if

$$b/c = c/d = d/a$$

then

$$(b/c)^2 = (c/d)(d/a) = c/a$$

or

$$c^3 = b^2a$$

Now, if $b = 1$ the cube root of a can be evaluated as long as c and d exist such that

$$c^2 = d \quad \text{and} \quad d^2 = ac \tag{10.3}$$

Omar Khayyam's great contribution was to discover the geometrical argument implicit in this algebra. If we think of c and d as variables and of a as a constant, then (10.3) are the equations of two parabolas with perpendicular axes and the same vertex. This is illustrated in Figure 10.8. The two parabolas, whose construction is explained in Example 10.3 later in this chapter, have the same vertex B, with axes AB ($=a$) and CB ($=b = 1$), and they intersect at E. In the rectangle BDEF, BF = DE = c and BD = FE = d. Since AB is a line segment and the point E lies on the parabola with vertex B and axis AB, the rectangle BDEF has the property that

$$(FE)^2 = AB \cdot BF \quad \text{or} \quad d^2 = ac \tag{10.4}$$

Similarly, for the other parabola with vertex B and axis BC,

$$(BF)^2 = CB \cdot BD \quad \text{or} \quad c^2 = bd = d \tag{10.5}$$

From (10.4) and (10.5) we get

$$c^3 = a \tag{10.6}$$

Therefore, DE = BF is a root of equation (10.6).

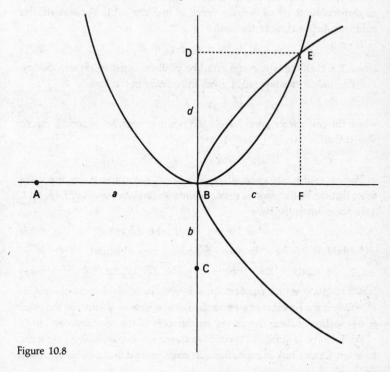

Figure 10.8

Table 10.1 A summary of Omar Khayyam's solutions of cubic equations. In each case one positive root was found.

Type ($a > 0, c > 0$)	Method
(1) $x^3 + c$	Intersection of two parabolas
(2) $x^3 + ax = c$	Intersection of circle and parabola
(3) $x^3 + c = ax$	Intersection of hyperbola and parabola
(4) $x^3 = ax + c$	Intersection of two hyperbolas
(5) $x^3 + ax + c = 0$	No positive roots

Applying similar reasoning, Omar extended his method to solve any third-degree equation for positive roots. He discussed nineteen types of cubic equation (expressed with only positive coefficients). Five of these could easily be reduced to quadratic equations. Each of the remaining fourteen he solved by means of conic sections. It is possible to classify these fourteen cubics, using modern notation, into five main types as in Table 10.1. To illustrate this reduction, take a cubic of the form

$$z^3 + pz^2 + qz + r = 0 \qquad (10.7)$$

where the coefficients p, q and r can be positive, negative or zero. Setting $z = x - p/3$ in equation (10.7) gives an equation of the form

$$x^3 + gx + h = 0$$

where the coefficients g and h are again positive, negative or zero. It can be shown that

$$g = q - (1/3)p^2 \quad \text{and} \quad h = (2/27)p^3 - (1/3)pq + r$$

There is evidence of Omar Khayyam's facility with other cubics, especially those that can be transformed to one of the standard types given in Table 10.1. Take for example the cubic

$$x^3 + ax^2 + b^2x + c = 0, \quad \text{for } a, b, c > 0 \qquad (10.8)$$

and substitute $2dy$ for x^2 to obtain the transformed equation

$$2dxy + 2ady + b^2x + c = 0, \quad \text{for } a, b, c, d > 0 \qquad (10.9)$$

Equation (10.9) is the equation of a hyperbola, while the transformation $x^2 = 2dy$ is the equation of a parabola. The abscissae or x values at the point of intersection of these two curves are the roots of the cubic equation (10.8).

Without the language of modern mathematics, in particular its symbolic notation, Omar's task of exposition was much more difficult. In his approach, *a, b, c* and *x* were line segments, and the problem was:

Given *a, b* and *c*, to construct a line segment *x* such that equation (10.8) holds.

Omar begins by declaring that a line segment cannot be constructed by using only a straightedge and compass; at some point in the construction conic sections must be introduced. His knowledge of conic sections was derived mainly from the *Conics*, the work of the Hellenistic mathematician Apollonius of Perga (*c.* 200 BC). This was one of the more difficult works of Alexandrian geometry. It is a measure of the level of sophistication of Arab mathematics in the tenth century that the *Conics*, together with Euclid's *Elements* and Archimedes' *On the Sphere and Cylinder*, were the three pillars of Arab geometry. (Omar's solutions for each type of cubic listed in Table 10.1 are too long and involved for us to discuss here. His own book, *Al-jabr w'al-muqabala* (Algebra, translated into English and edited by Kasir, 1931), gives an indication of the breadth of coverage of his approach to different cubics.)

Omar's achievement is typical of Arab mathematics in its application (in a systematic fashion) of geometry to algebra. While Omar made no addition to the theory of conics, he did apply the principle of intersecting conic sections to solving algebraic problems. In doing so he not only exhibited his mastery of conic sections, but also showed that he was aware of the practical applications of what was a highly abstruse area of geometry.

Also important was Omar's systematic classification of cubic equations, and his demonstration of a geometric solution for each type. Despite the constraints imposed by the character of the mathematical language of the time (he used either geometric magnitudes or numbers capable of geometric interpretation), the clarity of Omar's presentation is striking, and the number of cases he cannot demonstrate is relatively few. He was aware that sometimes there was more than one positive solution, sometimes none at all (for non-intersecting conic sections). His neglect of negative or imaginary roots is perfectly understandable given the mathematical climate of the time. But this does not imply that the methods he used were not adequate for the purpose of extracting negative roots. In fact, referring to Table 10.1, it can be shown that (i) the absolute value of the negative roots of type (5) is identical to the positive root of type (2), and vice versa; and (ii) that the absolute values of the negative roots of type (4) correspond to the positive roots of type (3), and vice versa. Indeed, Omar's geometrical methods were more comprehensive than the algebraic methods developed by the Italian algebraists, notably Girolamo Cardano and Niccolò Tartaglia. For example, Cardano was unable to solve certain cubics of the form $x^3 = ax + c$, type (4) in Table 10.1.

What Omar Khayyam and those who came after him failed to do was find an algebraic solution of cubic equations. In an earlier chapter we examined the numerical procedures the Chinese used to solve equations of any degree, but they showed little interest in the algebra of these solutions. A recent French translation of Sharaf al-Din al-Tusi's work (Rashed, 1986) contains an interesting application of what we would describe today as the calculus technique of testing for the existence of positive solutions to cubic equations. For an equation $x^3 + c = ax^2$ expressed in the form

$$x^2(a - x) = c \tag{10.10}$$

al-Tusi notes that whether it has a positive solution depends on whether the expression on the left-hand side reaches c. He is aware that, for any maximum value of x lying between 0 and a,

$$x^2(a - x) \leqslant (2a/3)^2(a/3)$$

We can easily see with the help of elementary calculus that a (relative) maximum occurs for x in equation (10.10) when $x_0 = 2a/3$. But there is no indication as to how al-Tusi found this value. He proceeds to make the following inferences:

if $4a^3/27 < c$,   no positive solution exists

if $4a^3/27 = c$,   only one positive solution ($x = 2a/3$) exists

if $4a^3/27 > c$,   two positive solutions (x_1 and x_2) exist

where $0 < x_1 < 2a/3$ and $2a/3 < x_2 < a$. However, there is no evidence that al-Tusi actually found the two positive solutions (x_1 and x_2).

Arab algebra and its influence on Europe

Tracing the lines of Arab influence on European mathematics is a difficult task at the best of times, although there are one or two lines of which we are reasonably certain. Al-Khwarizmi's *Algebra* is generally recognized, through its Latin translations, as having been highly influential in the development of European algebra. Abu Kamil (c. AD 900), popularly known as the 'Egyptian Calculator', wrote a commentary on al-Khwarizmi's work in which he systematically treated the fundamental rules of algebraic operations and solution of equations, including non-linear simultaneous equations. This work influenced Fibonacci, whose impact on medieval European mathematics cannot be overstated. There was a third Arab mathematician and scientist whose geometric theory of the solution of equations, particularly as applied to problems in optics, had a direct impact on Europe – al-Haytham. It is one of the ironies of history that the works of Thabit and Omar, two of the greatest

Arab algebraists who turned to geometry for rigorous derivations of results, were less well known than those just mentioned.

Not until the 1850s did Omar's work begin to be mentioned in the standard Western histories of mathematics, when Woepcke's translation of his *Algebra* appeared, though Kasir (1931, pp. 6–7) produces evidence of European interest in Omar's work from over a hundred years before. And there is no evidence that Thabit ibn Qurra may have directly influenced the development of mathematics in Europe, yet there are pieces of circumstantial evidence which are quite suggestive. The translations by Gherardo of Cremona (c. 1175), through which more Arab science entered Europe than in any other way, drew heavily on Thabit's works. Furthermore, we may question the originality of John Wallis's *Treatise on Angular Sections* (1685), in which there are echoes of the works of Ptolemy (c. AD 100) and of Thabit's generalization of the Pythagorean theorem. (Thabit's work in this area will be discussed in a later section.) It is important to keep an open mind about such possible transmissions, about which more may emerge from detailed research on how the Arab world influenced medieval Europe. It was not all that long ago that the Arabs were dismissed as mere custodians, or at best pale imitators, of Greek science and philosophy.

ARAB GEOMETRY

Throughout the Islamic world are to be found buildings decorated with intricate geometric designs. These are a common feature of Islamic art, which has an ornamental tradition since the Islamic religion has generally proscribed the portrayal of living things. Such superb craftmanship in various media, including wood, tile and mosaic, would have required considerable geometric skills in construction. Around the year 950, Abul Wafa wrote a book entitled *Kitab fi ma yahtaj ilayh al-sani min al-amal al-handasiyya* (On Those Parts of Geometry Needed by Craftsmen), which provides a number of constructions, many of which can be achieved with just a straightedge and compass. However, it was in the construction of conic sections with these basic implements that the Arabs made their distinctive contribution.

In *Rasm al-qutu attalata* (On Drawing the Three Conic Sections) by Ibrahim ibn Sina (d. 946), a grandson of Thabit ibn Qurra, there are detailed instructions on how to construct a parabola and an ellipse, and three different ways of constructing a hyperbola. To illustrate, let us consider the constructions of a parabola and a hyperbola. (The reader may like to try drawing these figures with just a rule and compass to appreciate the sheer ingenuity and geometric 'sense' displayed here.)

EXAMPLE 10.3 *To construct a parabola.*
Method (see Figure 10.9) Draw a line AB, and construct a perpendicular CE cutting AB at D. On the line segment DB, mark a number of points G, F, ... Next, construct circles with diameters AB, AF, AG, . . . which intersect CE at H and L, J and M, K and N, . . . respectively. Through H and L, draw lines parallel to AB, and through B draw a line parallel to CE. Let these lines through H, L and B meet at P and S. Similar lines drawn through J, M and F, and through K, N and G, intersect at points Q and T, and R and U, respectively. Ibn Sina provides a proof that all such points of intersection lie on a parabola. The parabola has turning point at D, axis AB and parameter AD.

EXAMPLE 10.4 *To construct a hyperbola.*
Method (see Figure 10.10). AB is a line segment which is also the diameter of a semicircle, centre O. Produce AB in the direction of B. Choose

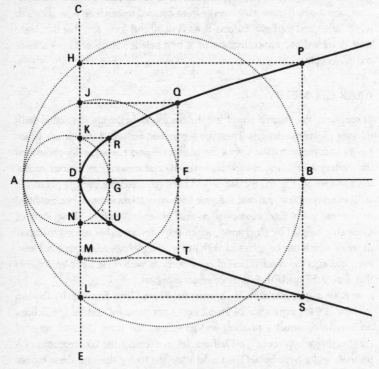

Figure 10.9 Ibn Sina's construction of a parabola.

Figure 10.10 Ibn Sina's construction of a hyperbola.

points C, D, E, . . . and through them construct tangents of the semicircle: CH, DI, EJ. From points H, I, J construct lines parallel to AB such that HR = HC, IQ = ID, JP = JE. Ibn Sina shows that points M, N, P, Q, R, . . . , B lie on a hyperbola.

It is difficult to establish the origins of the mathematician considered by many to have been al-Khwarizmi's natural successor. The difficulty arises from a characteristic of Arabic writing: sometimes letters are distinguished, not by their different shapes, but by the location of a dot near the letter. So whether this mathematician was al-Karkhi (which places his origins in Iraq) or al-Karaji (which places his origins in Persia) depends on whether the dot was placed above the relevant Arabic letter or below it; both versions are recorded. Whatever his origins, we shall refer to him as al-Karaji (i.e. one who was born in the town of Karaj, near Tehran). His many contributions include rules of operations with exponents, the solution of equations of higher degree, and an elaboration of the 'Indian calculation'. He also translated the first five books of the Alexandrian mathematician Diophantus. Al-Karaji gives an interesting geometric construction based on a result which has strong algebraic under-pinnings.

EXAMPLE 10.5   *Construct a circle whose area is equal to a given fraction (1/n) of the area of a given circle.*

Method (see Figure 10.11)   AOB is the diameter of the given circle, with centre O. Draw TA perpendicular to AB. Extend BA to C such that CA = $(1/n)$AB. Construct a circle with CB as diameter. The point at which this circle cuts TA is denoted by D. AD is then the diameter of the required circle, equal to $(1/n)$AB, so the area of this circle, C_2, is $1/n$ of the given circle, C_1.

Figure 10.11

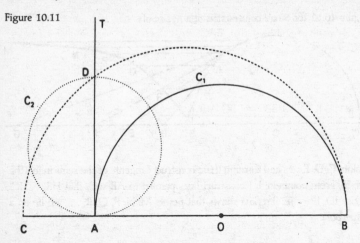

Proof   The proof follows easily from two propositions of Euclid's on intersecting chords (*Elements*, Book II, Propositions 12 and 13) which give

$$CA \cdot AB = DA^2 = (d/n)\,d, \quad \text{where } AB = d, \quad CA = (1/n)AB$$

Therefore $DA = d/\sqrt{n}$. Now, the circle with diameter AB is C_1 and the circle with diameter $AD = d/n$ is C_2. So the ratio of the areas of C_1 and $C_2 = d^2 : d^2/n = 1 : 1/n$.

Thabit ibn Qurra's generalization of the Pythagorean theorem

In a letter to a friend Thabit expressed disappointment with an existing (so-called Socratic) proof of the Pythagorean theorem because it applied only to isosceles right-angled triangles. He then proceeded to give three results, of which the third is a generalization of the Pythagorean theorem applicable to all triangles, whether right-angled or not. We shall look at the third result, which Sayili (1960, p. 35) has described as an 'important contribution' to the history of mathematics.

Consider Figure 10.12, which is constructed in the following manner. From the vertex A of a triangle ABC drop lines intersecting the base BC at B' and C' and forming angles AB'B and AC'C respectively, each of which equals angle BAC. We wish to show that

$$AB^2 + AC^2 = BC(BB' + CC')$$

Thabit ibn Qurra provides no proof, except to say that it follows from Euclid.

A reconstruction of the proof that makes use of similar triangles is as follows. It can easily be shown that BAC, BAB' and CAC' are similar triangles. Therefore

$$BC:AC:AB = AC:CC':AC' = AB:AB':BB'$$

Then

$$AB/BB' = BC/AB, \quad \text{so} \quad AB^2 = (BC)(BB') \tag{10.11}$$

and

$$AC/CC' = BC/AC, \quad \text{so} \quad AC^2 = (BC)(CC') \tag{10.12}$$

Adding equations (10.11) and (10.12) gives Thabit's generalization of the Pythagorean theorem:

$$AC^2 + AB^2 = BC(BB' + CC')$$

Figure 10.12 shows an obtuse triangle, with the angle at A greater than a right angle. Thabit also considers an acute triangle, for which B' and C' lie outside BC, but for which the above proof (with minor modifications) still applies, and the Pythagorean right-angled triangle, for which B' and C' coincide at D.

Thabit's work on this theorem was discovered as late as 1953 in the library of the Aya Sofia Museum in Turkey. However, it made its first appearance in European mathematics in 1685 when John Wallis's proof of the theorem was published in his *Treatise on Angular Sections*. It is a reasonable conjecture that Wallis was aware of Thabit's work in this area, since he was sufficiently acquainted with Arab mathematics to carry out a translation into Latin of al-Tusi's work on the parallel postulate. Indeed, Scriba (1966, p. 67) is of the opinion that the *Treatise on Angular Sections* is based on Thabit ibn Qurra's generalization of the Pythagorean theorem and Ptolemy's work in this area.

Figure 10.12 Thabit's generalized Pythagorean theorem.

ARAB TRIGONOMETRY

As with so many other areas of mathematics, the Arabs selected Hellenistic and Indian concepts of trigonometry and combined them into a distinctive discipline that bore little resemblance to its precursors. It then became an essential component of modern mathematics. We shall consider three aspects of Arab trigonometry:

1 the introduction of six basic trigonometric functions, namely sine and cosine, tangent and cotangent, secant and cosecant;
2 the derivation of the sine rule and establishment of other trigonometric identities; and
3 the construction of highly detailed trigonometric tables with the aid of various interpolation procedures.

Introduction of trigonometric functions

Basic to modern trigonometry is the sine function. It was introduced into the Arab world from India, probably through the famous Indian astronomical text *Surya Siddhanta*. This was one of the texts brought to the court of al-Mansur during the eighth century by a diplomatic mission from Sind. We saw in an earlier chapter that there were two types of trigonometry: one based on the geometry of chords (see Figure 9.1) and best exemplified in Ptolemy's *Almagest*, and the other based on the geometry of semi-chords (see Figure 9.2), which was an Indian invention.

In this scheme, the length of the semi-chord AM that corresponded to the semi-angle α at the centre of the circle (of radius 3438', where each minute was a unit of length equal to 1/60 of the length of 1° of arc on the circle) was given at intervals of 3°45': effectively, a sine table. The only difference between this table and a modern one is that it gives the Indian sine, or *jya*, of the angle α:

$$\text{jya}\,\alpha \;=\; r\sin\alpha \;=\; 3438\sin\alpha$$

From the tenth century onwards, starting with the work of Abul Wafa, the Arabs brought the sine function closer to its modern form by defining it in terms of a circle of unit radius, although it remained defined for an arc of a circle rather than the angle subtended at the centre.

The etymology of the word 'sine' is instructive for it shows what can happen as a result of imperfect linguistic and cultural filtering. The Sanskrit term for sine in an astronomical context was *jya-ardha* ('half-chord'), which was later abbreviated to *jya*. From this came the phonetically derived Arabic word *jiba*, which, following the usual Arab practice of omitting vowels, was written as *jyb*.

Early Latin translators, coming across this apparently meaningless word, mistook it for another word, *jaib*, which had among its meanings the opening of a woman's garment at the neck, or bosom; *jaib* was translated as *sinus*, which in Latin had a number of meanings, including 'fold' (in a toga), 'bosom', 'bay' and, indeed, 'curve'. And hence the present word 'sine'.

There are two other functions which the Arabs may have derived from the Indians. *Kojya* (i.e. $r \cos \alpha$, or OM in Figure 9.2) and *ukramajya* (i.e. $r \operatorname{vers} \alpha = r(1 - \cos \alpha)$, or MC in Figure 9.2) were trigonometric functions commonly used in Indian astronomy during the period of contact between the Indians and the Arabs. But the tangent and cotangent functions are of Arab origin.

During the ninth century the Arab astronomer al-Hasib examined the length of the shadow of a rod of unit length horizontally mounted on a wall when the Sun was at a given angle to the horizontal. It is easily shown (Figure 10.13(a)) that the length s of the shadow on the wall can be calculated as

$$s = \sin \alpha / \cos \alpha = \tan \alpha$$

where α is the angle of elevation of the Sun above the horizon. The length t of the shadow cast by a vertical rod (see Figure 10.13(b)) is

$$t = \cos \alpha / \sin \alpha = \cot \alpha$$

Figure 10.13

The secant and cosecant functions seldom appeared in Arab trigonometric tables, though there are scattered references to them in the works of Abul Wafa and al-Tusi. The secant and the cosecant were known respectively as the 'hypotenuse of the shadow' (the distance from the top of the horizontal rod to the tip of the shadow in Figure 10.13(a)) and the 'hypotenuse of the reversed shadow' (the distance from the top of the vertical rod to the tip of the shadow in Figure 10.13(b)). A long-standing trigonometric tradition based on shadow lengths is found in both Indian and Arab mathematics.

Derivation of trigonometric relationships

Abul Wafa's work on trigonometry contains more than a systematic treatment of the six functions. In his *Zij almagesti* he gives a rule for calculating the sine of the sum of two arcs and the sine of their difference when each of them is known:

> Multiply the sine of each of them by the cosine of the other, expressed in sixtieths, and we add the two products if we want the sine of the sum of the two arcs, but take the difference if we want the sine of their difference.

Expressed in modern notation, this rule becomes the familiar

$$\sin(\alpha \pm \beta) = \sin\alpha\cos\beta \pm \cos\alpha\sin\beta$$

The reference to sine and cosine functions expressed in sixtieths shows that calculations were carried out in sexagesimal fractions. Abul Wafa provides a proof of this result in terms of arcs of a circle of unit radius (Berggren, 1986, pp. 136–8).

The sine rule in its modern version is attributed to Nasir al-Din al-Tusi. The rule may be stated in the following way: given any triangle ABC (see Figure 10.14),

$$\frac{b}{c} = \frac{r\sin B}{r\sin C}$$

where r is taken to be 60 units. Al-Tusi provides a proof for this rule (Berggren, 1986, pp. 138–9) and proceeds to consider how the result could be used to calculate the dimensions of a triangle, given a knowledge of different combinations of angles and sides. For example, knowing the values of one angle (say B) and two sides (b and c), the other angle (C) can be calculated by first using the rule given above and then looking up the angle in a sine table. This would immediately yield the other angle (A), and the sine rule could then be used to obtain the length of side a. The sureness with which al-Tusi tackles a variety

Figure 10.14

of problems using the sine rule is an indication of the maturity of thirteenth-century Arab trigonometry.

Construction of trigonometric tables

Early Arab interest in trigonometry was triggered by their discovery of sine tables in the Indian *Siddhantas*. They soon realized that trigonometric calculations, whether applied to astronomy or to geometry, required detailed and accurate tables, and they proceeded to construct tables that were more accurate than any before. Al-Hasib (*c.* 850) constructed the first sine and tangent tables at intervals of 1°, accurate to three sexagesimal (five decimal) places. Subsequent work concentrated on reducing the intervals and increasing the accuracy of these tables. Thus, in the works of the astronomer-king Ulugh Beg in 1440, there are tables for the two functions at intervals of 1/60 of a degree correct to five sexagesimal (nine decimal) places. The computation necessary to produce such a table is quite breath-taking. For each of the 90 degrees there would be 60 entries, making a total of 5400 entries.

The calculation of the sine of 1° (assuming, for simplicity, a unit radius) was itself a considerable undertaking. We know from Abul Wafa's work in the tenth century that the procedure was to apply the formula for the sine of the difference of two arcs, namely sin (72° − 60°), which would give sin 12°. The choice of 72° and 60° was deliberate, and quite revealing about the level of sophistication of Arab trigonometry, since from the sides of a regular pentagon and of an equilateral triangle inscribed in a circle it is possible to work out the required values of the sines of the angles mentioned to any degree of accuracy.

To illustrate this, Figure 10.15 shows a triangle ABC whose base angles are each 72° and whose third angle is therefore 36°. It can be established that

$$2\cos 36° = 2\cos 72° + 1 \tag{10.13}$$

From the familiar expansion for $\cos(A + B)$ and the result $\sin^2 A + \cos^2 A = 1$, equation (10.13) can be expressed as

$$2c = 2(2c^2 - 1) + 1, \quad \text{where } c = \cos 36°$$

Therefore $4c^2 - 2c - 1 = 0$, the solution of which is

$$c = \frac{2 \pm \sqrt{4 + 16}}{8} = \frac{1 + \sqrt{5}}{4}, \quad \text{for } c > 0$$

or $\cos 36° = \frac{1}{4}(1 + \sqrt{5})$.

Now, to calculate $\sin 72°$, since $\sin 72° = \cos 18°$, we set $c = \cos 18°$. Then

$$2c^2 - 1 = \frac{1 + \sqrt{5}}{4}$$

and

$$c^2 = \frac{5 + \sqrt{5}}{8}$$

Figure 10.15 Equation (10.13) is established by drawing the line AD bisecting angle CAB, then dropping perpendiculars from A and D.

so

$$\cos 18° = \sin 72° = \sqrt{\frac{5 + \sqrt{5}}{8}}$$

The fact that $\sin 60° = \sqrt{3}/2$ was known to the Indians as early as AD 500 probably explains the early Arab acquaintance with the result. The difference between the calculated values of $\sin 72°$ and $\sin 60°$ is taken and used with the expansion

$$\sin(72° - 60°) = \sin 72° \cos 60° - \sin 60° \cos 72°$$

to obtain the value of $\sin 12°$. Next the half-angle formula* is applied to yield, successively, $\sin 6°$, $\sin 3°$, $\sin 1\frac{1}{2}°$ and $\sin \frac{3}{4}°$, and then some linear interpolation procedure is applied to the last two values to give an estimate of $\sin 1°$.

Different types of interpolation procedure were experimented with, particularly when it was recognized that while linear interpolation would work well over small intervals, where the growth was uniform, it was not appropriate for large intervals, or for the upper bounds of a tangent function, where the value of the tangent of an angle approaches infinity as the angle approaches 90° (i.e. the function has a vertical asymptote at 90°). This recognition was already implicit in Indian mathematics before it came into contact with the Arabs. One of the greatest of the Arab astronomers, ibn Yunus, who lived during the first half of the tenth century, had devised a second-order interpolation procedure and used it in constructing his sine table. New ground was broken by al-Kashi. In his book *Risala al-watar wa'l-jaib* (Treatise on the Chord and Sine) he approached the problem of obtaining an accurate estimate of $\sin 1°$ in a different manner, devising an iterative procedure involving the solution of cubic equations. His method highlights some of the similarities between the numerical methods used for this purpose by the Chinese and the Arabs.

The approximation to $\sin 1°$ was based on two pieces of information known to Arab mathematicians at least three centuries before al-Kashi's time. (For simplicity we shall use base 10 here, not the sexagesimal base with which al-Kashi worked. Also, we shall use the modern sine function, whereas al-Kashi's sine function is 60 times the modern one.) The two relationships are, first, that for any given angle α,

$$\sin 3\alpha = 3 \sin \alpha - 4 \sin^3 \alpha$$

---

*The half-angle formula follows from the identities $\sin^2 x + \cos^2 x = 1$ and $\cos 2x = \cos^2 x - \sin^2 x$. Thus

$$\sin x = \sqrt{1 - \cos^2 x} \quad \text{and} \quad \sin \tfrac{1}{2}x = \sqrt{\frac{1 - \cos x}{2}}$$

so that

$$\sin 3° = 3 \sin 1° - 4 \sin^3 1° \qquad (10.14)$$

Second, by the method discussed earlier, we can calculate that

$$\sin 3° = 0·052\,335\,956 \qquad (10.15)$$

Combining equations (10.14) and (10.15), and denoting the unknown value $\sin 1°$ by x, we get the cubic equation

$$x^3 - 0·75x + 0·013\,083\,989 = 0$$

To solve this cubic equation al-Kashi used an iterative procedure which, expressed in modern terms, is as follows. To solve an equation of the form $x = f(x)$, choose an arbitrary value x_0 as a first approximation to the root. Then, by using the relation $x_n = f(x_{n-1})$ for $n = 1, 2, 3, \ldots$ a sequence of values $x_1, x_2, \ldots$ is obtained which approximates more and more closely to the solution, irrespective of what was chosen as x_0, as long as $\lim x_n$ exists. In modern numerical analysis this procedure is known as 'fixed point' or 'direct' iteration.

By this method al-Kashi computed the value of $60 \sin 1°$ correct to nine sexagesimal (sixteen decimal) places − a remarkable exhibition of computational skill, even by today's standards. The reader who wishes to know more about al-Kashi's method will find Aaboe (1954) illuminating. Values of sines for $\frac{1}{2}°$, $\frac{1}{4}°$, $\frac{1}{8}°$, and so on were obtained by applying the half-angle formula; other fractions or finer divisions were achieved by applying some appropriate interpolation formula. Variants of al-Kashi's method were used by other astronomers, mainly working in Samarkand, including his patron Ulugh Beg, who paid him the tribute quoted earlier. An iterative procedure similar to al-Kashi's was used by the German astronomer Johann Kepler (1571–1630).

THE ARAB CONTRIBUTION: A FINAL ASSESSMENT

In Chapter 1 we contrasted modern Eurocentric attitudes to the Arabs' contribution with the seminal role they played in transmitting mathematics to Western Europe, setting the stage for the development of modern mathematics. It should be clear from the present chapter that the traditional view of the Arabs as mere custodians of Greek learning and transmitters of knowledge is both a partial view and a distorted one. We have seen how original were their contributions to algebra and trigonometry, and how crucial was the role they played in bringing together two different mathematical strands − the algebraic and arithmetic traditions so evident in the mathematical cultures of Babylonia, India and China, and the geometric traditions of Greece and the Hellenistic world. The intertwining of these strands had already begun with later

Alexandrian mathematicians such as Heron, Diophantus and Pappus, who had absorbed much of their mathematics from Babylon and Egypt, but there remained the constraints imposed by the straitjacket of Greek mathematical tradition. It was left to the Arabs to bring together the best of both traditions. In doing so, they provided us with an efficient system of numeration, in which calculations were no longer tied to mechanical devices, an algebra which was both practical and rigorous, a geometry which was no longer an intellectual pastime, and a trigonometry freed from its ties to astronomy to become indispensable in fields as diverse as optics and surveying.

The Arab approach to mathematics was no doubt helped in earlier years by the existence of a creative tension between the 'algebra people' and the 'geometry people', best exemplified by al-Khwarizmi and Thabit ibn Qurra, respectively. Each group remained open to influences from the other group, as shown by al-Khwarizmi's geometric approach to the solution of quadratic equations and Thabit's discovery of a rule for generating amicable numbers. As Arab mathematics developed, work on 'pure' geometry, such as attempts to prove or modify Euclid's parallel postulate, continued alongside the development of skilful numerical methods for extracting roots and solving higher-order equations. Indeed, the main reason why modern mathematics moved away so substantially from the spirit and methods of Greek mathematics was the intervention of the Arabs. Perhaps, if the Arab lessons had been absorbed earlier, and if the work of the chief figures of Arab mathematics such as Omar Khayyam and Thabit ibn Qurra had been better known than they were, then the period of painful transition and the repetitious nature of some medieval European mathematics could have been avoided altogether.

Both history and religion in Europe conspired to stem the flow of ideas from the Arab world at a time when Europe was rousing herself from her long slumber and taking her first confident steps into the realm of ideas. Increasingly, Europe was exposed only to the Greek vision as represented by various translations into Latin and other languages from Arabic texts. In a search for her roots she bypassed her Arab and non-European heritage and homed in on Greece and Rome. Greece thereby became the fount of her intellectual and cultural heritage, while for her religious roots she looked towards Rome and Byzantium. Here there had eventually been rewoven a synthetic Christianity from some of the various strands into which doctrinal controversies had split the original Pauline faith, although the result was in some ways far removed from its Eastern and Judaic origins. The history of the last five hundred years has tended to strengthen these ties, partly as a consequence of European dominance and partly under the pressure of 'classical' scholarship, much of which regards Greece as the sole source of knowledge and culture. In the late 1960s, a series of programmes entitled 'Civilisation' was shown on British television. It dealt exclusively with the cultural and intellectual heritage of Europe. By implication, no other society or culture came within the purview of this description.

In mathematics, the glorification of ancient Greece during the Renaissance led to a concentration on Hellenistic texts. The Arabs also admired the Hellenistic contribution, particularly in geometry. One wonders, especially with Thabit ibn Qurra, whether some of the time and effort he spent on translating Greek works might have been put to better use in developing his own, very promising algebra. Nevertheless, the Arabs remained open to other influences as well.

There is no denying that the Greek approach to mathematics produced some remarkable results, but it did hamper the subsequent development of the subject. The strengths of the Greek approach have been discussed extensively; any standard textbook on the history of mathematics deals with this, so there is little point in going over the issue again. But the limiting effect of the Greek mode of thought is another matter. The Greek preoccupation with geometry until the infiltration of Babylonian and Egyptian influences in the later Hellenistic period was a serious constraint. Great minds such as Pythagoras, Euclid and Apollonius spent much of their time creating what were essentially abstract, idealized constructs; how they arrived at a conclusion was in some ways more important than any practical significance. There were in fact two different geometries coexisting at the time: the 'pure' geometry of the Greeks, whose validity was determined wholly by its internal consistency and coherence, and the 'applied' geometry of other mathematical traditions, whose validity was judged solely by its ability to describe physical reality. (It is interesting to speculate what a Euclid who had absorbed the arithmetic and algebra of the Babylonians and had sympathy with their analytic/algebraic approach to geometry might have created with his particular brand of deductive reasoning.) Apollonius' *Conics* seemed to be a product of a Greek abstract geometry in need of no further refinement. Only with the emergence of the Arabs were the works of the period rescued and given a new direction. However, the pioneers of modern mathematics in the post-Renaissance period found themselves compelled to undergo a sometimes painful distancing from the Greek geometric approach their predecessors had too readily espoused, unleavened as it was with the Arab spirit.

We conclude this chapter by returning to the question of transmissions, which we have considered in several previous chapters. The impact of the

Arabs on the intellectual life of Europe is better chronicled than most other cross-cultural influences. The spread of Indian numerals, the growth of algebra, the introduction of trigonometry, the dissemination of Greek geometry and the Arab extensions to it – these are all well authenticated and recognized in the more recent histories of mathematics. The possible transmission of certain techniques through third parties for which written records are non-existent or incomplete is more problematic. We have looked at several examples, including numerical solutions of higher-order equations by Horner's method and Thabit's rule for generating amicable numbers (sometimes credited to Fermat and Euler). As a further instance, it is now apparent that Nicolaus Copernicus (1473–1543) owed a considerable debt to the Arab mathematician-astronomer Nasir al-Din al-Tusi (1201–74), some of whose ideas are incorporated in the Copernican model of the solar system (though it was Copernicus rather than al-Tusi who put the Sun at the centre, thus reviving an idea which dates back to the Greek Aristarchus of Samos in the third century BC). There is clearly a need for further examination of known medieval European sources, and a search for other archival material, especially in Arab and Ottoman sources.

Apart from transmissions to the West, there are two other links which require further elaboration.

1 There is the whole question of possible transmissions of mathematical ideas between the Arab world and China. We have remarked in this chapter on how the Chinese methods of solving numerical equations of higher order may have influenced Arab mathematics; at the end of Chapter 7, on Chinese mathematics, we briefly considered the likelihood of Arab trigonometry and arithmetic having reached China. In looking for channels along which such transmissions could have occurred, we must take into account the political and social climate of the first half of the second millennium AD. There is evidence, particularly from the time of the Sung dynasty, of political and cultural contacts between the two societies. The Mongol empire stretched across a good part of Central Asia. We have seen how receptive some of the Mongol rulers were to scientific ideas from lands they had conquered. Hulegu Khan and Ulugh Beg were not only patrons but practitioners of science. The intriguing question remains: did they fulfil the same role as the caliphs of Baghdad by encouraging cross-cultural contacts between their scientists? Here again, more research needs to be done before we can provide any definite answers.

2 Indian mathematics and astronomy absorbed much from Arab and Persian sources; astronomy was the main beneficiary. In 1370 Mahendra Suri, the astronomer at the court of Firoz Tughlaq, published his work *Yantraraja*, which introduced Arab and Persian astronomy into the Sanskrit *Siddhanta* tradition.

This flow of astronomical ideas, as well as instruments, continued into the seventeenth century, providing the basic materials for those training in the Ptolemaic system. However, attempts to synthesize the two systems, which had such promising beginnings with Mahendra Suri, proved unsuccessful. Over a period of time there developed two distinct schools in Indian mathematical astronomy: the old Sanskrit school and the new Islamic *madrassa* school. Occasionally they came together, usually under the patronage of an enlightened ruler such as the Mughal emperor Akbar, or Rajah Jai Singh of Rajasthan. The latter left as his monument the large masonry astronomical instruments at Delhi, Jaipur, Ujjain and Varanasi (Benares).

This is the end of our story. We have travelled around the world in search of our 'hidden' mathematical heritage, and in the rich tapestry of early human experience we have discovered mathematics in bones, strings and standing stones. No society, however small or remote, has ever lacked the basic curiosity and 'number sense' that is part of the global mathematical experience. The need to record information that gave birth to written language also brought forth a variety of number systems, each with its own strengths and peculiarities. And yet if there is a single universal object, one that transcends linguistic, national and cultural barriers, and is acceptable to all and denied by none, it is our present set of numerals. From its remote beginnings in India, its gradual spread in all directions remains the great romantic episode in the history of mathematics. It is hoped that this episode, together with other non-European mathematical achievements highlighted in this book, will help to extend our horizons and dent the parochialism that lies behind the Eurocentric perception of the development of mathematical knowledge.

Reflections

1 The History of Mathematics: Alternative Perspectives

Labels

The responses of some critics to the first edition of this book have helped to confirm my belief that words or labels in common use need careful scrutiny. It had already been pointed out that terms such as 'Dark Ages' and 'Renaissance' are peculiarly European concepts of little relevance to the rest of the world. Also, words such as 'ancient', 'medieval' and 'modern' are of doubtful provenance when applied to other histories. It is, however, in the labelling of geographical areas that the distortions could, potentially, take on grotesque proportions.

Consider the term 'European continent'. A Eurasian peninsula has been elevated to the status of a continent, equal in importance, if not superior, to the rest of the continent combined. The Mercator projection may have contributed to this perception, with its visually distorted image exaggerating the northward bounds to make 'Europe' look larger than the whole of Africa, and enormous compared to the other Eurasian peninsula, India. The special status accorded to Europe in the standard histories of the world has strengthened the notion, shared by many Europeans and their overseas descendants, that they played a starring role in the Eurasian theatre of world history. The resulting categories such as European/Non-European, West/East, Europe/Asia have tended to reinforce this Eurocentric illusion.

It is precisely because the tendency to enhance Europe through the use of labels that any project involving the writing of a balanced history should carefully address the question of labels. Unfortunately, these labels have existed for so long that they have acquired legitimacy through usage. Fairly early in writing this book, I was faced with the problem of finding a subtitle for it. With the hardback edition, the publisher took the decision out of my hands and provided the subtitle *Non-Western Roots of Mathematics*. The Penguin paperback version replaced it with *Non-European Roots of Mathematics*, which at least had the benefit of confronting the widely held view that mathematics began in Greece – a part of Europe. However, both titles fail to avoid the implication that a large part of the world's population is defined by *not* being something.

A further problem was in the choice of labels to identify different mathematical traditions. Labelling traditions by names of regions such as India and China, or of historical cultures such as the Islamic cultures between AD 800 and 1500 or Greek/Hellenistic cultures between 600 BC and AD 300, has both a spatial and a temporal justification. If we use the linguistic marker, descriptions of Greek, Chinese and Arabic mathematics have legitimacy, although there appears to be a certain inconsistency implied in the first, and an incompleteness implied in the last, of the descriptive terms.

When Alexander appointed his general, Ptolemy I, to rule Egypt, he also appointed Seleucus I to rule Mesopotamia. Historians who were eager to discount the African influence called the science of Ptolemaic Egypt 'Greek', while they continued to label the science and mathematics of the Seleucid period as 'Mesopotamian'. For example, in the case of astronomy, Neugebauer (1969, p. 97) writes: 'Early Mesopotamian astronomy appeared to be crude and merely qualitative, quite similar to its contemporary, Egyptian astronomy . . . Only the last three centuries BC [have] furnished us with texts [from the former] . . . fully comparable to the corresponding Greek systems [of the latter] . . .'

Outstanding among these so-called Greek systems were those of Alexandria, which Neugebauer labels 'Greek' rather than 'Egyptian', unlike the description of Seleucid astronomy as 'Mesopotamian', although Mesopotamia at that time was also under Greek rule. In this edition we shall continue to retain the label 'Greek' to describe all mathematical works written in Greek, irrespective of their geographical origins. However, a distinction will be drawn between 'Greek' and 'Hellenistic' (or 'Alexandrian') in providing a geographical reference. Thus, Alexandrian mathematics would be firmly situated in Egyptian mathematics.

The label 'Arabic mathematics' is a clumsy and problematic construction that neglects the rich source of non-Arabic mathematical texts, especially those written in Persian, Syriac and Hebrew. In place of 'Arabic' other labels are available, such as 'Islamic', 'Muslim', 'Middle Eastern' and 'Semitic'. However, the terms 'Middle Eastern' and 'Semitic' were quickly rejected for their geographical and conceptual imprecision.

In an illuminating discussion, Hodgson (1977, pp. 56–61) examined labels favoured in academic studies relating to Islam. He found that the label 'Islamic' is often used interchangeably to describe a religion, as well as a society and the culture associated with that religion, with the former usage now becoming increasingly preferred to the latter. At the height of Islamic civilization in the latter part of the first millennium AD, the two groups (i.e. Muslims by religion and Muslims by culture) were not co-extensive. Among the notable mathematicians and scientists of the day there were those who professed Christianity, Judaism, Zorastrianism and other faiths. Hodgson recommends the terms 'Islamdom' (following the analogy of 'Christendom') and 'Islamicate' (following the analogy of 'Italianate'), the former to identify a society in which Muslims, and their faith, were socially dominant, and the latter an adjective to describe a particular feature of the first. Thus, the mathematical pursuits of Islamdom become 'Islamicate' mathematics. This would appear to be an attractive option to adopt, despite the archaic nature of the first term and the novelty (and, therefore, general ignorance) of the second. The label adopted in the first edition to describe Islamicate mathematics was 'Arab' mathematics, which, as some critics pointed out, has in more recent times assumed an ethnic and nationalistic gloss. I had hoped that the footnote on p. 5 would serve as a warning against such ethnic privileging. Unfortunately, this was not borne out by some of the reactions to the first edition of this book. So the problem remains: the 'Arabs' then were seldom 'ethnic' Arabs, but included Iranians, Tadjiks and Moors. However, they all used the Arabic language, just as their European contemporaries used the Latin language. Therefore the use of the term 'Arab' as a descriptive term for Islamicate mathematics is retained in the new edition despite the misgivings.

The characterization of the mathematics of the Indian subcontinent raises other problems. Would it not be appropriate, following the argument in the previous paragraphs, to describe this mathematical tradition as 'Sanskrit' mathematics or 'Hindu' mathematics since most of the texts were written in some form of Sanskrit, and most people involved in the creation or dissemination of the subject were 'Hindus' according to the present usage of that term?

Until recently, the term 'Hindu' had never been one that the inhabitants of the Indian subcontinent ever applied to themselves. Historically, it derives from the Sindhu river, now more widely known as the Indus. The label 'Hindu', when used to describe a religious category, is of relatively recent origin. It arose from a number of factors:

1 the search for a religious identity to oppose Christian evangelism introduced during British rule;

2 the reaction to the British Imperial doctrine of 'divide and rule' which pitted the Hindus against the Muslims; and

3 the romantic perceptions of some regarding the spirituality and other-worldliness of the Hindu.

It became a political category only after the partition of the Indian subcontinent in 1947. And, as a social category, the term could well be used as an all-embracing description of the sects, life styles, agreements and disagreements found in the geographical boundaries between the Hindu Kush mountains in the north and Kanya Kumari (Cape Comorin) in the south. In that sense, the term encompasses everything and, at the same time, nothing — as do certain schools of 'Hindu' philosophy. And in the same sense, all inhabitants of India, sharing common life styles and values, may be labelled Hindus.

It is, therefore, tempting to use the term 'Hindu' mathematics to describe the mathematical tradition of the subcontinent. Eurocentric historians of mathematics often do so, for example in the continuing references to the 'Hindu–Arabic' numerals. It was decided not to use the label 'Hindu' in this book for reasons similar to those stated for the rejection of the label 'Islamic' mathematics, notably because of the strong religious and political dimension of its usage today. As is evident from Chapters 8 and 9, the mathematical traditions of the Indian subcontinent arose from different sources, including Buddhist, Jaina and, later, Muslim groups. While most seminal texts were written in Sanskrit, there were important works in local languages, notably those from the states of Kerala and Tamilnadu in South India, and Persian and Arabic in North India. For these reasons, we chose the label 'Indian' to describe the mathematical activities of this region. The other labels, such as 'Chinese', 'Egyptian', 'Babylonian' and 'Mayan', in the book are consistent with this usage and are uncontroversial.

The Hegemony of European Mathematics

A central theme of this book, seen by some as its principal strength and by others as either an irritating irrelevance or even a fatal weakness, is the critique of the widespread acceptance of the hegemony of a Western version of mathematics, following from the assumption that mathematics was largely a Greek and European creation. The two tactics used to propagate this myth are

1 *omission and appropriation* and,

2 *exclusion by definition*.

The first has been explored in sufficient detail in Chapter 1. The second needs further elaboration.

A Eurocentric approach to the history of mathematics is intimately connected with the dominant view of mathematics, both as a socio-historical practice and as an intellectual activity. Despite evidence to the contrary, a number of standard histories view mathematics as a deductive system, ideally proceeding from axiomatic foundations and revealing, by the 'necessary' unfolding of its pure abstract forms, the eternal/universal laws of the 'mind'.

The concept of mathematics outside the Graeco-European praxis was very different. The aim was not to build an imposing edifice on a few self-evident axioms, but to validate a result by any suitable method. Some of the most impressive work in Indian and Chinese mathematics, such as the summations of mathematical series or the use of Pascal's triangle in solving higher-order numerical equations, the derivations of infinite series for π, or 'proofs' of the Pythagorean theorem, involve visual demonstrations that are not formulated with reference to any formal deductive system. The Indian view of the nature of mathematical objects, like numbers, is based on a framework developed by Indian logicians (and linguists), and differs at the foundational level from the set-theory universe of modern mathematics.

The view that mathematics is a system of axiomatic/deductive truths inherited from the Greeks, and enthroned by Descartes, has traditionally been accompanied by the following cluster of values that reflect the social context in which it originated:

1 an *idealist* rejection of any practical, material(ist) basis for mathematics: hence the tendency to view mathematics as value-free and detached from social and political concerns;

2 an *elitist* perspective that sees mathematical work as the exclusive preserve of a high-minded and almost priestly caste, removed from mundane preoccupations and operating in a superior intellectual sphere.

Mathematical traditions outside Europe that did not accept this cluster of values have therefore been dismissed on the grounds that they are dictated by utilitarian concerns with little notion of rigour, especially those of proof. Any attempt at excavation and restoration of non-European mathematics is a multifaceted task: confront historical bias, question the social and political values shaping the mathematics, and search for different ways of 'knowing' or establishing mathematical truths in various traditions. Chapter 5 has a short discussion of 'proof' in the Egyptian and Babylonian traditions. I have written elsewhere (Joseph, 1994c, 1997b) on the same subject in relation to the Indian tradition. The references are found in the expanded bibliography at the end of this book.

Trajectories and Transmissions

There have been criticisms of the three trajectories of mathematical development (i.e. the classical, the modified and the alternative) outlined in Chapter 1. The main purpose of introducing these trajectories was to bring out differing *perspectives* on the origins of mathematics. No claims are made for the completeness or balance in marshalling the historical evidence underlying the discussion. Sicily, as a staging post in the spread of mathematical ideas from the East, was of less consequence than Spain, despite the appearance to the contrary in Figure 1.3. The role of Jund-i-shapur as a centre of learning and scholarship remains speculative compared with the role of Baghdad or Toledo. In the case of the alternative trajectory (see Figure 1.3), the role of centres in the Maghreb from the beginning of the twelfth century to the end of the fifteenth century AD, discussed in the works of Djebbar (1981, 1985, 1990, 1997), was ignored in the first edition. A brief treatment of the subject is found in the 'Reflections' on Chapter 10 (pp. 422–3). However, it should be remembered that the main thrust of the argument in the case of the alternative trajectory was how mathematics spread into Europe through the Arab intervention. The role of the Maghreb in this dissemination process is, as yet, unclear.

Since the first edition of this book, we are no more closer to gathering further definitive evidence of transmission of mathematical knowledge to Europe. A number of historians of mathematics now admit that there were some remarkable mathematical ideas emanating from China, India and the Islamic world, but many would argue that Europeans in later centuries were unaware of this work and carried out their exploration of new ideas independent of these earlier efforts. It is interesting, in this context, that a strong point of contention has been the somewhat tentative suggestion, in the first edition, of possible transmission of seminal ideas in modern mathematics from India, China and the Arab world.

With hindsight, the author regrets not having expressed these ideas of transmission more forcefully. The area of greatest promise of new discoveries relates to the transmission from the Arab world to Europe. Cumulating circumstantial evidence now strongly supports the thesis that significant ideas – notably in trigonometry, non-Euclidean geometry, number theory and combinatorics – were transmitted from Islamdom to Christendom, through Arabic and Hebrew texts, to contribute to the development of modern mathematics. The absence of a tradition of attribution during this period makes our task of tracing the transmission more difficult. However, there are some intriguing possibilities raised by Katz (1997).

In twelfth-century Muslim Spain, a book by Jabir ibn Aflah on spherical geometry was translated into Latin and Hebrew. A method of solving triangles on the surface of a sphere, discussed in Jabir's book, appeared in a book by Regiomontanus (1436–1476), an influential European mathematician of his time. The Italian algebraist Girolamo Cardano (1501–1576) noted the close similarity between the passages in the two books. The trigonometry contained in Jabir's book also makes an appearance in Copernicus's seminal text, *De revolutionibus*. In Chapter 10 (on Arab mathematics) other examples are quoted, notably the influence of Nasir al-Din al-Tusi's (1201–1274) geometry on the Italian Girolamo Saccheri's (1667–1733) attempt to prove the parallel postulate; and Thabit ibn Qurra's (826–901) formula for finding amicable numbers, which was proposed and used by the French mathematician Pierre de Fermat (1601–1665), who may have come across a Latin translation of an Arab or Hebrew text from the twelfth to the fourteenth century, containing ibn Qurra's formula.

Katz also discusses another example of a possible transmission, starting from a Hebrew text of Levi ben Gerson (1288–1344) containing the basic formulae for finding permutations and combinations that reappear in one of Cardano's manuscripts, and then an almost carbon copy in Marin Mersenne's (1588–1648) classic book on music theory. It is interesting, in this context, that almost fifteen hundred years before Mersenne's book, the basic formulae are found in Indian mathematics through the work of the Jaina school (*c.* 300 BC), with the link between combinatorics and music theory explored by Pingala around the same time. A discussion of the Indian work is found on pp. 253–5 in Chapter 8. There is also a brief discussion of mathematics from Hebrew sources in the 'Reflections' on Chapter 10 (pp. 419–22).

An area worthy of further exploration is the possible transmission westwards of the remarkable work in Kerala on infinite series, discussed in Chapter 9. The two connecting links need further study – one may have been through the Arabs. We know that the works of ibn al-Haytham (*fl.* AD 1000), the great Arab scientist, particularly on geometric series, were studied at certain *madrassah* (Muslim 'schools') in close proximity to centres of the Kerala school of mathematics and astronomy. So, was it possible that, through the medium of the Arabs, some of this work moved west?

However, a more important connection was the possible role of the Jesuits and the Portuguese: there is evidence that Matteo Ricci, the Jesuit astronomer and mathematician who is generally credited with bringing European sciences to the Chinese, spent almost two years in Cochin, South India, after being ordained in Goa in 1580. During that time he was in correspondence with the Rector of the Collegium Romano, the primary institution for the education of those who wished to become Jesuits. The Jesuits of that time were not merely priests, but also scholars who were very knowledgeable in science and mathematics. In fact, if you wanted to be trained as a mathematician in Italy at that time, you couldn't do better than go to a Jesuit seminary. For a number of Jesuits who followed Ricci, Cochin was a staging post on the way to China. Cochin was only 70 km from the largest repository of astronomical manuscripts in Trissur (Trichur) from where, two hundred years later, Whish (discussed on pp. 287–8) and Heyne, two of the earliest Europeans who reported on the work in Kerala, obtained their manuscripts. The Jesuits were expected to submit regularly a report to their headquarters in Rome, and it is a reasonable conjecture that some of the reports may have contained appendices of a technical nature which would then be passed on by Rome to those who understood them, including notable mathematicians. Material gathered by the Jesuits was scattered all over Europe: at Pisa, where Galileo, Cavalieri and Wallis spent time; at Padua, where James Gregory was engaged in mathematical studies; and at Paris, where Mersenne, through his correspondence with Galileo, Wallis, Fermat and Pascal, acted as an agent for the transmission of mathematical ideas. We need to examine these links closely, for at the moment there is only circumstantial evidence, but it is gratifying that research in this area has already started in India.

Ethnomathematics and the Inclusion Criteria

Since the publication of the first edition, there has been an increase in interest in ethnomathematics, or the study of mathematical concepts in their cultural context, often within socially cohesive and small-scale indigenous groups. Within the compass of mathematics as stated on p. 3, the emphasis is on how structures and systems of ideas involving number, pattern, logic and spatial configuration arose in different cultures. This view has had to contend with the strongly entrenched notion that mathematics, having originated in some primitive unformed state, advanced in a linear direction to the current state of modern mathematics and will continue to grow in that direction. A mathematical system that emerges in a culture removed from this 'mainstream' would then be perceived as a mere distraction of little relevance to the ideas and activities supported by modern mathematics.

A telling criticism of the first edition of *The Crest of the Peacock* is that it implicitly subscribed to this 'linear' view, being 'epistemologically based on the idea of direct literal translations of non-western mathematics to the western tradition' (Eglash, 1997, p. 79). Page 12 spells out the broad criteria used in determining the scope of the coverage of the first edition of this book. In the second edition, the coverage is extended to include areas in the Pacific and elaborate further on the mathematical activities in the African and American continents.

2 Mathematics from Bones, Strings and Standing Stones

Counting Systems and Numeration: The Pacific Dimension

The study of worldwide systems of numeration as they occur in natural languages has had a chequered career. A rich source of information on the 'exotic' languages and customs from the published literature of the explorers, administrators and missionaries form the core of the data now available on non-European counting systems. Initiated originally by those interested in linguistics and anthropology, the subject is now of only marginal interest to these groups. Yet a growth of interest on the part of the historian or philosopher of mathematics has not matched the waning interest on the part of the linguist or the ethnographer.

There are various reasons for this lack of interest. For the Western philosopher, the study of natural language number systems seems to have little or no relevance to the understanding of the nature of number, an abstract philosophical concept derived from ancient Greeks and independent of linguistics and cultural vagaries. Historians of mathematics tend to concentrate on the origins of written numbers because of their uneasiness about straying into territories where culture and language interpose. In any case, they prefer to work with written records, even if these records are mainly confined to those in European languages. The occurrence of ideas relating to numbers as existing among 'primitive' tribes, if considered at all, are dealt with in a dismissive fashion.

This lack of interest in societies outside the usual ambit of historians and philosophers has had some unfortunate consequences. Societies that lack a tradition of historical documentation constitute the majority of world languages. Hence, a great part of humanity's numeration and counting practices are ignored. Also, those societies whose numerical practices are considered in historical writing tend to fall into a small range of counting types, leading to an overwhelming bias in discussions on systems that concentrate on base 10. Allowance may sometimes be made for irregularities introduced by vestiges of base 12 (English and German) or base 20 (French and Dutch). An unfortunate consequence of this concentration is the susceptibility to the reductionism fallacy that sees the human response to the need to enumerate the world as unvarying across time and culture.

The renewed interest, since the 1960s, in documenting mathematical ideas from non-Western societies is reflected in recent histories of numbers, notably the works of Flegg (1983), Ifrah (1985) and Menninger (1969). However, much of the material on which these studies were based was drawn from nineteenth-century reports of the agents of European colonization. Regions, such as Melanesia, Polynesia, Micronesia and Australasia, that account for more than one-quarter of the existing world languages have been neglected. This situation has now changed dramatically with the monumental 21-volume study *Counting Systems of Papua New Guinea and Oceania* by Glendon Lean (1996). The study, which remains unpublished, is a valuable guide to natural language counting systems used by 900 out of 1200 linguistic groups in Papua New Guinea, Irian Jaya, Solomon Islands, Vanuatu, New Caledonia, and parts of Polynesia and Micronesia. The study relating to Papua New Guinea and Irian Jaya, the other half of the island of New Guinea, is especially useful since nowhere in the world is the diversity of cultures and languages so marked as on this island. This study will form the main basis of the discussion here.

To present a coherent framework for a discussion of the large database available from Lean's study, a system of classifying the number systems is required. The classification adopted on pp. 41–4 is found in the older literature and is built around the descriptive term 'base'. This classification is unsatisfactory for various reasons. It is particularly inadequate for examining 'mixed' systems of counting which may, for example, have present elements of base 2, base 5 and base 10 at the same time. Also, Lean's study indicates that groups within Papua New Guinea, sometimes in close proximity to each other, differ in significant ways in their counting practices according to the importance that they attribute to the enumeration of objects, as well as the type of objects counted, and the circumstances in which the counting takes place.

Laycock (1975) introduced a more satisfactory fourfold classification of number systems. The first consists of a 'body part' tally system where number representation is based on the body and entails counting body parts according to a conventionally defined order. Consider the Oksapmin, a group found in the West Sepik Province of Papua New Guinea, as an example. Saxe (1982) describes their counting system in the following terms: starting with the thumb of a hand, counting proceeds along the fingers of that hand, so that the little finger would be five; proceeding down that hand, so that seven is the forearm along the upper periphery of the body; to the face, with 12 being the nearest ear to the hand that has been counted; to the nearest eye (13); the nose (14); the other eye (15); the other ear (16); and down along the other side of the body to the little finger on the other hand (27). If necessary, one could repeat the counting process again to reach 54 on the second count. A specific number is referred to by its associated body part.

Such a system of number representation has been used to describe a range of diverse activities, whether counting a set of objects (e.g. number of pigs), measuring the length of an object (e.g. a bow) or establishing the location of a house in relation to other houses on a path. And the system alters to take account of changing socio-economic circumstances. Saxe describes the ingenious modifications in counting made by the Oksapmin with the introduction of the Australian 'shilling and pound' currency (1 pound = 20 shillings) where, rather than using all 27 parts of the body for enumeration, an individual stopped at the inner elbow of the other side of the hand (20) and called it one 'round' or one pound. When the count continued, the individual began a new 'round' starting with the thumb of the first hand. In 1975, when Papua New Guinea became independent, a new currency in the form of kina and toea (100 toea = 1 kina, 200 toea = 2-kina note) was introduced. This did not lead to an abandonment of the old system of counting money. Many older Oksapmin people resorted to a 'translation' of kina and toea into pounds and shillings, and then continued the count by referring to the 10-toea coin as one shilling and the 2-kina note as one pound.

A second type of number system, according to Laycock, is one that has two to four discrete number words and a matching 'base' to carry out the tally. For example, in the case of an indigenous Australian group, the Gumulgal, counting proceeds as *urapon* (1), *ukasar* (2), *ukasar-urapon* (3), *ukasar-ukasar* (4), *ukasar-ukasar-urapon* (5), indicating counting by twos (or a base-2 system).

A third type is the 'quinary-vigesimal' system that has a 5, 20 cyclic pattern and may employ fingers and toes as an aid to tallying. For example, in the Melanesian language Sesake, counting proceeds as *sekai* (1), *dua* (2), *dolu* (3), *pati* (4), *lima* (5), *la-tesa* (6), *la-dua* (7), . . . *dua-lima* (10), . . . *dua-lima dua* (20). The system has a cycle of 5 where numbers for six and seven use the roots for one and two respectively, and a super-ordinate cycle of 20 where 20 is two-fives twice.

The final type is a decimal system, normally with no reference to 'body parts' and having six to ten discrete number words. In the Micronesian language Kiribati, counting proceeds as: *tenuana* (1), *uoua* (2), *tenua* (3), *aua* (4), *nimua* (5), *onoua* (6), *itiua* (70), *wanua* (8), *ruainua* (9), *tebwina* (10), *tebwina-ma-tenuana* (11), . . . *vabui* (20), *vabui-ma-tenuana* (21), . . . *tebubua* (100), . . . *tenga* (1000), . . . *tebina-tenga* (10 000), . . . *tebubuna-tenga* (100 000). Here we have a straightforward base-10 counting system with no association to body parts. In Kiribati, number words also vary according to the object being counted. Thus the number word for nine is *ruaman* when counting animals, *ruakai* when counting plants, *ruai* when counting knives, *ruakora* when counting baskets, and *rauawa* when counting boats.

Numeration relating to time has not been a significant influence on the development of counting systems in the Pacific. Time is usually reckoned in units determined by lunar and seasonal cycles. Precision of reckoning is not very important and in only two activities would reckoning time be significant: agriculture and ceremonial events. The Mae Enga, a group in New Guinea, have a lunar 'calendar of events' consisting of 12 *kana* (or 'garden' months) with which they monitor agriculture and other activities. The name of each *kana* is indicative of the activity undertaken that month. A month is allocated for each activity, such as planting of specific crops, harvesting or preparing the 'garden' for planting, as well as undertaking trading trips or engaging in fighting. It is not clear whether the Mae Enga ever kept a record of the passage of time as measured by their calendar, such as the Ishango Bone from Central Africa.

The recording of ceremonial events does not require anything as elaborate as a calendar. Among the Kewa, the body count system is sufficient to track the occurrence of ceremonial dances. According to Johnson (1997, p. 659) if a cycle of dances were to begin in eleven months' time, this would be counted as *komane roba summa* (elbow). If the next occurrence of the dance were three months later, this would be counted as *pesame roba suma* (shoulder). The third occurrence, six or seven months later, would be counted as *rigame robasuna* (between the eyes). This sequence of dances would continue until the climax of the feast, when pigs were killed. This happened on the month represented by the division of the wrist and the finger. A number of these festivals required a strong numerical sense on the part of the participants since a person's prestige was measured by the quantity of pigs, shells or any other 'currency' given away as gifts.

The Ishango Bone: A Lunar Calendar of Events?

Although I would now underplay the mathematical significance of the Ishango Bone, it remains fascinating as to how carved notches on a piece of bone can represent so many possibilities. Many cultures have based their lives on the moon and its cycles. A lunar calendar treats time in segments of 28 days. After many moons even the best memory is unable to correlate an event and the moon within which it occurred. Subdividing a month by phases of the moon roughly corresponds to four 'weeks': new moon to first quarter, first quarter to full moon, full moon to last quarter, last quarter to new moon. Some means of defining longer periods of time was necessary before the year could be delineated. A flood-based year of Egypt correlated with the movement of the star Sirius was probably the earliest approach to the study of the solar year. The solar year became (and remains) the basic unit of time, but the discovery of the solar year came at least 10 000 years after the discovery of the lunar month.

The notches of the Ishango Bone open up other interesting conjectures. The epoch of the Ishango Bone was around the same period when women were supposedly the temporal and spiritual leaders of their clans. Since a woman's menstrual cycle mimics the phases of the moon, would it be too fanciful to argue that the markings on the bone represent an early calendar of events of a ceremonial or ritual nature superimposed on a record of the lunar/menstrual cycle constructed by a woman? After all, among the Siaui of the Solomon Islands in the Pacific, a menstruating female is described as 'going to the moon'.

There is another interpretation. The Ishango Bone may have been a precursor of writing. In that case, writing originated not in drawing figures or in attempting to record speech but in storing numerical information. The rows of notches became 'graphically isomorphic' to the recorder's counting numbers. In this and subsequent chapters, there are illustrations of counting systems with forms of recording where the iconic origin of the dash or stroke is the human digit (i.e. the finger or toe). These strokes are graphically isomorphic with the corresponding words used for counting. It makes little difference whether we 'read' the sign pictorially as standing for so many fingers held up, or in 'script' as standing for a certain numeral.

This line of argument could be extended to the proposition that the invention of writing, or the division of writing and drawing as separate functions, occurred when the graphic representation of number shifted from a 'token-iterative' system such as that which appeared on the Ishango Bone to 'type-slotting' in later systems of recording. A token-iterative sign-system may be of the form 'goat, goat, goat, goat, goat', or 'goat, another, another, another, another', whereas a type-slotting system requires a sub-language handling messages of the form 'goat, five'. A token-iterative produces a list as long as the number of individual items recorded, while a type-slot list provides no meaningful information from counting the number of marks it contains. The progression from recording five goats by means of one 'goat' sign followed by five strokes to recording the same information by means of one 'goat' followed by a second sign indicating the number 'five' is a progression that has already crossed not only the boundary between pictorial and script representation but to the beginnings of written numerals. We take up this theme in the 'Reflections' on Chapter 4 (pp. 381–3), when discussing early Sumerian notation.

Further Reflections on African Mathematics

Most books on the history of mathematics ignore Africa, especially areas south of the Sahara. After a cursory treatment of Egyptian mathematics, which in any case is often not considered African, Africa disappears from the reckoning. The publication of Claudia Zaslavsky's *Africa Counts: Number and Pattern in African Cultures* seems to have made little impression on the dominant Eurocentrism of the practitioners. Yet a recognition and evaluation of Africa's contribution to mathematics is important, and not only for reasons of restoring a historical balance to the subject. The consideration of African mathematics reminds us of what is often forgotten: mathematics is a pan-cultural phenomenon which manifests itself in a number of different ways, in counting and numeration systems, in games and leisure pursuits, in art and design, in record-keeping and metrology. Our very definition of mathematics has to be broadened to include activities such as counting, locating, measuring, designing, playing, explaining, classifying and sorting. Further, a search for the origins and nature of African mathematics takes us on an adventure of establishing interdisciplinary connections: between mathematics and cosmology; between mathematics and philosophy; between mathematics and technology; between mathematics and linguistics; and between mathematics and cross-cultural psychology. Gerdes (1995, pp. 148–50) has given a list of references to recent studies which attempt to establish these connections.

In Chapter 2 we considered, in a piecemeal fashion, the mathematics of the African continent. Apart from Egyptian mathematics, discussed in Chapter 3, there was a discussion of the Ishango Bone and the Yoruba system of numeration, and a mention in passing of the counting system of the Zulus. The geometry of African art and design, the mathematical 'ingredients' of games and puzzles, and the implicit mathematics of certain aspects of African astronomy were ignored. Some illustrations of the type of work done in this area may be illuminating. Ascher (1988) discusses the cultural background and mathematical properties of the continuous graphs drawn by the Booshong and Tshokwe who live in the area where the borders of Angola, Zaire and Zambia meet. In a general discussion of archaeoastronomy, Aveni (1981) draws parallels between cultures in the tropics which appear to have adopted a horizon-and-zenith approach to the sky, as opposed to the approach based on the celestial pole (now marked by Polaris) and the ecliptic/celestial equator, which is more familiar to those from the temperate climates. As a result, navigators tended to use stars on the horizon instead of compass directions. The use of the star compass seemed a characteristic of cultures as geographically far removed as the inhabitants of the Caroline Islands in the Pacific, the Maya of Central America, the Mursi of Ethiopia and the Bambara of the Sudan. Crowe (1971, 1975, 1982) discerns complex geometrical patterns in the designs on the smoking pipes of Begho (Ghana), on the textiles and woodcarvings of the Bakuba (Tanzania) and in Benin art. Eglash (1995) considers an example of fractals (scaling in street branching, recursive rectangular enclosures, circles of circular dwellings, etc.) in the layout of settlements of the Mokoulek in Cameroon. Gerdes (1986, 1988a, 1995) has written voluminously on the 'frozen' geometry found in the material artefacts of different groups from southern Africa, and his 1995 publication provides a good overview of the burgeoning literature in this area in recent years. Zaslavsky's classic text, originally written in 1973, with a new edition that came out in 1999, still remains a valuable source of reference on the geometry implicit in African art and design.

There are two other aspects of African mathematics that are often ignored.
1 The enforced diaspora of the Africans resulting from slavery was destructive of existing mathematical traditions, yet games such as mancala (a game based solely on strategy akin to mathematical reasoning, and found today with minor variations in many parts of Africa) were taken

in their earlier forms to the Caribbean and the American continents.

2 The skills in drawing and design, and a rich tradition of mental arithmetic, were also taken over in slave ships.

The African Diaspora Mathematics: The Case of Thomas Fuller

In discussions of people with extraordinary powers of mental calculation, there is sometimes a mention of Thomas Fuller, an African, shipped to America in 1724 as a slave at the age of 14. He was born somewhere between present-day Liberia and Benin. Late in his life, his remarkable powers of calculation made him an example for the abolitionists to demonstrate that blacks are not mentally inferior to whites. After his death, Fuller became a source of interest for psychics and psychologists, where the latter, even when denying blacks of mental abilities, supported the notion of the *idiot savant*. This was not borne out by those who met him. They remarked on Fuller's general self-taught intelligence and decried a system that prevented him from obtaining formal education.

On his death in 1790, an obituary contained the following passage:

> Though he could never read or write, he had perfectly acquired the art of enumeration . . . He could multiply seven into itself, that product by seven, and the products, so produced, by seven, for seven times. He could give the number of months, days, weeks, hours, minutes, and seconds in any period of time that any person chose to mention, allowing in his calculation for all leap years that happened in the time; he could give the number of poles, yards, feet, inches, and barley-corns in any distance, say the diameter of the earth's orbit; and in every calculation he would produce the true answer in less time than ninety-nine men out of a hundred would produce with their pens . . . He drew just conclusions from facts, surprisingly so, for his (limited) opportunities . . . Had his opportunity been equal to those of thousands of his fellow-men . . . even a Newton himself, need not have been ashamed to acknowledge him a Brother in Science. (*Columbian Centennial*, 29 December 1790)

Our new understanding of the ethnomathematics of his birthplace allows us to claim that when Thomas Fuller arrived in 1724 Virginia, he had already developed his calculation abilities based on his indigenous traditions. The existing evidence for this claim is not conclusive. However, Bardot's 1732 account of the numerical abilities of the inhabitants of Fida (on the coast of Benin) may be of relevance:

The Fidasians are so expert in keeping their accounts, that they easily reckon as exact, and as quick by memory, as we can do with pen and ink . . . [This] very much facilitates the trade the Europeans have with them.

Thomas Clarkson backed this up in 1788, writing:

It is astonishing with what facility the African brokers reckon up the exchange of European goods for slaves. One of these brokers has perhaps ten slaves to sell, and for each of these he demands ten different articles. He reduces them immediately by the head to bars, coppers, ounces . . . and immediately strikes the balance. The European, on the other hand, takes his pen, and with great deliberation, and with all the advantage of the arithmetic and letters, begins to estimate also. He is so unfortunate often, as to make a mistake; but he no sooner errs, than he is detected by this man of inferior capacity, whom he can neither deceive in the name or quality of his goods, nor in the balance of his account.

Gerdes and Fauvel (1990) wrote an interesting account of Thomas Fuller; and recent research in West Africa by Shirley (1988) shows the survival of a numerate tradition, placing a high value on mental calculations among a mainly illiterate population, from whose ancestors captive slaves such as Fuller were transported to the American continents.

Native Americans and their Mathematics

Introduction

It is difficult to estimate precisely how many native Americans were present in the two continents when the Europeans 'discovered' them in 1492. A recent estimate (Denevan, 1992, p. 244) puts the number at 54 million. The question as to where they came from has fascinated scholars since Columbus's time. It is now generally agreed that the distribution of their population in the Americas, together with their physical appearance, and the underlying unity in some aspects of their cultures across the Americas, implies a common Asian origin. An interesting lacuna in all native American cultures may help to establish a lower bound for the date of their migration from Asia to the Americas. In no culture in the pre-Columbian America is there any evidence of the use of the wheel for transport. A probable date for the first use of the wheel for this purpose in Central Eurasia is the beginning of the sixth millennium BC. Further, the early Native Americans and the Asians had certain technologies in common: these included the smelting of bronze, the casting of gold, silver and copper, as well as the arts of weaving and dyeing. These facts together would indicate that the migration was likely to have occurred not long before 12 000 BC.

There is yet another singular feature in the early history of the native Americans: of all the American cultures, the Maya of Central America seem to exhibit in their arts, folklore and myths a clearer historical memory of their Asian origins. Among them one finds sculptures of elephants, a species not found on the American continents. The Maya are perceived as being among the most 'advanced' of all Native American cultures in terms of their mathematics, astronomy and technology, and are one of the few that possessed a written language. This leads to the conjecture that the Mayan may have been the prototype of the culture that was transferred from Asia to America during the migration. Other American groups, as they roamed across the Americas, splintered away and forgot their origins, whereas the Maya remained closer to their roots.

In Chapter 2, we considered the mathematics of three Native American cultures, namely the Maya, the Aztec and the Inca, in varying detail. As far as we know, the Maya and the Aztec were the only Native Americans who had written mathematics, but they were not, by any means, the only cultures to have mathematics in their societies. It is often difficult to find evidence of mathematics in cultures where such knowledge is transmitted orally.

It would perhaps be useful to elaborate on two aspects of the Inca and the Mayan mathematical traditions not discussed in the chapter: how the Inca calculated, and the social context in which Mayan mathematics was produced and disseminated.

The Inca Abacus

We noted in the chapter that, apart from the scheme of knotted cords called the *quipu* that the Inca used to record numbers and keep accounts, they also had an abacus. The Inca abacus may have been used for computations before recording numerical information on the *quipu*. The fact that they did so with little error is supported by the observations of the Spanish Jesuit, José de Acosta, quoted on p. 40. Acosta pointed to similarities between the customs of the Inca and those of the peoples of Siberia, but made the incorrect inference that the Inca came from Asia about two thousand years before the Spanish Conquest of Mexico.

However, exactly how the Inca abacus was used in computations still remains unclear. The discussion of Wassen's interpretation on p. 40 remains problematic. I am particularly grateful to Ramón Glez-Regueral of Madrid for an interesting interpretation of the passage quoted earlier from Acosta, as well as the drawing of the abacus in Figure 2.2(b) on p. 33. In the drawing (if we count only the black dots in each cell of the drawing), the numbers represented may be arranged as a Fibonacci sequence of the form 0, 1, 1, 2, 3, 5, 8, 13, 21, . . . Let me quote from a personal communication from Ramón Glez-Regueral (1980):

> Don't you see here a trace of Fibonacci before Fibonacci? It has been mathematically proven that Fibonacci, Lucas and Golden Mean number systems are possible. All of them use a binary alphabet (0, 1) . . . The Inca minds had not yet been 'contaminated' by any preconceived algorithmic ideas. Just as you observed (in *The Crest of the Peacock*) the mental algorithms developed in the most varied manner by ancient civilizations, and then adopted generation after generation. Maybe the Inca brain jingled some little Fibonacci bells . . . Maybe the Fibonacci approach could help in solving the Inca abacus riddle . . .
>
> The decimal system found in *quipus* is just a way of writing numbers before writing words. But sometimes humans use a system for arithmetic and a different system for putting their results on paper, papyrus or cords.

There are other intriguing aspects of Inca arithmetic that merit further investigation. Did the Inca use a base 4 or base 5 system of numeration? How would one explain the absence of fractions in Inca numeration? Does their likely use of remainders instead indicate that they had one of the earliest examples of a modular arithmetic system?

The Social Context of Mayan Mathematics

Evidence relating to pre-Columbian Mayan civilization comes from three main sources:
1 four screen-fold books called codices;
2 a large number of stone monuments; and
3 thousands of ceramic vessels.

The best account of the Mayan culture around the time of the Spanish conquest comes from a Franciscan priest, Diego de Landa, who recorded the history and traditions of the Mayan people around 1566. Piecing together these different strands of evidence, Closs (1995), and others, have constructed an account of the social context in which mathematical sciences flourished during the period between the fourth and tenth centuries AD. This note is based on their findings.

At the top of the educational pyramid was a hereditary leader who was both a high priest (*Ahau-Can*) and a Mayan noble. Under him were the master scribes, priests as well as teachers and writers ('engaged in teaching their sciences as well as in writing books about them'). Mathematics was recognized as such an important discipline that depictions of scribes who were adept at it appear in the iconography of Mayan artists. Their mathematical identity was signified by the manner in which they were depicted: either with the Mayan bar-and-dot numerals coming out of their mouths, or with a number scrolls carried under their armpit. Closs (1995, pp. 8–9) discusses a Mayan vase that portrays two seated scribes with opened codices bound in jaguar skin. One of the scribes has, under his armpit, a scroll containing numerical data. The location of the scroll with numerical data under the armpit would indicate that it was a status symbol. It supports Landau's observation that not all the scribes understood the *katuns* (calendrical computation), and those who did enjoyed special prestige.

There is an interesting illustration on another Mayan vase of about the same period: at the centre of the illustration is seated a supernatural figure with the ears and hooves of a deer, attended by a number of human figures, including a kneeling scribe mathematician from whose armpit emanates a scroll containing the sequence of numbers 13, 1, 3, 3, 4, 5, 6, 7, 8 and 9. At the top right-hand corner of this illustration there is the small figure of a scribe who looks female, with a number scroll under her armpit indicating that she is a mathematician and, possibly, the one who painted the scene and wrote the text on the vase. She is described as *Ah T'sib* ('the scribe'). Preceding this text is a glyph that has not been deciphered but which could be her name. Once the name is deciphered, and if the scribe is female, we may have the name of one of the earliest known female mathematicians. The existence of female mathematician-scribes among the Maya is further supported by another depiction found on a ceramic vase examined by Closs (1992). The text on this vessel contains the statement of the parentage of the scribe in question: 'Lady Scribe Sky, Lady Jaguar Lord, the Scribe'. Closs (1995, p. 10) adds: 'Not only does she carry the scribal title at the end of her name phrase but she incorporates it into one of her proper names, an indication of the importance she herself places on that reality.'

3   The Beginnings of Written Mathematics: Egypt

The Scribe in Egyptian Society

Existing mathematical texts come mainly from two different periods in Egyptian history:

1 the first half of the second millennium BC; and
2 the period of Greek and Roman domination ending around the middle of the first millennium AD.

There were two main sources: table texts and problem texts. An example of the table texts is the $2/n$ table discussed on pp. 68–9. A problem text dating from around the middle of the second millennium reads: 'A pyramid. Its side is 140 [cubits] and its slope is five palms one finger [per cubit]. What is its height?' Note that a cubit is divided into 7 palms and each palm into 4 fingers.

These texts were the product of a professional group of scribes, with their clearly defined hierarchy. An interesting glimpse into professional rivalry is shown in a text in which one scribe taunts another: 'You come here and [try] to impress me with your official status as "the scribe and commander of a work gang". Your arrogance and boastful behaviour will be shown up by the following problem: "A ramp, 730 cubits [long] and 55 cubits wide, must be built, with 120 compartments filled with reeds and beams. It should be at a height of 60 cubits at its peak, 30 cubits in the middle, a slope of 15 cubits with a base of 5 cubits. The quantity of bricks required can be obtained from the troop commander." The scribes are all assembled but no one knows how to solve the problem. They put their faith in you and say: "You are a clever scribe, my friend! Solve [the problem] quickly for your name is famous . . ." Let it not be said: "There is something he does not know." Give us the quantity of bricks required. Behold, its measurements are before you; each of its compartments is 30 cubits [long] and 7 cubits [wide].' It is not known whether the 'arrogant' scribe solved the problem.

It is clear in this instance that the task set for the scribe was a problem in practical mathematics. A number of other problems had little connection with real life. The teacher scribes were simply showing their student scribes how to apply certain procedures correctly. In general, the scribes were not 'mathematicians', as we understand the term today. A scribe was either an instructor or an accountant. If he was an instructor he was expected to teach 'advanced' calculations to his students. If he was an accountant he had to work out labour requirements, food rations, land allocation, grain distributions and similar matters for his employers, who were either government officials or wealthy private individuals. It is usually an accountant scribe who is depicted on frescoes walking a few paces behind his master.

Egyptian Numeration

It is normally a feature of courses in the history of mathematics for the Egyptian system of numeration to be introduced to show how time-consuming and cumbersome it was compared with the number system used today. There is also a tendency to compare the Egyptian system unfavourably with the Sumerian system (discussed in Chapter 4) because of the absence of a so-called place-value notation. What is often ignored is the role of the Egyptians in introducing the idea of cipherization (briefly discussed on p. 47), 'a decisive contribution to the development of numeration . . . and in every way comparable in significance to that of the Babylonians in adopting the positional principle' (Boyer, 1944, p. 157).

The introduction of cipherization essentially replaced tally marks with abstract symbols for numerals. Tallies were tedious because they had to be repeated until the count reached the desired value. The hieroglyphic numerals introduced abstract symbols for powers of 10, but these symbols had to be repeated and counted. However, as early as the Fourth Dynasty, the hieratic numerals came into use (Kemp, 1991, p. 83).

Hieroglyphic numerals use different symbols for each power of 10 up to 10 million. These symbols are grouped and repeated as needed to represent a magnitude. On the other hand, the hieratic system, by using simple symbols or ciphers for 'each of the first nine integral multiples of the integral powers of ten', produced 'decimal cash-register cipherization', like an old-style cash register that sends up a flag for each decimal place (Boyer, 1944, p. 158). The gain in economy and speed was impressive: a representation of 57, say, requiring 12 hieroglyphs, needed only two hieratic symbols (see pp. 62–3).

It is interesting, in the context of the debate about the influence of Egyptian mathematics on Greek mathematics, to note that the earlier Attic numerals, adopted in Greek cities around 500 BC, used the same repetitive principle as the Egyptian hieroglyphic numerals (Flegg, 1989, p. 88). But when the Attic numerals gave way to the Ionian numerals, the same cipherization principle of the Egyptian hieratic numerals came into operation. Boyer (1944, pp. 166–7) has argued that the Ionian numerals were based on Egyptian hieratic numerals which, in turn, may have had an influence on the development of the Hindu-Arabic (or Indo-Arabic) numerals.

The absence of zero is a shortcoming of Egyptian numeration that is often referred to in standard histories of mathematics. It is clear that an absence of zero as a place-holder is perfectly consistent with a number system such as the 'Egyptian cash-register decimal system'. However, in two other senses it may be argued, as Beatrice Lumpkin (2000) has done, that the concept of zero was present in Egyptian mathematics. First, there is zero as a number. Scharff (1922, pp. 58–9) presents a monthly balance sheet of the accounts of a travelling royal party, dating back to around 1770 BC, which shows the expenditure and the income allocated for each type of good in a separate column. The balance of zero, recorded in the case of four goods, is shown by the *nfr* symbol that corresponds to the Egyptian word for 'good', 'complete' or 'beautiful'. It is interesting, in this context, that the concept of zero has a positive association in other cultures as well, such as in India (*sunya*) and among the Maya (the shell symbol).

The same symbol appears as a zero point in a series of drawings for some Old Kingdom constructions. For example, in the construction of Meidun Pyramid, it appears as a ground reference point for integral values of cubits given as 'above zero' (going up) and 'below zero' (going down). There are other examples of these number lines at pyramid sites, known and referred to by Egyptologists early in the century, including Borchardt, Petrie and Reiner, but not mentioned by historians of mathematics, not even Gillings (1972), who played such an important role in opening up the treasures of Egyptian mathematics to a wider public. About 1500 years after Ahmes, in a deed from Edfu, there is a use of the 'zero concept as a replacement to a magnitude in geometry', according to Boyer (1968, p. 18). Perhaps there are other examples waiting to be found in Egypt.

Egyptian Fractions

Nowadays, we would write a non-integer number either as a fraction (2/7) or a decimal (0·285714). From very early times, the ancient Egyptians (as far as we can tell from surviving documents) wrote it as a sum of unit fractions. Thus, while a number like 1/7 in hieroglyphic notation consisted of seven vertical lines crowned by a halo, the number 2/7 was written, as a sum of unit fractions, as $2/7 = 1/4 + 1/28$. Note that the same unit fraction was not used twice in one representation (i.e. $2/7 = 1/7 + 1/7$ was not allowed). This system of representing fractions became known as the system of Egyptian fractions.

There have been different views as to why the Egyptians followed this style of fractional representation. A traditional and widely accepted view, even today, is that the representation reflected the notational or conceptual limitation of their number system. To many of us, brought up on fixed-base representations of numbers, it is difficult to imagine a situation that would favour the Egyptian usage. One could, of course, contrast the 'exactness' of the Egyptian representation with the 'approximate' nature of fixed-base expansions of the neighbouring Sumerians, whose system is discussed in Chapter 4. We could marvel at the way that Egyptian scribes performed completed arithmetical operations. They would take $16 + 1/56 + 1/679 + 1/776$, find $2/3$ of it as $10 + 2/3 + 1/84 + 1/1358 + 1/4074 + 1/1164$, then add $1/2$ of it and $1/7$ of it and show that it all adds up to 37!

The Egyptian preference for 'exactness' has some interesting historical parallels with the preference of their students, the Greeks, for geometry over symbolic arithmetic. The Egyptian style may have contributed to the Pythagorean number mysticism and the number theory that grew out of it. For example, the notion of a 'perfect number', which is equal to the sum of its proper divisors, would now seem to many as esoteric. But perfect numbers have certain practical uses in working with Egyptian fractions. And the subject provides, even today, a source of mathematical puzzles and problems in abstract number theory.

There are two separate aspects of Egyptian fractions that are worth exploring. One is the choice of variable 'base' in determining the denominators that constitute the expansion of the original fraction. For example, while the Babylonians, using their fixed base 60, expressed the 'approximate' expansion of the fraction $1/7$ as

$$1/7 \approx 8/60 + 34/60^2 + 17/60^3$$

the Egyptians required the exact expansion

$$1/7 = 1/14 + 1/21 + 1/42$$

The second aspect relates to the Egyptian preoccupation with fractions of unit numerators. Various explanations have been offered for this preoccupation. It may have been derived, as mentioned on p. 64, from a binary approach to integer arithmetic so that a number N is expressed in the form

$$N = c_0 2^0 + c_1 2^1 + c_2 2^2 + \dots, \text{ where the coefficients } c_i \text{ are either 0 or 1.}$$

Extending this representation to a fraction p/q we get

$$p/q = c_1/A + c_2/B^2 + c_3/C^3 + \dots, \text{ where the coefficients } c_i \text{ are either 0 or 1.}$$

Now putting $A = B = C = \ldots = 2$ would not allow exact expansions. So the answer lies in using variable denominators A, B, C . . . .

The puzzle still remains: what practical purpose was served by the Egyptian unit fractions? One reason for expanding rational fractions is to facilitate the comparison of different quantities. For example, if you had to choose between being paid $1/7$ of a bushel of corn and $13/89$ of a bushel, which should you take? In terms of the Babylonian arithmetic the two quantities could be expressed in base 60 as

$$1/7 \approx 8/60 + 34/60^2 + 17/60^3$$
$$13/89 \approx 8/60 + 45/60^2 + 50/60^3$$

It is clear from inspection that while the first terms on the right-hand side of both approximations are identical, the second term of $13/89$ is larger than the second term of $1/7$.

An estimate of the relative magnitude of the two fractions using the Egyptian approach is cumbersome:

$$1/7 = 1/14 + 1/21 + 1/42$$
$$13/89 = 1/15 + 1/30 + 1/60 + 1/178 + 1/267 + 1/356 + 1/445$$
$$+ 1/534 + 1/890 + 2/267 + 2/445$$

Note that the last two terms in the expansion for $13/89$ could be converted to unit fractions if we had a $2/n$ table extending up to $n = 445$. (The $2/n$ table given in the Ahmes Papyrus and shown in Table 3.1 of Chapter 3 provides only odd values of n up to 101.) At some point, the two quantities have to be reduced to expansions which have a common denominator for a comparison to be made.

Conversion to Unit Fractions

A number of well-known mathematicians over the years have looked at this problem, and invented different ways of doing this conversion process. Each of these methods has advantages and disadvantages in terms of the complexity of the Egyptian fraction representations it produces and the amount of time the conversion process takes. Only the Gillings procedure was discussed on p. 69. There are still some unresolved problems about whether some of the processes generated by the procedure finish or whether they get into an infinite loop. An interesting approach is that of Milo Gardner, which has the advantage of placing the discussion in a historical context. The discussion that follows owes a great debt to him and his work.

One of the earliest unsolved number theory problems is one from ancient Egypt, dating to about 4000 years ago. Prior to 2000 BC, Egyptian fractions followed a binary structure, discussed earlier, using a notation known as Horus-eye. Horus was the son of Osiris, who was killed by his brother Seth. In revenge, Horus fought Seth and killed him, losing an eye in the fight. The broken eye parts were later fixed by the god Djehuti (whose name the Greeks mispronounced as Thoth). Around the time of the appearance of the Ahmes Papyrus (*c.* 1850 BC) there was a significant shift in representation with the replacement of the Horus-eye hieroglyphic fractions with a hieratic composite form:

$$\frac{2}{pq} = \frac{2}{(p+1)} \left(\frac{1}{q} + \frac{1}{pq} \right), \text{ where } p \text{ and } q \text{ are prime, with } p > q$$

Much of the recent academic debate on Egyptian fractional representation has tended to concentrate on three departures from the above formula in the case of the $2/n$ table contained in the Ahmes Papyrus, namely the particular cases of $2/35$, $2/91$ and $2/95$. Gardner suggests that in the case of $2/35$ and $2/91$, a well-known pattern formed from the product of the arithmetic mean and the harmonic mean of p and q may be detected, i.e. if $A = \frac{1}{2}(p + q)$ is the arithmetic mean and $H = 2pq/(p + q)$ is the harmonic mean, then the decomposition is given by

$$\frac{2}{AH} = \frac{2}{pq} = \left(\frac{1}{p} + \frac{1}{q} \right) \frac{2}{(p+q)}$$

Substituting the values $p = 7$ and $q = 5$ for $2/35$, and $p = 13$ and $q = 7$ for $2/91$, will give the same decompositions for these fractions as in the $2/n$ table.

There still remains the exception $2/95$. Gardner suggests the following algorithm in this case:

$$\frac{2}{pq} = \frac{1}{q} = \left(\frac{1}{a} + \frac{2a - p}{ap} \right)$$

where a is a highly divisible number around two-thirds of the value of p.

If $p = 19$, $q = 5$ and $a = 12$, chosen such that the two sets of divisors $(2, 3)$ and $(4, 1)$ of a sum to $5 \le 2a - p$, then

$$2/19 = 1/12 + (3 + 2)/(12 \times 19)$$
$$= 1/12 + 1/76 + 1/114$$

So

$$2/95 = (1/5)[1/12 + 1/76 + 1/114] = 1/60 + 1/380 + 1/570$$

It is an interesting feature of the $2/n$ table in the Ahmes Papyrus that, for all n that are prime numbers, the above rule for the decomposition of $2/19$ is adhered to without exception. For example, the decompositions of 2/17 and 2/47, given in Table 3.1, are

$$2/17 = 1/12 + (4 + 3)/(12 \times 17) = 1/12 + 1/51 + 1/68$$
$$2/47 = 1/30 + (10 + 3)/(30 \times 47) = 1/30 + 1/141 + 1/470$$

An interesting question: was the famous sieve of Eratosthenes (*c.* 250 BC), a method of finding prime numbers, anticipated 1500 years earlier?

For other values of *n*, the Egyptians, over a long period, developed other patterns to obtain exact, short and small denominator unit fraction series. Two thousand years later, in the Akhmin Papyrus (*c.* AD 400), and in even more recent Greek and Arab work, interest in unit fractions persisted with increasingly sophisticated criteria applied for the selection of the table entries. As late as AD 1202, Fibonacci's *Liber abaci*, which introduced the Indo-Arabic numerals to Europe, contains tables for conversion from common fractions to unit fractions. However, Fibonacci used a different method (Diophantine equations) to the one employed by the Egyptians.

The whole episode of Egyptian fractional representation is instructive in a number of ways. In most discussions of Egyptian mathematics, after a condescending rendering of Egyptian numeration, the treatment of fractions is seen as the product of a peculiar mentality or of a declining mathematical culture waiting to be rescued by a superior one (see e.g. Neugebauer, 1962). Nevertheless, the fact remains that, despite attempts by number theorists of the eminence of J. J. Sylvester and Paul Erdös, they have been unable to improve upon (or even duplicate) the optimal decompositions of the 2/*p* and 2/*pq* series found in the Ahmes Papyrus and the Egyptian Mathematic Leather Roll. It would seem that our view of the superiority of algorithmic numbers has clouded our historical imagination. It is time that we ceased to look at the 2/*n* tables as merely an aid to calculation, and saw it as one of the earliest and most remarkable examples of number theory in action. And it is also time to reconsider the comparison between the Babylonian and Egyptian methods of numeration, usually to the advantage of the former, since this assessment is a reflection of our ease with an algorithmic view of fractions (with its approximations and 'rounding up') whose logical ancestor was the Babylonian system. As Milo Gardner states in a personal communication:

> The history of the Egyptian fraction system was closely associated with inheritances of live animals and other units that could not be subdivided. The problem of partitioning herds of animals is an example where endless calculating loops associated with the decimal system 'round off' had to be

avoided. Do you recall Ian Stewart's 'Camel Problem', published by *Scientific American*? The problem was passed down from Arabic sources, but clearly [originated in] ancient Egypt. Also recalling that Horus-eye binary fractions were limited by simple looping, as our binary computers suffered for many years, ancient scribes did look for a methodology that found full units (without using algorithms, equations and roots) . . . Seen in this way, and in others (like the use of prime numbers, 'red auxiliary numbers', lowest common multiples), the Egyptians did provide a strong foundation for number theory to flower amidst the Greeks.

The Relationship of Egyptian to Greek Mathematics

Over several years, both on the Internet and in academic journals, there has been a lively debate on the historical relationship of Greek to Egyptian science in general, and mathematics in particular. The start of this debate in recent years may be traced to Martin Bernal's (1987, 1992, 1994) claim that classical Greek culture was significantly influenced by ancient Egyptian civilization, partly through Egyptian colonization of parts of Greece. And this influence was also reflected in the debt that Greek science owed to its Egyptian counterpart. A debate that began between Palter (1993) and Bernal (1994) has spilt over into journals and popular magazines. An interesting intervention by Beatrice Lumpkin (2000) was not published at the time of writing; I am grateful to her for giving me the opportunity to read her contribution, which has influenced the discussion that follows.

The arguments relates to two claims made by Bernal, and rejected by Palter:

1 there were scientific elements in Egyptian medicine, mathematics and astronomy long before there was any Greek science at all; and

2 Egyptian medicine, mathematics and astronomy significantly influenced the development of corresponding Greek disciplines.

In relation to mathematics, we know from the evidence presented in Chapter 3 that the Egyptians certainly knew how to solve various kinds of problems, from solving linear equations to calculating the area of – let us assume – the surface of a hemisphere. What we are uncertain about is *how* the Egyptians discovered the methods they used in solving the problems. Presumably, at the very least, there was some 'scientific' underpinning their methods, although not necessarily based on reasoning from explicit axioms.

On the other question, of whether Egyptian mathematics influenced Greek mathematics, it may be possible to give a more definite 'yes'. We have already established a similarity in mathematical thinking between the Egyptians and the Greeks in number theory. After all, many of the ancient Greek sources acknowledge this influence. Not only is Pythagoras supposed to have studied in Egypt, but so are Thales, the supposed father of Greek geometry, and Oenopides. Herodotus, Heron of Alexandria, Diodorus Siculus, Strabo, Socrates (through Plato), and Aristotle all say that geometry was first invented by the Egyptians and then passed on to the Greeks. The question always seems to be, in this regard, what we mean by 'geometry'. If by 'geometry' we mean an axiomatic treatment with theorems and proofs in the style of Euclid, then it is clear that this was a Greek invention. But mathematicians have always known that, in general, one does not discover theorems by the axiomatic method. One discovers theorems by experiment, by trial and error, by induction, etc. Only after the discovery is there a search for a rigorous 'proof'. Srinivas Ramanujan's mathematics, discussed in the Preface to the First Edition (pp. xii–xiv), is a good example of the discovery and the 'proof' being undertaken by different persons and at different times. And it would seem clear that when the Greeks declared that the Egyptians invented (or discovered) geometry, it is the results that they had in mind and not necessarily the method of proof. Conceding that the Greeks learnt various geometrical results from the Egyptians takes nothing away from Greek creativity. They were simply doing what mathematicians have always done: building on the results of their predecessors.

4   The Beginnings of Written Mathematics: Babylonia

Introduction: Fleshing out the History

Along the Fertile Crescent between the Tigris and Euphrates rivers emerged the first cities occupied by the people who had originally migrated from the region of present-day Armenia, near the Black and Caspian Seas. By 3500 BC, the population pressure was such that the naturally irrigated flood plains could no longer sustain the basic needs of the inhabitants, especially since, unlike the case of Egypt, the flooding occurred somewhat erratically. The rivers were not navigable, making the city-states culturally and economically isolated from one another. The whole region was wide open and flat, lacking in natural defences, making it vulnerable to external invasions. Thus the physical environment of Mesopotamia influenced both the economy and the habitat of its inhabitants.

The first city-states, such as Uruk, Lagash, Ur and Eredu, developed in Sumer, the most fertile region of Mesopotamia. They competed vigorously with one another for land, resorting to war at the slightest provocation. As a result, for the first time there emerged empires, unions of city-states achieved through coercion or persuasion, or both, often with a single city dominating the others. The history of Mesopotamia after 3000 BC is the history of one empire succeeding another, with each developing its own 'style' of dominance and survival.

The Akkadian empire (*c.* 2375–2225 BC) was the creation of Sargon (2371–2316 BC), who, initially taking advantage of internal dissension in Sumer, conquered most of the Mesopotamian river valley during his reign of fifty-six years. To retain his hold, Sargon arranged the marriage of his daughter, Enheduanna, the High Priestess of the Akkadian religion, to the former king of Sumer (the High Priest of the Sumerian religion). It was hoped that if two peoples worshipped the same gods they were unlikely to go to war with one another. Enheduanna set herself the task of writing a text containing the liturgy and rituals from both religions. This text became the earliest known writing by a woman anywhere in the world.

About 2000 BC, the Akkadian empire was overrun by the Amorites (or Babylonians) who swept down into Mesopotamia from the northern land of Nimrod (present-day upper Iraq). Hammurabi built the city of Babylon as his capital. He arranged for a written legal code, which was among the first in history. His approach to safeguarding his power was two-pronged:

1 increase the prestige of the king; and
2 promote the use of organized law.

The royal prestige was safeguarded by centralizing the seat of power: locating the palace and temple within the same group of buildings, so enabling the ruler to perform the same set of rituals to underpin temporal and secular power. An audience with the king was no different from an audience with a god in a temple. To create an organized system of law, Hammurabi employed legal experts to collect, codify and disseminate laws across his kingdom. His new Code of Law was carved onto pillars, situated in every city of his empire. All citizens could call upon the protection of Hammurabi's justice; the unity and stability of the empire would thereby be assured by popular support. Hammurabi's dynasty was a high point in Mesopotamian history and a period of flowering of mathematical and astronomical achievements. Indeed, as indicated in Chapter 4, much of the evidence on accomplishment within both disciplines belongs to this period. The Old Babylonian empire lasted about 700 years, finally broken down by internal disorder and weakness of later rulers.

The Assyrian empire (c. 900–600 BC) that followed may be described as 'Nazis of the Ancient World'. The rulers held their empire together through terror, proudly displaying the severed heads and flayed skin of conquered enemies. Like many tyrannies, they were adept in introducing new technology of war, such as iron weapons and horse-drawn chariots. Their power was based on fear, terror and superior military skill. But once the conquered people got over their fears and gained the new technology, they rebelled. The Assyrian capital, Nineveh, was finally destroyed in 612 BC. The Assyrian epoch is marked by relative stagnation in practically all scientific activities, with the possible exception of astronomy.

The Chaldean empire (c. 600–550 BC) tried the 'restoration' approach. Nebuchadnezzar (605–562 BC) conquered a little, but built a little and spent a lot. He conquered Israel and brought a substantial section of its population to Babylon as slaves. He developed Babylon, which had fallen into bad times under the Assyrians, including restoring the famous Hanging Gardens. The New Babylonian empire came into existence and there was a revival of what we would now call algebra. Nebuchadnezzar's approach was, on the whole, successful until the arrival of the Persians.

In his campaigns between 550 and 530 BC, Cyrus united the Persians and the Medes, then moved his show on the road to Mesopotamia. The Persian 'style' was tolerance and benevolence to all. They respected local customs and traditions, thereby gaining many supporters. The Jews were allowed to go home; Cyrus even gave them money to rebuild their smashed temple. Many cities just opened their gates to Cyrus and asked to be part of his empire. The Persians gave the region uniform coinage, shared technology, trade and roads, and asked only for allegiance and taxes in return. There was considerable work on astronomy during this period. Cyrus and his successors ruled in peace for over 200 years, until they were conquered by the Macedonian Alexander, and in 312 BC the Seleucid dynasty was established. But this is another story which we will not pursue further, except to add that work on astronomy and algebra continued apace to merge into the swelling stream of Hellenistic mathematics and astronomy.

The Material Basis of Sumerian Mathematical Culture

The empires that grew out of the city-states of Sumer required a large bureaucracy to carry out their wishes. From the middle of the third millennium this bureaucracy began recruiting scribes. A scribe was a member of a specialized profession, trained in special schools where, increasingly, the curriculum was dominated by applied mathematics. Even when empires collapsed, the two lasting legacies that the scribes had helped to create over the centuries remained: a method of systematic accounting and the introduction of a place-value number system. Both innovations were a result of the ability of scribal schools to respond effectively to the increasingly sophisticated demands made by the administrative apparatus engaged in collecting taxes, conducting land surveys, supervising large-scale building programmes and ensuring the supply of young men for wars.

The centralized Sumerian states collapsed, in part at least, because of the weight of its top-heavy bureaucracy. This led to the emergence of Hammurabi and the Old Babylonian period. A new economic, social and ideological order asserted itself. Instead of large-scale agriculture or craft workshops, often owned by the ruler, the emphasis changed to small-scale enterprises owned by private individuals. This decentralization spread into many spheres, including the occupations of scribes. Scribes were no longer small cogs in the large wheel of government. They could be found writing letters for private individuals or tutoring children in private homes. The school for scribes probably continued with the old curriculum for a long time, teaching the student scribes accounting, surveying and other administratively useful topics. The high-status jobs were still to be found in the state bureaucracy; however, one begins to discern from the clay tablets of the period that a new mathematics was developing: a mathematics that was no longer purely utilitarian. This was a period of great interest in second- and higher-degree equations (discussed on pp. 108–14). The Plimpton Tablet (pp. 115–18) belongs to this period. Even the scribal schools were caught up in this wave of new thinking. There are signs of mathematics developing as a separate discipline, loosened from the coat-tails of narrow utilitarian pre-occupations.

This period of Mesopotamian history, however, came to an end. For about a thousand years, 'pure' mathematics took a back seat, to be re-established during the Chaldean period, when there was once again a resurgence of mathematics. But that is another story. However, what this note seeks to establish in a limited fashion is that mathematical development, howsoever defined, was shaped by institutions such as scribal schools, which in turn were products of the material and social forces driving the society.

The Origins of Mesopotamian Numeration

From about 8000 BC, a system of recording involving small clay tokens was prevalent in the Near and Middle East. Tokens were small geometric objects, usually in the shape of cylinders, cones and spheres. They were first identified in societies evolved from a life based on hunting and gathering to one based on agriculture, like the Ishango in Central Africa. The earliest tokens were simple in design: they stood for basic agricultural commodities such as grain and cattle. A specific shape of token always represented a specific quantity of a particular item. For example, 'the cone . . . stood for a small measure of grain, the sphere represented a large measure of grain, the ovoid (a roughly egg-shaped solid with one end being more pointed) stood for a jar of oil' (Schmandt-Besserat, 1992, p. 161). Two jars of oil would be represented by two ovoids, three jars by three ovoids, and so on. Thus the tokens not only became an abstraction for the things being counted, but also constituted a system of great specificity and precision.

With the development of city-states and the emergence of empires came a more complex economic and social structure, reflected in both the diversity and the standardization of tokens. This increased the scope for record-keeping and commercial contracts in a way that counting using pebbles or twigs could not do. A collection of tokens could represent a future promised transaction or, stored in a temple or palace, a record of a past transaction. Both contracts and archives required secure methods of preserving groups of tokens. The Sumerians devised two main systems of storage: stringing the tokens on a piece of cord and attaching the ends of the strings to a solid lump of clay marked with a security seal called a *bulla*; or storing the tokens inside a clay envelope bearing impressions of the enclosed tokens for identification purposes. 'For reasons we do not know, plain tokens were most often secured by envelopes and complex tokens by *bullae*' (Schmandt-Besserat, 1992, p. 110).

Of the two systems, the practice of storing tokens in clay envelopes was more significant for the development of mathematics. The last step in the evolution of tokens was a merging of the two systems of *bullae* and envelopes. Simple tokens were pressed into a solid lump or tablet of clay to make marks. Only the clay tablet was then kept. Within a couple of hundred years, this new system was also being used for the complex tokens, but here, because of their complicated shapes and designs, the image of the token did not transfer satisfactorily onto the clay. This new system, in place by about 3000 BC, afforded greater ease of use and storage, at the price of a certain loss of security. These pressed or drawn marks on the clay tablets were the beginnings of the Babylonian numeration system.

From about 3000 BC, among the Sumerians, tokens for different goods began appearing as clay tablets, represented by different symbols, and multiple quantities were represented by repetition. Thus three units of grain were denoted by three 'grain marks', five jars of oil were denoted by five 'oil marks', and so on. The limitations of such a system became evident with the increasing complexity of Sumerian economic life: a confusing proliferation of different styled tokens to be learnt and the tedium of representing large magnitudes. Recording the sale of five jars of oil or of a limited range of commodities was a simple affair, but an increase in the quantity and range of commodities was a different matter. Temple complexes, such as the temple of Inanna at Uruk, were large-scale enterprises, dealing in considerable quantities of goods and labour. A new system of recording and accounting needed to be devised.

The first great innovation, as we saw earlier in the 'Reflections' on Chapter 2, was the separation of the quantity of the good from the symbol for the good: that is, to represent three units of grain by a symbol for 'three' followed by a symbol for 'grain unit' in the same way that we would write three goats or three cows or, even more generally, 3 l or 3 km.

Whereas we use the same number signs, regardless of their metrological meaning (the 'three' for sheep is the same sign as the 'three' for kilometres or jars of oil), the Sumerians resorted to a wide variety of different symbols. Nissen *et al.* (1993) have identified around sixty different number signs, which they group into a dozen or so metrological systems. For example, the Sumerians used one system for counting discrete objects, such as animals, and other systems for measuring areas or volumes. Each system had a collection of signs denoting various quantities.

Any metrological system contains a number of different sized units with fixed conversion factors between them, so that, for example, there were 12 inches in a foot and 3 feet in a yard. Just as in our old weight and measure systems, Sumerian metrology featured all sorts of conversion factors, although it is notable that they were all simple fractions of 60.

The symbolic representation of numbers in the early Sumerian system is shown on p. 96. A parallel system used for counting most discrete objects, a single object, such as a sheep, cow or fish, was shown by a small cone. Ten cones equalled one small circle, six small circles equalled one big cone, ten big cones equalled a big cone with a circle inside it, six of those made a large circle and ten large circles became a large circle with a small circle inside. This last unit, whose symbol is not shown, equaled 36 000 base units. The choice of these shapes was probably dictated by design considerations: a circle and a 'cone shape' could be easily made by a stylus pressing on the clay, either vertically for the circle or at an angle for the cone.

In the early stages, however, there were different systems of numerical representation in Mesopotamia, depending on what was being measured. For a short period, a 'bisexagesimal' system operated with conversion factors of 10, 6 and 2, so that the symbol for the largest quantity, this time a large circle containing two small circles, represented 7200 base units. Yet another system was used for measuring grain capacity: the conversion factors were 5, 10 and 3, so that the largest unit, a large cone containing a small circle, was worth 1500 base units. To add to the confusion, a single sign could be used in several systems to denote different multiples of the base unit. In particular, the small circle could mean 6, 10 or 18 small cones, depending on context and the system in use.

Gradually, over the course of the third millennium, these signs were replaced by cuneiform equivalents so that numbers could be written with the same stylus that was being used for the words in the text. The final step in this story, occurring probably some time around the end of the third millennium, was the introduction of a sexagesimal place-value system. The number of signs was reduced to just two: a vertical wedge derived from the small cone often used for the base unit, and a corner wedge, derived from the small circle. A detailed account of this innovative system is contained on pp. 97–102.

The Plimpton 322 Clay Tablet

The nearly 4000 year old lump of clay called Plimpton 322 is undoubtedly the most famous mathematical tablet from Mesopotamia. Since its decipherment in 1945, it has been discussed in the 'Babylonian' chapter of almost every general history of mathematics, and in a number of specialist works as well. There have been two main approaches to interpreting the tablet:

1 it is a sophisticated listing of the so-called Pythagorean triples; or
2 it is a remarkable trigonometric table (two thousand years before the apparent development of angle measurement in Alexandria).

In the first edition, I chose to dwell rather narrowly on the first interpretation.

But more recent research opens up other possibilities: Plimpton 322 is no longer a freakish anomaly in the history of early mathematics, but one of the peaks of Babylonian mathematical culture.

To understand the significance of Plimpton 322 it may be necessary to examine the 'anthropology' of Babylonian mathematics, notably how the people of this area approached mathematical problems and what role these problems played in the wider society. It should be remembered that the tablet contains a table of fifteen rows and four columns. The usual assumption is that the first column contained all numbers beginning with '1'. Then, it is easily seen that each number in the column is a perfect square. Also, subtracting 1 from each of these numbers gives a perfect square. Consider, for instance, the number in row 11 of Table 4.3 (p. 116). The number 1;33,45 represents $1 + 33/60 + 45/3600 = 1 + 9/16 = 25/16$, which is the square of $5/4$. One less than $25/16$ is $9/16$, the square of $3/4$. The second and third entries in this row represent these fractions: 45 represents $45/60 = 3/4$ and 1;15 represents $1 + 15/60 = 5/4$.

The relationship would still hold if 45 represented 45 and 1;15 represented 75, in which case these two entries are proportional to the fractions. Considered as a table of Pythagorean triples, the second column (b) is one side of a right-angled triangle or the width of a rectangle. In this interpretation, the third side of the triangle (the fifth column, h) does not appear in the fragment of the tablet that we have at present. However, the first column has been interpreted as

$$(d/h)^2 = 1 + (b/h)^2$$

This interpretation of the Plimpton Tablet views each first column entry, in modern terms, as the square of the secant of an angle between sides d and h of a right-angled triangle with successive angles roughly $1°$ apart. In other words, the above equation may be written as

$$\sec^2 2 = 1 + \tan^2 2$$

So we have here the first table of a trigonometric function consisting of squares of secants for $45°$ down to $30°$: the earliest case of degree measurement by about two thousand years. This is a strong claim, but is not backed up by supporting evidence from other sources of Babylonian mathematics.

6 Ancient Chinese Mathematics

The Versatility and Primacy of Chinese Numeration

Throughout the history of China, mathematics was referred to as the 'art of calculation' (*suan chu*). From the first evidence of mathematical activity in China, consisting of numeration signs on tortoise shells and cattle bones (known as oracle bones) from the Shang dynasty, an interest in calculation was matched by the invention of a positional number system which lent itself easily to calculation. The rod numeral system that came later was economic, both in terms of the symbols used and the space taken up in its written form. It may be remembered (pp. 141–3) that, in this system, each of the first nine numbers is represented by no more than five rods. By letting a single rod represent the number five, the numbers from six to nine are simpler to represent than by the use of roman numerals. The introduction of the principle of place value made it easier to carry out difficult arithmetical operations. The early development in China of an algebra of negative numbers as well as of decimal fractions were both by-products of the rod numeral system.

Addition, subtraction, division and multiplication of fractions were familiar to the Chinese from the early Han dynasty (206 BC to AD 220) onwards. To divide by a fraction, both numbers were multiplied by the denominator of the fraction and then rules similar to those of today were used. In adding fractions each numerator was multiplied by the denominators of the other fractions, and then all the denominators were multiplied by one another. The numerators were then added. This is very similar to the method we use today, except that we try to find the lowest common denominator (a number which each fraction denominator can divide evenly into) to make things easier. Subtraction was carried out by multiplying the numerator of each fraction by the denominator of the other fraction, subtracting the smaller of the two products from the greater and dividing the remainder by the product of the denominators. Of course, this method of subtraction ensured that one always ended up with a positive quantity. It is interesting, in this context, to note that the step-by-step procedures of multiplication and division using Indo-Arabic numerals, described by the Arab mathematicians al-Khwarizmi and al-Uqlidisi, are very similar to those contained in the earliest extant Chinese text (*Sun Tsu Suang Ching*) describing these operations with rod numerals. For this, and other reasons, Lam Lay Yong (Lam and Ang, 1987; Lam, 1992) concludes that our numeral system has its origins in the Chinese rod numeral system.

The Chinese did not need a bar, or any other device, to separate the numerator and denominator. In written form, the notation for three-fifths was ||| at the top and ||||| at the bottom. In spoken form, the numerator was referred to as 'son' and the denominator as 'mother'. The concept of operations with negative numbers posed no problem either. To distinguish between positive and negative numbers, as mentioned on pp. 146–7, different-coloured counting rods were used and, if written, a negative number had a diagonal line through its last column. With the rod numeral system deeply ingrained in their minds, the Chinese were able to develop the arithmetic and algebra of the *Chiu Chang Suan Shu* ('Nine Chapters'), discussed on pp. 156–76.

The Foundations of Chinese Mathematics

The fact that the concept of decimal fractions and negative numbers posed no problems raises certain interesting questions regarding foundational differences between early Chinese and Greek (Western) mathematics. To the Chinese, the primacy of arithmetical operations was their main concern. The notion of number as an atomic and indivisible entity, a characteristic of early Greek mathematics, did not trouble them. Neither did Chinese mathematics have to face the dilemma of the irrational, which had perplexed the Greeks.

There was a further difference between the Greek and Chinese mathematical traditions: while the Greeks were interested in formal logic from early times, the Chinese continued to use a form of 'dialectical logic'. This is best exemplified in the Chinese method of solving equations. A solution of an equation is assumed beforehand, and the answer is obtained to the desired accuracy by performing as many iterations as are necessary. As an illustration consider the solution offered to Problem 6.8 (p. 173). This involved the 'rule of double false position', one of the oldest methods of approximating the real roots of an equation, of which there are variants to be found in Babylonian mathematics (p. 106), in Greek mathematics (known as Heron's method), and in the Indian Bakhshali Manuscript (pp. 260–61).

For over a thousand years, the Chinese used this method, refining and extending it, passing it on to the Arabs, who in turn passed it on to the Europeans. The method still continues to have basic application in modern numerical analysis under the name of the chord or chain rule. Its Chinese antecedents are, however, less well known. The method would seem to be in line with the naturalistic philosophical tradition of *yin* and *yang* of China, in which there was little appreciation of ideal entities, and their associated ontological realm of *stasis*, found in the Greek tradition (Joseph, 1994c, 1994d).

Recreational Problems

There has been a general tendency to label problems and puzzles of unknown origin as 'Chinese'. Whether this is a reflection of the physical and cultural distance of China from the West, or an acknowledgement of the considerable debt owed by the world to Chinese technology, is a moot point. One example, a prototype for a popular pre-school child's puzzle today, involves the rearrangement of a set of wooden pieces (a square, a rhombus and five triangles of different sizes). Known to the Chinese as *Chi Chiao* ('Seven Subtle Shapes'), it became known in Europe as 'tangrams'. Its Chinese origin may have been related to the importance of 'dissection and reassembly' in Chinese geometry (p. 169), as well as the variety of geometrical forms used by Chinese builders in designing the lattice-work of windows. Other examples include *Lien Nuan Chuan* ('Ring of Linked Rings') which, as the 'Puzzle of the Chinese Rings', was first mentioned by the Italian algebraist Girolamo Cardano, and then by John Wallis in greater detail a hundred years later. Interest in this puzzle continued into the nineteenth century, when Gros discussed the puzzle using binary arithmetic notation, which has come to be known as the Gray code, and patented by Bell Laboratories in 1953 (Singmaster, 1994, p. 1570). The puzzle utilizes a related puzzle, *Che Chih Shu*, about the folding of paper, and was mentioned in a famous poem written around the beginning of the Christian era.

One of the most famous recreational problems began its journey from China, making its first appearance in *Suan Ching Shih Shu* ('Ten Mathematical Manuals'), which appeared in the seventh century AD. The 'hundred fowls' problem reads: 'A cock is worth 5 *qian*, a hen 3 *qian* and 3 chicks 1 *qian*; with 100 *qian* we buy 100 of them; how many cocks, hens and chicks are there?' (Martzloff, 1994, p. 96). The Indian version of the problem appears around AD 850 in the *Ganita Sara Samgraha* of Mahavira, and reads: 'Pigeons are sold at the rate of 5 for 3 *rupa*, cranes at the rate of 7 for 5 *rupa*, swans at the rate of 9 for 7 *rupa* and peacocks at the rate of 3 for 9 *rupa*. How would you acquire 100 birds for exactly 100 *rupa*?' The Arab mathematician Abu Kamil (*fl.* AD 900) gave his version of the same problem as part of his treatment of indeterminate equations. It finally appeared in Europe in *Liber abaci* (1202), the influential book by Fibonacci. The problem is by no means trivial. There are 26 solutions in all to the Indian version of the problem. Mahavira obtained four of the solutions, and when students were asked to write an algorithm for the solutions using a programmable calculator, the 'fowls' algorithm took 122 minutes to run on one type of calculator! (Joseph, 1994d, pp. 59–60, 159).

Another favourite recreational problem is the 'chessboard' problem. The man who invented the game of chess was asked by his grateful ruler to request anything, including half the kingdom, as reward. He requested that he be given the amount of grain that would correspond to the number of grains on a chessboard, arranged in such a way that there would be one grain on the first square, two on the second, four on the third, and so on up to the 64th square. The ruler initially felt insulted by what he thought was a paltry request but, on due reflection, realized that there was not sufficient grain in his kingdom to meet the request.

In its modern form, this recreational problem involves work with series, permutations and combinations, and probability. The traditional problem, whose origins are usually attributed to India, could equally have come from China. An Arab mathematician, Al-Biruni, worked out the total number of grains required as 18 446 744 073 709 551 615. How did he arrive at this number? He probably summed the geometric progression

$$1 + 2 + 4 + 8 + \ldots + 2^{63} = 2^{64} - 1$$

This problem entered European mathematics through Fibonacci's *Liber abaci*.

Another famous 'chessboard' problem (relating to the specifics of Chinese chess) is associated with a Thang monk named I-Hsing. Shen Kua, the author of *Meng Chi Pi Than* ('Dream Pool Essays') written in AD 1086, recounts the story of I-Hsing, who claimed that he could calculate the total number of possible positions. If the chessboard consisted of two rows and four pieces, denoted by (2, 4), there would be 81 different positions possible. For (3, 9) and (4, 16), the number of possible positions would be 19 683 and 43 046 721, respectively. When all rows and chess pieces are specified, the number would be of an order of magnitude impossible to calculate, although Shen Kua claims that it is possible to enumerate (at least theoretically) all possible changes and transformations occurring on the chessboard.

There are other problems such as the 'St Ives' problem from the Ahmes Papyrus of Egypt (pp. 79–80), the 'monkey and coconut' problem from ancient India and the 'Josephus' problem of discarding every ninth person from a circle of 30 persons until 15 remained, known in Europe from the ninth century AD. There are also the 'magic squares', which seem to thrive in many cultures for their mental, aesthetic and metaphysical attributes. All of them serve as historical markers, with great variations in the time taken for their development and transmission across cultures. They have frequently turned up ideas which have been creative in terms of further development of mathematics. As Singmaster (1994, p. 1570) reminds us:

The Königsberg bridge problem and Hamilton's Icosian Game have led to basic techniques of combinatorial optimization ... Knots have recently become of interest in molecular biology because DNA can form closed loops which may be knotted ... [But recreational problems also emphasize] our increasing realization that modern algebra and arithmetic may derive more from Babylonia, China, India and the Arabs than from Greece. Such easily understood examples are extremely useful in conveying the antiquity and the cultural diversity of the origins of mathematics, particularly to school students, school teachers and the general public.

7   Special Topics in Chinese Mathematics

The Influence of Chinese Mathematics

Three main cultural areas came under the influence of Chinese mathematics: Korea, Tibet and Japan. The circumstances under which the influence manifested itself, and the response to, and the assimilation of, this influence differ considerably between each of these areas.

Mathematics in Korea

In AD 682, under the Shilla dynasty, a system of mathematics education was established in Korea. Based on the Tang *Iludian*, a codified mathematics programme established during the Tang dynasty in China (AD 618–906), its purpose was to train professional mathematicians. It was a closely controlled programme which prescribed the number of mathematicians to be trained, the length of study and the content of the curriculum. The students could be enrolled at any age between fifteen and thirty, and the study lasted for nine years or longer. The full curriculum, according to Kim (1994, p. 112), consisted of elementary ('six-chapter arithmetic'), intermediate ('nine-chapter arithmetic') and advanced ('continuation techniques') components. The original Tang model, on which it was based, recruited students at a younger age, with the duration of study being seven years, and a more elaborate curriculum was offered, based on the 'Nine Chapters' and its commentaries. The Korean programme remained in place as the official programme until the end of the Choson dynasty in 1910. It influenced the system set up in Japan around AD 701, known as the *Taihorei*. China and Japan soon discontinued their programme of studies.

During the Choson dynasty in Korea (1392–1910), the programme was revised and strengthened, with the requirement that all bureaucrats had to undergo mathematical training. A class of bureaucrats, called the *chungin* ('middle men'), were put in charge of the technical civil service examinations. Mathematicians were recruited to become an increasingly important section of this class.

The *chungin* soon became an exclusive class, intermarrying among themselves and establishing an almost caste-like line of descent, where a father who was a mathematician would be succeeded by a son who was also trained in the same discipline. The discipline was further regimented during the reign of Sejong (AD 1419–50), when a Bureau of Mathematics and an Agency for Calendars were established at the same time as a system of awarding titles such as *Sanhak Paksa* (Doctor of Mathematics), *Sanhak Kyosu* (Professor of Mathematics) and *Sansa* (Mathematician).

The growing professionalism of the discipline was not, however, matched by greater creativity on the part of its members. We have an account of Hong Chong-ha, a professor of mathematics and a *chungin*, who wrote a text entitled *Kuilchip* ('Nine Chapters of Arithmetic in One'). He was born in 1684, into a family where his father, both grandfathers and a great-grandfather, as well as his wife's father, were all mathematicians! (Kim, 1994, p. 113). His book could well have been written three hundred years earlier. Its content was no different from the older Chinese methods of the original programme, and even retained calculation rods when, in China, they had already been replaced by the abacus.

The stranglehold that Chinese mathematics had on Korean mathematics was never significantly loosened. Even when new ideas came to China from Europe, they had virtually no impact on Korean mathematics. The intellectual climate and the institutionalization of mathematics as a discipline in Korea would seem to have hindered the understanding and assimilation of modern mathematics.

Astronomy in Tibet

Tibet has a living astronomical heritage influenced by both China and India. As a product of cultural mixing – a mixing of religions, languages and astronomical knowledge – Tibetan astronomy provides another model of how outside influences are fused in the development of science.

Tibetan astronomy (or *rtsis*) may be broadly divided into four branches: *nag-rtsis* ('black calculation'), *rgya-rtsis* ('Chinese calculation'), *skar-rtsis* ('star calculation') and *dbayabis-char* ('divination'). The first two branches are of Chinese origin while the last two are of Indian origin. The first and the third may be loosely interpreted in modern terms as astronomy while the other two are closer to astrology. We will concentrate on the Chinese components.

A popular work of the seventeenth century on *nag-rtsis* lists nine integral components of the subject, being a blend of Chinese natural philosophy — such as the theory of the five basic elements and the *yin—yang* principle — and of Chinese astrological practices that include the association of each year of a 12-year cycle, each month in a year being associated with a particular animal (e.g. rat, tiger, horse, sheep). There is also a description of the Chinese lunar zodiac which contains 28 lunar mansions or houses.

The *rgya-rtsis* is based on a Tibetan translation of an original Chinese text on the construction of a Shixian calendar, the last luni-solar calendar to be constructed before being replaced by the Gregorian solar calendar in China in 1912. A luni-solar calendar consists of 'short' months of 29 days and 'long' months of 30 days. The idea was to arrange 'short' and 'long' months so that the new moon would occur on the first day of each month and the full moon on the fifteenth day of that month. A solar calendar, on the other hand, requires an independent system of solar intervals which, in the case of a tropical year, consists of 12 equal intervals of time, with the interval centre being the middle point of each equal interval. Since the synodic month (based on the moon and so varying between 29 and 30·1 days) was always slightly shorter than an equal interval in a solar calendar, an interval centre would not occur in certain months. The month of non-occurrence was known as an intercalary month; the year in which an intercalary month occurs has 13 synodic months. This occurs roughly once in every three years.

The Tibetans have had a long tradition of constructing a luni-solar calendar and making adjustments for intercalary months; this went hand in hand with astronomical and calendar construction practices which originated in India. The co-existence of these different calendars used for different purposes, reminiscent of the Mayan practice discussed in Chapter 2, is one of the more interesting aspects of Tibetan astronomy.

The culmination of traditional Tibetan astronomy, however, was found in the works of Bu-ston Rin-chen grub (1290–1344). Bu-ston wrote an astronomical text called *Kas-pa dga'byed* which summarizes the subject-matter of Tibetan astronomy in seven chapters. Apart from the final chapter of the text, which contains a discussion of word numerals based on Indian texts (see the 'Reflections' on Chapter 8), mathematics (as opposed to 'calculation') is treated primarily as a tool for astronomical calculations. Indeed, there is no known Tibetan text which deals with mathematical work outside astronomy.

Mathematics in Japan

The introduction of Chinese mathematics into Japan took place during two periods: in the eighth and ninth centuries, and in the sixteenth and seventeenth centuries AD. In the first period, when 'Nine Chapters' was brought to Japan, the impact was limited – probably because of the lack of understanding of the text on the part of the Japanese. However, the text was preserved and a Japanese translation made, to be studied by a few scholars.

The introduction during the second period was a more creative encounter. Two important thirteenth-century texts found their way to Japan via Korea. An encounter occurred between two cultures which, despite other differences, shared the same mathematical language. This experience would be different from the Japanese encounter with Western mathematics that occurred a few centuries later. The language of one had to be translated and interpreted for the other.

We will approach the Sino-Japanese encounter through a case study of Seki Kowa (1642–1709). He is a central figure in Japanese mathematics in a number of ways and was mentioned briefly in Chapter 6 for his work on magic squares and the discovery of determinants. He is generally accepted as the greatest Japanese mathematician of his time. His role was crucial in giving final shape to the original and individualistic creation of Japanese mathematics, the *Wasan*, the importance of which is reflected in the fact that the word *Wasan* is sometimes used to describe the whole indigenous mathematical tradition of Japan.

Although Seki Kowa was strongly influenced by Chinese mathematics, it is difficult to determine precisely which Chinese works he had studied. However, he was a prolific writer, and a number of his publications are either transcriptions of mathematics from Chinese into Japanese, or commentaries on certain works of well-known Chinese mathematicians.

There is little known about the personal life of Seki Kowa, and doubts have been expressed regarding the accuracy of the name by which he is commonly referred to outside Japan. Shigeru Jochi and Rosenfeld (1997) suggest that the correct name is Seki Takakazu. He was born in Fujioka, the second son in a samurai warrior family, and began writing mathematical texts when he was employed as an auditor by Shogun Tokugawa Ienobu. In later life he became the landlord of a '300 person' village.

Seki Kowa's interests ranged widely over a number of fields of mathematics (15 according to certain Japanese historians of mathematics) from constructions of calendars and recreational mathematics (e.g. a solution to the Josephus problem mentioned in the 'Reflections' to Chapter 6) to magic squares and magic circles, and solutions of higher-order and indeterminate equations. He had two able disciples, the brothers Katahiro Takebe (1664–1739) and Kataakira (1661–1739). Between them they brought out an encyclopaedia of the mathematics of their forerunners, entitled *Taisei Sankyo* ('Large Account of Mathematics') which ran into 20 volumes. Seki Kowa died in Edo (now Tokyo) on 5 December 1708.

According to Shigeru and Rosenfeld (1997), Seki Kowa made contributions in advance of, or contemporaneous with, the works of European mathematicians of the same period: he discovered determinants ten years before Leibniz, extended the Chinese work on solving numerical equations of higher order using the Ruffini–Horner method, discovered the conditions for the existence of positive and negative roots of polynomials, did innovative work on continued fractions, and discovered the Bernoulli numbers a year before Bernoulli. In geometry, he calculated the value of π correct to nine places of decimals by applying an ingenious extrapolation to a polygon with 2^{17} sides. This work was extended by his disciple Takebe, who obtained an accuracy, remarkable for the time, of 31 places of decimals, from a formula equivalent to Taylor's expansion of $\sin^{-1}x$, 15 years before Euler. Other notable geometrical contributions included the calculation of the volume of a sphere using an original integral method called *enri*, calculations on conic sections and Archimedes' spiral. The last few years of his life were spent mainly on astronomical work, which is found in *Shiyo Sampo* ('Mathematical Methods of Computing Four Points on the Lunar Orbit') and *Temmon Sugaku Zatcho* ('Notes on Astronomy and Mathematics').

Many of Seki Kowa's works are remarkable considering the standard of Japanese work of that time. The subjects he took up were mainly those that interested Chinese mathematicians of the Song and Yuan dynasties. These included Yang Hui's *Suan Fa* ('Method of Computation'), Shen Kua's *Meng Chi Pi Than* ('Dream Pool Essays'), Chu Shih Chieh's *Szu Yuen Yu Chien* ('Precious Mirror of the Four Elements'), Li Yeh's *Tshe Yuan Hi Ching* ('Sea Mirror of Circle Measurements') and Chin Chiu Shao's *Shu Shu Chiu Chang* ('Nine Sections of Mathematics').

As an illustration of the way that Seki Kowa put his own stamp on the

Chinese work, consider the first and the last works given in the above list. Yang Hui's work on magic squares indicates that the Chinese interest in the subject had both mathematical and philosophical significance. As mentioned on pp. 149–150, Chinese philosophers believed that the smallest magic square (*Lo shu*) had mysterious power. Chinese mathematicians were constrained by this 'magical' element to confine themselves to only those squares of certain sizes. Seki Kowa was not under any such constraint: his interest in the subject was purely mathematical. As a result, the work on this topic carried out by Seki Kowa and other Japanese mathematicians after him ranged widely to include larger squares, as well as other shapes such as circles, cubes and spheres.

One may detect a similar disposition when it comes to Seki Kowa's approach to the problem of solving indeterminate equations. In China, indeterminate equations arose from astronomical studies, in particular to compute 'accumulated years from an initial epoch' (pp. 207–8). Shigeru and Rosenfeld (1997) maintain that it is likely, as in the case of magic squares, that Seki Kowa would have ignored the metaphysics contained in Chin Chiu Shao's works and concentrated instead on the mathematics. However, Seki Kowa's work on indeterminate equations does not represent any significant advance on *Shu Shu Chiu Chang*.

There is an area in which Seki Kowa had a far-reaching impact on Japanese mathematics: in his book *Jinko-ki*, he invented a form of notational algebra (*endan jutsu*) which helped him to understand and decipher the solutions of higher-order numerical equations that Chin Chiu Shao had obtained using the *tian yuan* notation (pp. 199–204). Seki Kowa proceeded to develop this approach, known as the *Wasan*, to study parabolas, hyperbolas and the spirals of Archimedes. He applied this approach to calculate a determinant of order 5 in his work *Kaifukudai-no-ho*, published in 1683. Seki Kowa's disciples took up his work and developed the Japanese mathematical tradition of *Wasan* to a significant degree, climaxing in the work of Yoshiro Kurshima (*d.* 1757) and Ryohitsu Matsunaga (*d.* 1744). For further details of the strengths and weaknesses of the *Wasan* tradition, the publications of Murata (1975, 1980, 1994) are particularly valuable.

8 Ancient Indian Mathematics

Mathematics in Harappa

The argument for Indian mathematics beginning in Harappa is not based on any direct evidence: nothing like the Mesopotamian clay tablets or the Egyptian papyri exist to testify to its origins in other centres of early civilizations. However, the elaborate constructions excavated there cannot be understood without attributing knowledge of a number of geometrical propositions relating to the shapes and mensuration of rectilinear figures and circles.

Notwithstanding these conjectures, an important source of evidence, the written script, has so far thrown no light on this subject. It remains unread, despite years of ingenious attempts to do so. The script poses certain problems which were not present in the case of other ancient scripts. The Harappan writing is only available in the form of specific types of object principally seals made of steatite (soapstone), each seal, on average, containing a text of only five graphemes (or signs). No bilingual or multilingual text, such as the Rosetta Stone for Egyptian script, is available. Also the language, or the language family, of the Indus script is unknown, although the common assumption made in the past was that it is some form of proto-Dravidian language and had 'disappeared' sometime before the middle of the second millennium BC. There is, apparently, a long hiatus between this disappearance and the emergence of the so-called historical period of the Indian subcontinent (the Vedic period), thereby causing a big 'hole' in the chronology of Indian mathematics right from its inception.

There have been a number of attempts to read the inscriptions ever since a substantial collection of them became available around the 1920s. Many of the early attempts were phonetic interpretations based on unverifiable *a priori* linguistic attributes and other speculations sometimes involving mythological elements of Hindu traditions. A notable 'objective' attempt was that of Hunter (1934), who carried out a positional and functional analysis of the signs of the Indus script and suggested methods for splitting the texts into certain sign combinations which constituted 'words', irrespective of their linguistic attributes. Following Hunter's work, more recent investigations have involved detailed structural analysis of the texts with the aim of classifying the signs or sign combinations into linguistic units, such as root morphemes, attributes and other grammatical suffixes, and then read the texts phonetically, adapting a form of Dravidian as the underlying language.

Any fresh approach to the deciphering of the Indus script needs to take account of three distinctive features that were identified in earlier studies. First, there are rich structural regularities in the texts which make them distinct from other ancient writings. Second, the texts occur in almost all cases on seals, so that the purposes of these seals become a matter of some importance. Finally, a closer examination should be made of the nature and significance of a number of animal and other motifs, named by archaeologists as 'field symbols', that occur on many of the seals together with the writing. And the examination of all three features should be made within the historical context of the emergence of the Harappan culture and its aftermath. Some recent studies (Jeganathan, 1993; Kak, 1989, 1990; Subbarayappa, 1993) are noteworthy attempts to incorporate these features and look at the seals with fresh eyes.

It is generally accepted that the Indus seals are records of administration and of internal as well as external trade. Indus seals have been found in sites of West Asia, indicating commercial contact between the Harappans and the neighbouring areas. There is also general agreement among archaeologists of the existence of an efficient and centralized administration, governing the vast area that constituted the Harappan culture, ensuring a degree of uniformity whether it was in the construction of houses and public amenities, or the promotion of arts and commerce or other activities.

What do the inscriptions on the seals mean? To attempt an answer, some features of the Harappan culture need to be highlighted. The geographical spread of this culture makes it highly unlikely that the language was the same throughout the length and breadth of that culture. Even the Mesopotamian civilization, which was contained in a much smaller area than that of the Harappan culture, had regional languages. India has had a multilingual culture from ancient times. There have been marked variations not only in spoken but also in written languages. As for a common script, whatever may have been the situation in other ancient civilizations, the picture could well be different on the Indian subcontinent. Even the earliest extant written records from the first half of the first millennium BC used two scripts — Kharosti and Brahmi — as vehicles of the same language.

Are we therefore justified in assuming that there was one language throughout the Harappan culture — in the urban centres such as Mohenjo-Daro, Harappa, Chanhu-Daro, Lothal and Kalibangan, and in the rural settlements — spanning a total area of 1.2 million square kilometres? Yet the assumption of a similar literary script is necessary since the seals, and other inscribed objects found in different parts of the Harappan culture, have more or less identical forms with a noticeable uniformity of their own. Such continuity is more plausible in the case of a well-established system of numerical notation. Subbarayappa (1993) compares the Indus signs to Brahmi, Kharosti and the Chinese 'oracle-bone' numerals and finds a number of similarities between these numeral forms. This finding has important implications for examining the origins of our decimal numerals. Future work based on longer written records found in more recent excavations is needed to verify other conjectures linking the Indus script with the Brahmi script, in the case of Kak (1989), or with the Dravidian counting etymology, in the case of Jeganathan (1993).

Vedic Geometry

In Chapter 8, we examined in some detail one of the earliest written records of Indian geometry, the *Sulbasutras*. Little is known about the authors of these texts, except that they were not merely scribes but also priest-craftsmen performing tasks including design, construction and maintenance of sacrificial altars. Their work may be traced to earlier texts, notably the *Satapatha Brahmana*, which is about three thousand years old. Despite its obscurities and archaic character, this text is valuable for an early discussion of the technical aspects of altar construction.

An important section in *Satapatha Brahmana* deals with constructing altars for a 12-day Agnicayana ceremony to mark the passage of time. The ceremony is carried out on an open ground divided into two sections:

1 The *Mahavedi* (Great Altar) was laid out as an isosceles trapezium with the two parallel sides of 30 *prakrama* and 24 *prakrama* and altitude 36 *prakrama*. A *prakrama* is about 0.8 metres. The choice of these dimensions may have been dictated by a calendrical consideration: to reconcile the discrepancy of the two calendars in use at the time – the lunar and the solar – by choosing a nominal year of 360 days which is four times the sum of the above dimensions. The *Mahavedi* contained within it a falcon-shaped brick altar (*vakrapraksa-syena*) shown in Figure 8.2 on p. 227. There were other constructions in the *Mahavedi* of functional and ritual significance, but of little mathematical interest.

2 The *Pracinavamsa* was a smaller rectangular section that lay to the west of the *Mahavedi* and contained three fire altars consisting of *Garhapatya* (of circular shape, symbolizing the Earth), *Dakshinagni* (of semicircular shape, representing space) and *Ahvantya* (of a square shape, representing the sky). These three altars had to be of equal area of one square *purusha* (approximately 5 square metres). Seidenberg (1962, 1983) has an interesting

discussion of the ambiguities in the Vedic texts relating to equivalence of area as well as the philosophical underpinnings of such a requirement. The last of the fire altars mentioned (the sky altar) was laid out in five layers, with the first representing the Earth, the second being the joining of Earth and space, the third space, the fourth representing the joining of space and sky, and the fifth the sky. We have already discussed in Chapter 8 how the need to maintain equivalence of areas between different shaped altars raised a number of geometric problems, the solutions of which led to early Indian geometry.

Another problem that led to some interesting mathematics related to ensuring the precise distance and relative positions of the three fire altars. The general requirement was that *Dakshinagni* should lie south of the line joining the other two fire altars and be at a distance from *Garhapatya* of one-third the distance between the other two fire altars. A discussion of how this was solved is given in Joseph (1996b, 1996c).

The early Vedic texts describe how equivalence is established between phenomena through numbers as well as through areas. A starting point is the centrality of the number 360 in the Vedic calendar and philosophy. Parallels were drawn between human anatomy and planetary motions. Thus, the *Caraka Samhita*, an early medical text, counts the total number of *asthi* (bones, teeth, nails, hard cartilage) in the human body as 360, obtained from considering the 308 bones of the new-born baby (before they fuse into a smaller number of 206 in the adult), 32 teeth and 20 nails, where each of these *asthi* is associated with a day of the year. The parallels between the nominal year (360 days) and man (*purusha*) are carried further in *Satapatha Brahmana* where the basic falcon-shaped altar of 7.5 square *purusha* or 108 000 square *angula* (1 *purusha* equals 120 *angula*) is linked to 10 nominal years (1 nominal year equals 360 days) or 108 000 *muhurta* (1 *muhurta* equals 48 minutes).

Kak (1993) has argued that the concept of equivalence is of central importance in interpreting Vedic astronomical knowledge. The design of altars contained astronomical codes to be deciphered. For example, the circular *Garhapatya* fire altar that symbolized Earth (or the womb) was constructed with 21 ordinary bricks and had an area of 1 square *purusha*. If the basic falcon-shaped altar, having an area of 7.5 square *purusha*, corresponded to 360 days, then 1 square *purusha* would be equivalent to 48 days. The augmentation of the basic falcon-shaped altar by one square *purusha* at a time was to be seen as a correction to make the altar correspond closer to an actual year (366 days, or 372 *tithi* or lunar days). A *nakshatra* year was taken as the number of *nakshatra* or signs of the zodiac (27) multiplied by the number of months (12) which would give 324 days (or *tithi*). An additional 48 *tithi* as a correction were needed to get an actual year. However, this would mean an excess of 0·93761 *tithi* every year since the number of *tithi* in a solar year is 371·06239. So, by constructing a falcon-shaped altar of 95 layers, symbolizing a 95-year cycle, with each augmentation being 1 square *purusha* (or 48 *tithi*), starting from 7.5 square *purusha* and finishing with 101.5 square *purusha* at the end of that cycle, the practice of a major adjustment every 95 years to the calendar by 90 *tithi* (or 3 lunar months) made sense. Such a correction implied that the length of the solar year is $372 - (90/95) = 371·05263$ *tithi* which corresponds to 365·24675 days. Compare this value to the present-day estimate for the tropical year of 365·25636 days, and we are struck by the accuracy of estimates made about three thousand years ago.

With similar ingenuity, Kak proceeds to unravel the astronomical codes contained in the Vedic texts, notably in the *Rigveda* and *Satapatha Brahmana*. The distances of the Moon and the Sun from the Earth were expressed as 108 times the diameters of these heavenly bodies, respectively. The values obtained today from using modern instruments are 110·6 for the Moon and 107·6 for the Sun.

Restoring Historical Continuity to Early Indian Mathematics

A common view is that Indian mathematics originated in the service of religion. Since time immemorial, the argument goes, the needs of religion have determined not only the character of Indian social and political institutions, but also the development of scientific knowledge. Astronomy was developed to help determine the auspicious day and hour for performing sacrifices. The 49 verses of *Jyotisutra* (the *Vedanga* that contains astronomical information) gave procedures for calculating the time and position of the Sun and Moon in various *Nakshatra*. A strong reason in Vedic India for the study of phonetics and grammar was to ensure perfect accuracy in pronouncing every syllable in a prayer or sacrificial chant. The construction of altars and the location of sacred fires had to conform to clearly laid-down instructions about their shapes and areas if they were to be effective instruments of sacrifice.

There is a danger, however, that the magico-religious beliefs surrounding the Vedic rituals can be over emphasized when considering the origins of Indian mathematics. We have discussed the role played by the *Agnicayana* ceremony in generating geometrical concepts and techniques found in the *Sulbasutras*. The rituals associated with the construction of fire altars may be looked at from two standpoints. The first is that the beliefs connect the shapes of altars with specific desires to be fulfilled by their use in the sacrifices. The second is simply technological: how exactly does one construct the altars with specific shapes and sizes, and by using a specific number of bricks, or how does one vary their shapes without affecting their size or area?

It may be argued that the geometry of the *Sulbasutras* had little to do with the first standpoint. Thus, for example, whether a falcon-shaped altar ensures for the beneficiary heaven or the annihilation of enemies has little relevance to the problem of constructing it to conform to a certain size and shape. As a matter of fact, these problems would remain even if the purpose of the construction was to create an ornamental structure in a garden. In other words, the geometry developed in the *Sulbasutras* was required basically to solve technological problems of construction. It is this geometry placed in a social context that should be of primary interest to the historian of mathematics.

Once the *Sulbasutras* are seen primarily as manuals for technicians, the question then arises as to where and when the practical knowledge relating to brick technology was acquired. References to bricks are conspicuous by their absence from the most sacred and earliest of Vedic literature, the *Rigveda Samhita*. When they do make an appearance in a recension (*Tattiriya Samhita*) of a later *Veda*, the *Yajurveda Samhita*, bricks are viewed as marvelous and mysterious entities. In *Taittiriya Samhita*, there is a reference to bricks as 'milch cows' (a ready source of gain). In *Yajurveda Samhita*, there are exhortations that 'tiles or potsherds' found in ruins, of what were probably Harappan cities, should be gathered for ritual purposes. It is, therefore, likely that the priests were acquainted with the burnt bricks from the same sites and would in the course of time invest them with magico-religious properties. In one of the last revisions of *Yajurveda* appears the *Satapatha Brahmana* in which a discussion of the conducting of the *Agnicayana* is accompanied by a short discourse on the construction of brick altars of various shapes and sizes. While the discussion lacks the sophistication of the *Sulbasutras*, it is clear that knowledge of brick technology from the Harappan culture had percolated into the Vedic rituals. Staal's (1978) conclusion that 'it is certainly reasonable to suppose that knowledge of the techniques for firing bricks was preserved among the inhabitants of the subcontinent even after the Harappa civilization had disappeared' is particularly apposite. So, the geometry embodied in the *Sulbasutras* should be viewed as the outcome of a long and sophisticated tradition of brick technology inherited from the Harappan civilization. If this is correct, the first and earliest of the discontinuities in the chronology of Indian mathematics has been filled with the assistance of bricks.

Numeracy in India

Fascination with numbers has been an abiding characteristic of Indian civilization, not only large numbers but very small ones as well. Operations with zero attracted the interest of both Bhaskaracharya (*b.* 1114) and Srinivas Ramanujan (1887–1920). In an elementary class that Ramanujan attended, the teacher was explaining the concept of division (or 'sharing') through examples: between three children, each child would get one banana. Similarly, the share would be one banana if four bananas were shared among four children, five bananas among five children, and so on. And when the teacher generalized this idea of sharing out *x* bananas among *x* boys, Ramanujan asked whether, if *x* equalled zero, each child would then get one banana! There is no record of the teacher's reply. Ramanujan explained later to his school friends that zero divided by zero could be anything since the zero of the denominator may be any number of times the zero of the numerator.

In *Ramayana*, the great Indian epic, there is a description of two armies facing each other. The size of the larger army led by Rama is given as follows in a seventeenth-century translation of the epic by Kottayam Kerala Varma Thampuran:

> Hundred hundred thousands make a Crore
> Hundred thousand crores make a Sankhu
> Hundred thousand sankhus make a Maha-sankhu
> Hundred thousand maha-sankhus make a Vriundam
> Hundred thousand vriundams make a Maha-vriundam
> Hundred thousand maha-vriundams make a Padmam
> Hundred thousand padmams make a Maha-padmam
> Hundred thousand maha-padmams make a Kharvam
> Hundred thousand kharvams make a Maha-kharvvam
> Hundred thousand maha-kharvams make a Samudra
> Hundred thousand samudras make a Maha-ougham.

The importance of number names in the evolution of the decimal place-value notation in India cannot be exaggerated. The word-numeral system was the logical outcome of proceeding by multiples of ten. Thus, in an early system, 60 799 is denoted by the Sanskrit word *sastim* (60), *sahsara* (thousand), *sapta* (seven) *satani* (hundred), *navatim* (nine ten times) and *nava* (nine). Such a system presupposes a scientifically based vocabulary of number names in which the principles of addition, subtraction and multiplication are used. It requires:

1 the naming of the first nine digits (*eka, dvi, tri, catur, pancha, sat, sapta, asta, nava*);

2 a second group of nine numbers obtained by multiplying each of the nine digits in **1** by ten (*dasa, vimsat, trimsat, catvarimsat, panchasat, sasti, saptati, astiti, navati*); and

3 a group of numbers which are increasing integral powers of 10, starting with 10^2 (*sata, sahasara, ayut, niyuta, prayuta, arbuda, nyarbuda, samudra, madhya, anta, parardha . . .*).

In forming the words of the second and third groups of numbers, the multiplicative principle applies, as in the example quoted: 60 000 is *sastim-sahasari*. The additive principle is employed when the numbers from the first and second groups are used: for example, 27 is *sapta-vimsati*. The subtractive principle may apply occasionally and in a limited way, as for example *ekanna-catvarimsat* indicates $20 - 1 = 19$, where *ekanna* means 'one less'.

To understand why word numerals persisted in India, even after the Indian numerals became widespread, it is necessary to recognize the importance of the oral mode of preserving and disseminating knowledge. An important characteristic of written texts in India from times immemorial was the *sutra* style of writing, which presented information in a cryptic form, leaving out details and rationale to be filled in by teachers and commentators. In short pithy sentences, often expressed in verse, the *sutras* enabled the reader to memorize the content easily.

As a replacement for the older word-numeral system that consisted of merely names of numbers, a new system was devised to help versification and memory. In this system, known as *bhuta-samkhya*, numbers were indicated by well-known objects or ideas. Thus, zero was *sunya* (void) or *ambara akasa* ('heavenly space' or ether) or . . . , one was *rupa* (Moon) or *bhumi* (Earth) or . . . , two was *netra* (eyes) or *paksha* (waxing and waning of the Moon) or . . . , three by *kala* (time: past, present and future) or *loka* (heaven, Earth and hell) or . . . , and so on. With multiple words available for each number, the choice of a particular word for a number would be dictated by literary considerations. This form of notation continued for many years in both secular and religious writings because it was aesthetically pleasing and offered an easier way of remembering numbers and rules.

There were two major problems with the *bhuta-samkhya* system. First, there was an 'exclusionist' element in that to decode the words for their numerical values required considerable familiarity with the philosophical and religious texts from which the correspondences were established in the first place. Second, at times the same word stood for two or more different numbers, where some writers had their own preferences when it came to choosing words to correspond to numbers, as, for example, when *paksa* was used for 2 as well as 15, and *dik* for 8 and 10 as well as 4.

A third system of numerical notation originated with Aryabhata (*b.* AD 476). In his *Aryabhatiya*, he introduced an alphabetical scheme for representing numerals, based on distinguishing between classified (*varga*) and unclassified (*avarga*) consonants and vowels. The *vargas* fall into five phonetic groups: *ka-varga* (guttural), *ca-varga* (palatal), *ta-varga* (lingual), *ta-varga* (dental) and *pa-varga* (labial). Each group has five letters associated with it, and represented numbers from 1 to 25. There were seven *avargas* consisting of semi-vowels and sibilants representing numerical values 30, 40, 50, ..., 90. An eighth *avarga* was used to extend the number to the next place value. The ten vowels denoted successive integral powers of 10 from 100 onwards.

This form of representation, closer to the system that preceded *bhuta-samkhya*, has the advantage of brevity and clarity but the disadvantage of having limited potential for formations of words that are pronounceable and meaningful, both necessary requirements for easy memorization. For example, according to the Aryabhatan system the representation of the number of revolutions of the Moon in a *yuga* (57 753 336) is the unpronounceable and meaningless word *cayagiyinusuchlr*!

From a refinement of Aryabhata's alphabet-numeral system of notation emerged the *katapayadi* system which the founder of the Kerala school of astronomy, Varurici, popularized around the seventh century AD. In this system, every number in the decimal place-value system can be represented by words, each letter of the word representing a digit. A vowel not preceded by a consonant stands for zero, but vowels following consonants have no special value. In the case of adjacent consonants, only the last consonant has a numerical value. Number words are read from right to left so that the letter denoting the 'units' is given first, and so on.

This was a system devised to help memorization since memorable words can be made up using different chronograms. For example, if such a system is applied to English, the letters b, c, d, f, g, h, j, k, l and m would represent the numbers zero to nine. So would n, p, q, r, s, t, v, w, x and y. The last letter, z, denotes zero. The vowels, a, e, i, o and u are helpful in forming meaningful words but have no numerical values associated with them. Thus, the sentence: 'I love Madras' represents the numbers 86 and 9234. To take another example from Kunjunni Raja (1963, p. 123), the number 1 729 133 could be represented by *balakaltram saukhyam* (i.e. the (company) of a young woman is sheer happiness) or *lingavyadhir asahyah* (i.e. the demise of sexual virility is unbearable).

The close relationship between literacy and numeracy, implied by such varied systems of numerical notation, may find its roots in the way that Sanskrit developed in its formative period after its separation from other languages of the Indo-European family. A long tradition of oral communication of knowledge was a characteristic of that period and left a singular mark on the nature and transmission of knowledge, whether religious or scientific, in Indian culture. After many years, as Sanskrit became a written language, three kinds of scientific Sanskrit developed with varying degrees of artificiality: grammatical, logical and mathematical Sanskrit.

Mathematical Sanskrit remained the least artificial of the three, with the greatest artificiality found in the development of *grammatical* Sanskrit by Panini and Patanjali, followed five hundred years later by the *logical* Sanskrit of *Nyaya* that culminated a thousand years later in *Navya-Nyaya*. This has some important implications for a comparative study of the historical development of Indian and Western mathematics, according to Staal (1995). First, the chronological order of the development of artificial scientific languages in the West was a reversal of the Indian experience. In the West, the study of logic came after mathematics and linguistics was a late developer. In India, mathematical Sanskrit never quite became an artificial language, although it employed abbreviations and artificial notations outside Sanskrit as shorthand for practical procedures. And logical Sanskrit never became, like its Western counterpart, an important adjunct to 'mathematical philosophy'.

The Spread of Numeracy in India

A search for the social origins of numeracy must consider the everyday practices and institutions that make the numerals and operations with them familiar to the ordinary person. The structure of Indian mathematics education for all may have been set by a Jaina text, called *Sthananga Sutra*, dating back to about 300 BC. Of the ten topics treated in that text, the first two, *parikarma* (number representation and the four fundamental operations of arithmetic) and *vyavahara* (arithmetic problems, including the 'rule of three') came to be referred to as *patiganita* ('sand-calculation') and meant to be studied by all. The other eight topics were plane geometry calculations as carried out with a rope (*rajju*), mensuration of plane figures and solids (*rasi*), advanced treatment of fractions (*kalasavrna*), study of that which is unknown or algebra (*yawat-tawat*), problems involving squares and square roots (*varga*), problems involving cubes and cube roots (*ghana*), problems involving higher powers and higher roots (*varga-varga*), and permutations and combinations (*vikalpa*).

Apart from those who were taught at home, which was the usual practice for the higher-caste males, and for all females, all others attended schools. There are early British descriptions of indigenous village schools where emphasis on numeracy was an important part of the school curriculum. A report, submitted in 1838 by William Adam, of such schools in some districts of Bengal and Bihar (Dharampal, 1983) is quite illuminating. The period a student spent in an elementary school was divided into four stages. The first stage, when the child first entered school, seldom exceeded ten days. During that time the young child was taught 'to form letters of the alphabet on the ground with a small stick or slip of bamboo', or on a sand-board. The second stage, lasting from two and a half to four years, involved pupils being taught to read from and write on palm leaves. During the same period, the pupil was expected to memorize 'the Cowrie Table, the Numeration Table as far as 100, the Katha Table and the Ser Table', the last two being tables of weights and measures. To help them with this enormous task, different systems of word numerals were taught. The third stage, lasting between two to three years, was spent on improving their literary skills as practiced on plantain leaves, as well as completing the basic course on *patiganita*. In the fourth and final stage, lasting up to two years, pupils were expected to read religious and other texts, both at school and at home, undergo training in commercial and agricultural accounts, and compose letters and petitions. A few would continue their education in higher institutions where Sanskrit was the language of instruction and the teachers and students were predominantly Brahmins.

Apart from numeracy skills, *patiganita* consisted of all the mathematics needed for daily living. The *vyavaharaganita* included problems involving calculation of volumes of grains and heaps, estimating amounts in piles of bricks and timber, construction of roads and buildings, the calculation of the time of the day, interest and capital calculations, barter and exchange, and recreational problems. In modern terminology, this is practical mathematics, which included commercial mathematics. The authors who wrote texts on *patiganita*, such as the unknown author of the Bakhshali Manuscript, or Mahavira (*fl.* AD 850) or Sridhara (*fl.* AD 900), began with a review of arithmetic operations, though the extent and detail to which this was done varied with different texts – the earlier the text, the more detailed the treatment.

The level of numeracy in traditional Indian society was high, partly because of the manner in which numeracy was acquired and passed on, and partly because of the lack of any institutional, religious or philosophical inhibitions to the acquisition and practice of numeracy. Yet the absence of a commercial revolution in India meant that the social milieu which nurtured interest in matters scientific in Europe was missing. In particular, no artificial language evolved, and while notations were fun and intellectually distracting, they did little to advance science, which ultimately stagnated. And practical mathematics, the handmaiden of numeracy, continued to remain at the same level for about a thousand years, eventually to be submerged by the rise of Western mathematics. Even the remnants of indigenous numeracy that exist in subterranean occupations, such as astrology and traditional architecture, may soon become a historical memory.

9   Indian Mathematics: The Classical Period and After

The growing volume of research into medieval Indian mathematics has dented the common perception that mathematics in India after Bhaskara II made 'only spotty progress until modern times' (Eves, 1983, p. 164). In these notes, it is intended to extend the discussion of Kerala mathematics on pp. 286–94 by sketching in more detail the social landscape within which this mathematics occurred, and identify in greater detail the main actors who played their part in this episode of the development of Indian mathematics. We conclude with a brief survey of Indian mathematics from Persian and Arabic sources which, for a short period became a parallel stream to the traditional one already examined in Chapters 8 and 9.

Kerala Mathematics

The Context

Two powerful tools contributed to the creation of modern mathematics in the seventeenth century: the discovery of the general algorithms of calculus, and the development and application of infinite series techniques. When introduced to calculus one is usually told that the names normally associated with the development of the subject are Newton and Leibniz. The other lesser known stream, the discovery and applications of infinite series, is often ignored despite its importance in the development of modern mathematics.

The two streams tended to reinforce each other in their simultaneous development by each extending the range of application of the other.

The origin of the analysis and derivations of certain infinite series, notably those relating to the arctangent, sine and cosine, is not to be found in Europe, but in an area in South India which now falls within the state of Kerala. From a region of about a thousand square kilometres north of Cochin, and during the period between the fourteenth and sixteenth centuries, there emerged discoveries in infinite series that predate similar work of James Gregory, Newton and Leibniz by at least 200 years.

There are a number of questions worth asking about the activities of this group of mathematician-astronomers (referred to hereafter as the Kerala School), apart from those relating to the mathematical content of their work. They include specific ones relating to the social and historical landscape in which the Kerala School developed, as well as to the motivation behind their work.

There are five main texts that constitute the main evidence of the work of the Kerala School. They are: *Tantrasamgraha* ('A Digest of Scientific Knowledge') by Nilakantha (1444–1545), *Yuktibhasa* ('An Exposition of the Rationale') by Jyesthadeva (*fl.* 1530), *Kriyakramakari* ('Operational Techniques') by Sankara Variar (*c.* 1500–1560) and Narayana (*c.* 1500–1575), *Karanapaddati* ('A Manual of Performances in the Right Sequence') by Putumana Somayajin (*c.* 1660–1740) and *Sadratnamala* ('A Garland of Bright Gems') by Sankara Varman (1800–1838).

An important feature of these texts is their claim to have derived their main ideas from Madhava of Sangamagramma (*c.* 1340–1425). We have no knowledge of his family background, except that he belonged to a sub-caste of Brahmins known as the *Empranatiri* who were not originally Nambuthri Brahmins (the highest-ranking Brahmins in Kerala) but had over the years attempted to enter this group. His only surviving works are on astronomy. We know, from the reports of those who came after him, of Madhava's contribution to the development of Kerala mathematics. His fame rests on his discovery of the infinite series for circular and trigonometric functions, notably the Gregory series for the arctangent, the Leibniz series for π, and the Newton power series for the sine and cosine. There are also some remarkable approximations attributed to him based on incorporating 'correction' terms for these slowly converging series.

Madhava's distinguished student was Paramesvara. Born in 1360 into a Nambuthri Brahmin family of Vedic scholars in the village of Alattur, he was reputed to have learnt his mathematics and astronomy from Madhava. Alattur had become a famous centre of learning and scholarship some centuries earlier during the period of Jain and Buddhist dominance. Paramesvara wrote a number of commentaries including ones on Aryabhata's *Aryabhatiya*, on Bhaskara I's *Mahabhaskariya* and *Laghubhaskariya*, and on Bhaskara II's *Lilavati*.

The foundation laid by Paramesvara heralded the emergence of the major figure of Nilakantha Somayaji, born in 1443 into a Nambuthri Brahmin family of *Somatris* (those who performed the *soma* sacrifice) in Trikkantiyur. He stayed and studied in the house of Damodara, the son of Paramesvara, where he was probably taught by Paramesvara. His major works included a commentary on three chapters of *Aryabhatiya*, and the seminal text *Tantra-samgraha*.

The latter consists of eight chapters, containing 432 verses, dealing mainly with various topics connected with astronomical calculations, including the setting-up of a sundial, calculations of the meridian, of latitude, of declinations, etc., and the prediction of eclipses. In *Tantrasamgraha*, Nilakantha carried out a major revision to the Aryabhatan model for the inferior planets, Mercury and Venus, arriving at a more accurate specification of the equation of the centre for these planets than any other that existed in Arab or European astronomy before Kepler (born about 130 years after Nilakantha). In *Aryabati-yabhasya*, Nilakantha developed a computational scheme for planetary motion more efficient than that of Tycho Brahe in that it correctly takes account of the equation of centre and latitudinal motion of the inferior planets. This computational scheme implied a model of planetary motion in which the five planets (Mercury, Venus, Mars, Jupiter and Saturn) move in eccentric orbits around the Sun which, in turn, goes round the Earth. This model is similar to the one suggested by Tycho Brahe when he revised Copernicus's heliocentric model. It is significant that all astronomers of the Kerala School who followed Nilakantha accepted his planetary model (Ramasubramanian *et al.*, 1994).

The other works which he wrote late in life were either commentaries on his earlier texts, such as those on his *Chandrachayaganita* and *Siddhantadarpana*, the latter being a short work in 32 verses, dealing with certain important astronomical constants and the theory of epicycles; or works such as *Golasara*, a book in three chapters on spherical astronomy; or *Sundararaja Prasnottara*, a work no longer extant but mentioned elsewhere as giving answers to questions raised by a Tamil astronomer called Sundararaja. The last work is important because it provides rare evidence of the spread of the influence of Kerala mathematics and astronomy to other areas of South India.

One of Nilakantha's students was Chitrabhanu (*c.* 1475–1550). He was a Nambuthri and came from the Brahmin village of Sivapuram (situated in present-day Trissur). His work, *Karanamrta*, contained four chapters dealing with advanced astronomical calculations. He was also the author of *Ekamvimsati Prasnottara* ('Twenty-one Questions and Answers'), in which he offered solutions for each of a set of 21 pairs of simultaneous equations in two unknowns. The 21 pairs arose from taking, at a time, any two of the following seven quantities (*a* to *g*) given on the right-hand side of the following equations:

$$x + y = a; bx - y = b; xy = c; x^2 + y^2 = d; x^2 - y^2 = e; x^3 + y^3 = f; x^3 - y^3 = g.$$

The solutions to 15 of the 21 pairs ($_7C_2$) are fairly straightforward, while the remaining six are not.

A student of Chitrabhanu, Narayana (*c.* 1500–1575), completed one of the major texts of the Kerala School, *Kriyakramakari*. A commentary on Bhaskara II's *Lilavati*, it was begun by Sankara Variar (*c.* 1500–1560), a student of both Nilakantha and Chitrabhanu. The Variars were a group of non-Brahmin temple officials who assisted the Brahmin priests in their religious rituals. A number of them were skilled in astrology and many were fluent in Sanskrit. The text *Kriyakramakari* is important in the history of Kerala mathematics and astronomy for its detailed discussion of the works of earlier writers, some of which are not extant, and for providing rationale, and proofs for a number of earlier results.

Another student of Nilakantha was a Nambuthri Brahmin from the Alattur village (the birthplace of Paramesvara), Jyesthadeva. He was the author of the seminal text of the Kerala School, *Yuktibhasa*. There are at least three versions of this text, of which the Malayalam version became well known throughout Kerala. Based on Nilakantha's *Tantrasamgraha*, it is unique in Indian mathematical literature for giving detailed rationales, proofs or derivations of many theorems and formulae in use among the astronomer-mathematicians of that time.

A student of Jyesthadeva came from the Pisharoti community. They were not Brahmins but performed traditional functions as cleaners and suppliers of flowers and plants for the temple. They were also employed by some Nambuthri families to give instructions to family members on the calculation of the astrological calendar (*panchagam*) and on time reckoning. Achuta Pisharoti (*c.* 1550–1621) was a versatile scholar who made a mark not only in astronomy but also in literature and medicine. His major contribution is found in his work *Sputanarnaya*, where he introduced for the first time in Indian astronomy a correction called 'reduction to the ecliptic', around the same time as Tycho Brahe did in Western astronomy.

In Charles Whish's 1832 paper, mentioned in Chapter 9, appears the passage, 'The author of the *Karanapaddhati* whose grandson is now alive in his seventieth year was Putumana Somayajin, a Nambuthri Brahmana of Trisivapur [Trissur] in Malabar.' An influential work in the dissemination of Kerala mathematics and astronomy, not only in Kerala but also in the neighbouring areas of present-day Tamilnadu and Andhra Pradesh, *Karanapaddhati* was written in 1732, almost 200 years after Jyesthadeva's *Yuktibhasa*.

After Achuta Pisharoti, little in the way of original work was done, although the tradition of providing corrections and contributing to the preparation of the astronomical ephemeris for the daily needs of the faithful observers and practitioners continued for a long time. About one hundred years after *Karanapaddhati* was written came the last of the known texts of the Kerala School, *Sadratnamala*. The author of this book, Sankar Varman of Katattanad in North Kerala, belonged to the royal family of that area and was a contemporary of Charles Whish. *Sadratnamala*, written in 1823, contains many of the results of the Kerala School, but given without the rationales or derivations found in the earlier texts. Whish met him and described him as 'a very intelligent man and acute mathematician'. He died six years after Whish's article on Kerala mathematics and astronomy appeared in 1832.

The authors mentioned above form part of a tradition of continuing scholarship in Kerala over a period of 400 years, from the birth of Madhava in 1340 to the probable death of Putumana Somayajin in 1740. The current level of knowledge of source materials means that it is difficult to assign many of the developments to any particular person. The results should be seen as produced by members of a school spread over several generations.

The Social Background

To understand the context in which mathematics developed in Kerala, there is a need to take a careful look at the social landscape of medieval Kerala society in particular the structure of medieval Kerala society, the pivotal role of the Kerala temple, and the mode by which scientific knowledge was acquired and disseminated in medieval Kerala. Each of these topics could well provide sufficient subject-matter for a whole chapter. Instead, we shall confine ourselves to making a few observations.

It is clear from the discussion so far that the members of the Kerala School were mainly Nambuthri Brahmins. Within a mainly two-tier caste system in Kerala, consisting of Brahmins and Nairs, two institutions operated to strengthen and sustain the economic and social dominance of the Nambuthris to a degree not known elsewhere in India: a system of feudalism (the *janmi* system of land-holding) headed by the Nambuthris, and their control of vast areas of arable land owned by temples.

There were also certain social factors that strengthened the Nambuthri dominance over the Kerala society of that time. The Nairs practiced the *marumakkattayam* (matrilineal) system of descent without a formal institution of marriage. Sexual alliances between Nair women and Nambuthri men were permitted, indeed sometimes encouraged, with children of such unions remaining the sole responsibility of their mother's family. At the same time, the Nambuthris operated a system of patrilineal descent (*makkatayam*), with a form of primogeniture that allowed only the eldest son to inherit property and to marry Nambuthri women. The eldest son was also required by custom to provide for the material needs of his siblings, consisting of younger brothers and unmarried sisters (of whom there were a large number, given the operation of the system).

Even a cursory examination of the social background of the members of the Kerala School would indicate that although the majority were Nambuthri Brahmins, the founder, Madhava was not one. The pursuit of activities such as studying mathematics and astronomy did not traditionally confer high status, which was reserved for those who carried out ceremonial and ritualistic duties. The most notable member after Madhava, Neelakantha, belonged to the highest rank among the Nambuthris. His high social status arose from the fact that he was a *somayaji*, one of the select sub-castes among the Nambuthris who carried out the *soma* sacrifices. Yet he single-mindedly pursued his interests in mathematics and astronomy. There were other members of the Kerala School who were not even Brahmins. There was, for instance, Sankara Variar, where the name Variar indicated that he belonged to the *Ambilavasis*, a caste of temple servants, and similarly with Achuta Pisharoti. This would indicate that the Kerala School were a mixed group, probably brought together by their interest in mathematics and astronomy, undertaking pursuits that did not have great social status, a group that cut across caste lines, and a group that probably had strong contacts with the temple personnel.

Rajan Gurukkal's (1992) study of the medieval history of Kerala, and the importance of the temple culture, is particularly instructive in examining the activities of this unusual group. The temple fulfilled an important purpose as an institution for acquiring and disseminating scientific knowledge. It was an influential organization since it combined religious power with secular power, many of its members being powerful landlords in their own right. The temple served as a medium through which the Nambuthris asserted their power and kept other groups in check. There are clearly parallels between the Church in medieval Europe and the Temple in medieval Kerala.

Another aspect, delving into the background of the members of the Kerala School, shows that a number of the Nambuthri members of the Kerala School were younger sons. As mentioned earlier, the Nambuthris followed a strict system of primogeniture, so that all landed property went to the eldest son. There was an additional twist: only the eldest son could marry a Nambuthri woman. The younger sons never married but formed sexual partnerships, often of a transient nature, with Nair women. So we have a situation here of a group of Nambuthris, freed of all economic and family responsibilities, a truly leisured class with religious duties confined to a few, not very demanding rituals. Some in such a situation whiled away their time writing erotic poetry, but many were engaged in other, less demanding pursuits. But there were a few groups who pursued their interest in astronomy and mathematics consistently over a period of a few centuries, sustained by the institution of the *guru–chela* (teacher–disciple) relationship which was a characteristic of the Indian educational system.

While this explanation for the emergence and continued existence of the Kerala School would appear attractive, particularly to those who are disposed to seek major explanations of any Indian phenomenon in the caste system, such as Pingree (1981), it is somewhat unconvincing. First, it does not account for the presence and the role of non-Brahmins in the Kerala School. Second, this explanation ignores the symbiotic nature of the relationship between the traditional *jyotish* (astronomer-astrologer), who often came from the lowly Kaniyan caste, and the Nambuthri *jyotish* that has been observed elsewhere. Third, the *granthaveri* (or village records) of Kerala of this period contain ample evidence of the metrical precision of a number of artisans and craftsmen (such as the carpenter, the trader, the builder and the architect). These records show some awareness among these artisans of the developments taking place in astronomy and mathematics during that period. The *granthaveri* and temple records remain a good but relatively untapped source of information about the 'calculating people' of the period.

A study of the social context of Kerala mathematics has an unexpected bonus. There is a deeply entrenched notion, as mentioned in relation to Egyptian and Babylonian mathematics, that all non-European mathematics is utilitarian. A number of Indian scholars have fallen into the same trap. They search for the motivation behind Kerala mathematics in astronomy, navigation and other practical pursuits. One should never ignore the practical motivation.

After all, many of the members of the Kerala School were both mathematicians and astronomers. The texts of that period cover both subjects. However, a lot of the works on infinite series do not have any direct applications to astronomy. So what led them on in their pursuit of knowledge? I have this vision, of a group of pure mathematicians in Kerala between the fourteenth and sixteenth centuries, like Ramanujan, Hardy and Littlewood in Cambridge, indulging in their passion and probably boasting of the fact that the mathematics that they did was of no use to anyone! Some members of the Kerala School took delight in long and tedious calculations, such as the one reportedly undertaken by Madhava in calculating sine tables to 12 places of decimals. Such fascination with numbers and delights in calculation has been a characteristic of Indian mathematics over the ages, as we saw in Chapter 8.

The Motivation and Method: A Case Study

The motivation of the Kerala School may be traced to a verse in *Aryabhatiya* which explains how, for a given diameter, the circumference of a circle is calculated:

> Add 4 to 100, multiply by 8, and add 62 000. The result is
> approximately [?] the circumference of a circle whose diameter is
> 20 000. (Verse 10)

Certain historians of mathematics (notably G. R. Kaye and Morris Kline) have argued, partly on the basis of this quotation, that the Indians were not aware of the fact that π could never be exactly determined. Their confusion may have risen because of the mistranslation of the word *asana* as 'approximate' or 'rough value' as in the quotation. The word is subtler than that. What it conveys is the notion of 'unattainability', i.e. something that cannot be reached.

This is illustrated by a passage from Nilakantha's commentary on *Aryabhatiya*:

> Why is only the approximate value (of circumference) given here? Let me explain. [It is approximate] because the real value cannot be obtained. If the diameter can be measured without a remainder, the circumference measured by the same unit [of measurement] will leave a remainder. Similarly, the unit that measures the circumference without a remainder will leave a remainder when used for measuring the diameter. Hence, the two measured by the same unit will never be without a remainder. Though

we try very hard we can reduce the remainder to a small quantity but never achieve the state of 'remainderlessness'. This is the problem.

What the passage shows is that Nilakantha and others understood the 'irrational' nature of π. So the question arose as to what could be done as a result. The following passage from Sankara Variar and Narayana's *Kriya-kramakari* contains a strategy:

Thus even by computing the results progressively, it is impossible theoretically to come to a final value. So, one has to stop computation at that stage of accuracy that one wants and take the final result arrived at by ignoring the previous results.

Indian mathematicians, unlike their Greek counterparts, were not generally preoccupied with numbers as mathematical objects. Faced with irrationality, they attempted to arrive at as accurate an estimate as possible. And this is what *Kriyakramakari* recommends.

However, in applying infinite series to estimate the circumference, the Kerala mathematicians came across a serious difficulty: the special case of the Madhava–Gregory series (referred to as Euler's series on p. 290) converges very slowly. For example, summing the first 19 terms on the right-hand side of

$$\pi/4 = 1 - 1/3 + 1/5 - 1/7 + \dots$$

gives a highly inaccurate estimate of π as 3·194.

The problem was tackled by Kerala mathematicians in two directions: obtain

1 rational approximations by applying corrections to partial sums of the series; and

2 more rapidly converging series by transforming the original series.

The results are reported in *Yuktibhasa* and *Kriyakramakari*. As an illustration of 1 from the *Yuktibhasa*, consider the incorporation of the following correction term to the series given above:

$$F(n) = (n^2 + 1)/(4n^3 + 5n)$$

where n is the number of terms on the right-hand side.

When this correction was applied for $n = 11$, the implicit estimate of $\pi = 3\cdot141592527$ was obtained, correct to 7 places of decimals. And this interest in increasing the accuracy of the estimate continued for a long time, so that as late as the nineteenth century the author of *Sadratnamala* estimated the circumference of a circle of diameter 10^{18} as 314 159 265 358 979 324 — correct to 17 places!

There were also a number of effective transformations of the original series to improve the convergence of the original series. The details of the derivations of the rational approximations and different transformations used are discussed in Hayashi *et al.* (1989), Hayashi (1990) and Joseph (1996d). What the work of the Kerala School exhibits is a measure of understanding of the concept of convergence, of the notion of rapidity of convergence and an awareness that convergence can be speeded up by transformations. Similar work was found in modern mathematics only as late as the end of the eighteenth century.

An unusual aspect of the Kerala approach to the derivation of a number of infinite series is the use of the method of direct rectification. This is an interesting geometric technique different from the method of exhaustion used in Arab and European mathematics. In the Kerala case, you are subdividing an arc into unequal parts while in the Arab and European case there is a subdivision of the arc into equal parts. A different technique was used in Kerala not because the method of exhaustion was unknown to the Indians. Indeed, it is likely that Aryabhata preferred the octagon method rather than the hexagon method used by Greek and Arab mathematicians to compute his accurate estimate of the circumference of the circle. The method of exhaustion was probably avoided because the calculation involving working out the square roots of numbers at each stage of the calculation, a tedious and time-consuming task.

Indian Mathematics from Persian and Arabic Sources

The arrival of the great Arab scientist al-Biruni in India in 1018 as a prisoner in the invading army of Muhammad Ghaznavi was an important landmark in the scientific contact between the two cultures. During his enforced stay in India, which he profitably occupied by learning the language and culture of the country, al-Biruni translated from Arabic into Sanskrit Euclid's *Elements*, Ptolemy's *Almagest* and his own work on the construction of the astrolabe. None of these Sanskrit translations are extant today. However, al-Biruni's example was followed by Sultan Firus Tughlaq, who occupied the throne of Delhi between 1351 and 1388. He ordered the astronomical text *Brhatsamhita* of Varahamihira to be translated into Persian, while his court astronomer, Mahendra Suri, wrote *Yantraraja*, which introduced astronomical principles and practices from Central and Western Asia into India. Unfortunately, the ideas proposed in the book, welding together principles from the *Siddhantas* and the Persian–Arabic system, did not get a favourable reception. However, ideas from the West continued to flow into India, including those of Nasir al-Din al-Tusi (1201–74). His *Kitab Zij-Ilkani*, based on the observations made at the Marghah Observatory, became the model for future *zij* (astronomical tables) in the Islamic world. Two of his texts, one each in Arabic and Persian, became compulsory reading for students of astronomy in Indian *madrassahs* (mosque schools) and appear to have inspired an astronomer, Muhammad Jaunpuri, to question the validity of Ptolemaic planetary model.

Geometry has always been an integral part of astronomy. However, it has developed on different lines in different cultural areas. In India, as we saw in Chapter 8, it originated to serve the needs of rituals and cosmographic speculations. In Central and Western Asia, under Islamic rule, Euclidean geometry, which developed as a deductive science, became dominant. As a result, students of mathematics and astronomy, taught in Arabic-Persian language schools in India, required a thorough knowledge of Euclidean geometry before they took up the study of Ptolemy's *Almagest*, or Archimedes' *On the Sphere and Cylinder*, or Apollonius' *Conics*. The most widely used Arabic translation of Euclid's *Elements* was Nasir al-Din al-Tusi's *Tahrir Uqlidis*, which in turn went into a number of Persian translations. In 1732, seven centuries after al-Biruni's attempt to introduce Euclid to India, Jagannatha Samrata translated the Persian text into a Sanskrit version of the *Elements* entitled *Rekhaganita*. Only 100 years later, five chapters of Hutton's *Euclidean Geometry* were rendered from English into Sanskrit by Yogadyana Misra in Calcutta. By then, English-language education had become the norm in a number of schools in major cities of India.

The major differences between the two streams of mathematical activity, namely the one within the Sanskritic tradition, and the other within the Arabic–Persian tradition, is well brought out if one considers the research preoccupations of the two groups. The Kerala mathematicians, with their work on infinite series, were inspired by Aryabhata and his school. The Indian mathematicians working within the Greek–Arabic–Persian tradition, whose interests were primarily in Greek geometry and Ptolemaic astronomy, offered another model. An interesting illustration of the difference between the preoccupations of the two traditions is shown by the work of Ghulam-Hussain Jaunpuri (*b.* 1790), one of the notable Indian mathematicians from the Arabic–Persian tradition. In *Jame-i-Bahadur Khani*, composed in 1833, Juanpuri tackles the problem of trisecting an angle, one that has engaged mathematicians from the Greek–Arabic–European tradition over a long period of time, including such notable names as Archimedes, Al-Biruni, Thabit ibn Qurra and François Viète. Various approaches were tried out, including those involving conic sections, transcendental curves, circles and *neusis* (insertion of a static line). Jaunpuri used the last of these methods to achieve a construction that is sound and practicable. The details of his method are found in Rizvi (1983) and are of little importance within the context of this discussion. What would have been very unlikely is that such a problem would have engaged the interest of an Indian mathematician from the Sanskritic tradition. The two parallel traditions met in a few cases involving astronomy, but hardly ever on matters relating to pure mathematics. This lack of contact was a missed opportunity which has had considerable repercussions for the development of Indian mathematics. But that is another story.

10 Prelude to Modern Mathematics: The Arab Contribution

Changing Algebra: Al-Khwarizmi to Abu-Kamil to Al-Karaji

Around AD 830, Muhammad ibn Musa al-Khwarizmi composed the earliest known Arab treatment of algebra, beginning a preoccupation that continued for several centuries. His treatise, *Hisab al-jabr wa'l-muqabala*, began with a discussion of the algebra of first- and second-degree equations, proceeding to practical applications of this algebra to questions of mensuration and legacies. Different influences and sources went into the making of this treatise. The use of geometry to justify algebraic manipulation has been traced to Euclid's *Elements*, although Chinese and Indian mathematics, discussed in earlier chapters, contain similar strands. His treatment of practical geometry has similarities to an anonymous Hebrew text, *Mishnat ha-Middot*. This is not surprising now that it has been established that the Hebrew text was indebted to al-Khwarizimi's work in the first place (Sarfati, 1968). His interest in second-degree equations and the rhetorical style of his presentation show vestiges of both Babylonian and Indian influences.

In the opening pages of his book, al-Khwarizmi distinguishes between six types of algebraic equation, up to and including the quadratic, and then proceeds to systematically present solutions, some of them known since Babylonian times. However, what differentiates his treatment from the

Babylonian approach is the geometric justification for his algebra. Thus, after explicitly stating that '. . . it is necessary that we should demonstrate geometrically the truth of the same problems which we have explained in numbers', he proceeds to justify the solution to the quadratic expressed in modern notation as $x^2 + 10x = 39$ with two different geometrical constructions both yielding completion of the square. One of the geometric constructions is discussed on pp. 325–6.

A generation later, Abu-Kamil ibn Aslam (*c*. 850–930) wrote his own text, *Kitab fi al-jabr wa'l-muqabala*, based on al-Khwarizmi's work. In it, Abu-Kamil not only quotes directly from al-Khwarizmi, but also incorporates almost half of al-Khwarizmi's 40 examples into his work, just with numerical changes. However, the mathematical environment in which Abu-Kamil's thought developed had undergone significant changes, notably in that the Greek influence had become more manifest. Compare, for example, Abu-Kamil's proof of the solution of the equation $x^2 + 10x = 39$ with the proof of al-Khwarizmi. A detailed step-by-step comparison is found in Joseph (1994c, pp. 65–7) and will not be discussed here. What Abu-Kamil did was to use the underpinnings of Greek mathematics without destroying the concrete base of al-Khwarizmi's algebra to create an algebra based on practical realities. This approach remained unchanged in other commentaries on al-Khwarizmi's work until the middle of the tenth century.

The next change was a result of the impact of Abul-Wafa's commentary on Diophantus' *Arithmetica*, which introduced the Diophantine approach to the solutions of determinate and indeterminate equations. It is important, in this context, to distinguish between the approach to and the actual solution of indeterminate equations since it is likely that the Arabs had come across Indian and, possibly, Chinese work on the subject.

One mathematician whose work reflected this change in approach was Abu Bakr al-Karaji (al-Karkhi) (*c*. 1000). In *al-Fakhri*, al-Karaji solved indeterminate equations of degrees two and three in up to three unknowns by methods clearly influenced by the *Arithmetica* while retaining the geometrical algebra of al-Khwarizmi. At the same time, he rejected aspects of both approaches, in particular al-Khwarizmi's complete reliance on geometry for proof and Diophantus' yncopated notation. However, al-Karaji's innovations had only limited impact on Arab algebra, which continued to be dominated by the ideas of al-Khwarizmi.

It is interesting that, in the westward movement of algebra, it was mainly the twelfth-century Latin translations of al-Khwarizmi's text, by Robert of Chester and Gerard of Cremona, that shaped the mathematical environment of medieval Europe. The same applied in the case of Fibonacci, who came across al-Khwarizmi's work either from reading one or both translations, or learning algebra during his youth in North Africa, or during his travels as an adult in Egypt, Syria, Greece and Sicily. The fact remains that al-Khwarizmi's ideas on the theory of quadratic equations figured prominently in the fifteenth and final chapter of Fibonacci's influential book, *Liber abaci* (1202).

Following the example of Fibonacci's Babylonian/Indian/Greek/Arab ancestors, Fibonacci adopted a rhetorical style in stating the problems, solved the problems using algebra and provided a geometric rationale to justify his solutions. And, in conformity with the mood of the day, the algebraic treatment combined the practical with the theoretical in a mode of presentation which Europe retained for the next two centuries, until the needs of the emerging mercantile class in Italy eventually led to the replacement of Latin texts by texts written in the vernacular. The latter (known as *libri d'abaco*) retained vestiges of the Arab style under the cloak of authority of Fibonacci. However, the demands of the new commercial environment eventually led to a growing gap between the emerging popular textbook approach and the approach favoured by the Fibonacci model in the range of problems considered, in the type of justification presented, and in the language and form of presentation. Eventually, the need for problem-solving texts gradually diminished during the course of the fifteenth century, with the practical texts being supplanted by texts of a more theoretical nature. An early example of the new model was Luca Pacioli's *Summa* (1494), which aimed at a synthesis of the ideas and approach of the *libri d'abaco* while also drawing from the theoretical portions of works such as Euclid's *Elements* and Fibonacci's *Liber abaci*. Europe had at last outgrown its need for Arab algebra.

Mathematics from Hebrew Sources

The oldest mathematical text in Hebrew is the *Mishnat ha-Middot*. Of unknown authorship and uncertain date, it gave practical rules for mensuration, including that of the Tabernacle that the Jews constructed in the desert. There is a section on the solution of quadratic equations and the geometric analogues of these equations, which may have been a later inclusion since its composition was strongly influenced by al-Khwarizmi's algebra from a later period. It is only in the twelfth century that there emerged significant Hebrew mathematics in the persons of Abraham bar Hiyya (*c.* 1065–1145) and Abraham ibn Ezra (1092–1167).

Abraham bar Hiyya spent most of his life in Barcelona in Christian Spain, although it is likely that he was educated in the Muslim kingdom of Zaragossa. His major text, written in Hebrew and entitled *Hibbur ha-Meshiha we ha Tisboret* ('The Composition on Geometrical Measures'), has an extensive coverage that includes geometrical algebra akin to al-Khwarizmi's work, mensuration of plane figures and solids including the truncated pyramid, some trigonometry of chords and a lost fragment of Euclid's work on the 'division of figures'. The text was translated into Latin as *Liber embadorum* (1145) by Plato of Tivoli, and would play a significant role in the spread of mathematical knowledge into Europe.

The contribution of Abraham ibn Ezra is more difficult to assess. Born in Aragon and imbued with Arab culture and learning, he was a versatile scholar and poet. His mathematical reputation is essentially based on one book, *Sefer ha-Mispar* ('The Book of Numbers'). Written before 1160, it is one of the first expositions in Europe of the Indo-Arabic numerals and arithmetical operations performed with them. It was an important text in the diffusion of Indian arithmetic among the Hebrew readers. He also wrote another book, *Sefer ha-Ehad* ('The Book of the One'), that was highly prized by some Jewish scholars of the period for the 'mathematical hints' (or numerology) contained in it.

The role of the two Abrahams was to open up the availability of basic mathematical knowledge to Jewish communities who were either unfamiliar with Arabic sources or did not have direct access to these sources. A scientific language close to biblical Hebrew became the vehicle for the transmission of such knowledge. Translations of scientific and philosophical texts from Arabic into Hebrew gathered momentum. For example, Jacob Anatoli (*c.* 1194–1256), living in Italy under the patronage of Frederic II of Sicily, translated Ptolemy's *Almagest*, Ibn Rushd's *Compendium on Astronomy* and Euclid's *Elements* into Hebrew. Moses ben Samuel ben Yehuda of Montpellier (*fl.* 1240–83) carried out a number of translations of mathematical texts from Arabic, including works of Euclid, al-Farabi ('Commentary on the *Elements*'), Ibn al-Haytham ('Commentary on the *Elements*') and al-Hassar ('The Arithmetic'). Other major translators of mathematical literature from Arabic into Hebrew included Jacob ben Makhir of Montpellier (*c.* 1236–1305), Yehuda ben Solmon ha-Kohen (*b.* 1215) and Qalonymos ben Qalonymos (*b.* 1287). The activities of these and other translators meant that, by the middle of the thirteenth century, there were Hebrew versions of the major Greek texts of Euclid, Archimedes and Ptolemy, and works of Arab mathematicians, such as Abu Kamil ('Algebra') and Ibn al-Haytham ('Astronomy').

An outstanding example of the resulting creativity was Levi ben Gershom

(1288–1344), better known in Europe by his Latinized name, Gersonides. He was one of the most versatile Jewish scholars of the European Middle Ages, and wrote on philosophy, logic, mathematics and astronomy. His treatise *Ma'asheh Hoshev* ('Computer's Manual') contains studies of permutations, combinations and summations of series, as well as an introduction to arithmetic and algebra. He also wrote commentaries on the *Elements*, including a treatise on the 'parallel postulate', and on Thabit ibn Qurra's *Risala fi Shakl al-qatta* ('On the Secant Figure'). However, his most lasting contribution is contained in his book on astronomy, *Sefer Tekhunah*, which is preserved in both the Hebrew and Latin versions. A lengthy work, divided into 136 chapters, it is notable for its emphasis on astronomical observations and its critical assessment of Ptolemy and the Arab astronomer al-Bitruji. Using his own solar and lunar models, he calculated sine tables, spherical astronomical parameters, and solar and lunar mean motions and corrections. His practical bent led him to the invention of a transversal scale that helped to reduce the random errors introduced into observations when reading off minutes from the linear scale of an astrolabe calibrated in degrees. His writings on harmonic numbers (now only extant in Latin translation), combinatorial analysis and the geometry of Euclid have been studied in depth by Tony Levy (1996) for their Arabic sources, and for their impact on later work. As mentioned in the 'Reflections' on Chapter 1, there is the possibility of a transmission of the basic formulae for finding permutations and combinations from Gerson's *Computer Manual* to a Cardano manuscript which reappeared in Mersenne's classical book on music theory.

In this short survey of mathematicians from the Hebrew tradition, we have omitted discussion of one of the most influential figures in Jewish intellectual history, Moses Maimonides (*d.* 1204). His early education in Andalusian Spain introduced him to the scientific treatises of fellow Andalusians, Jabir ibn Aftah and al-Mu'tamir ibn Hud. He wrote in both Arabic and Hebrew, and his notable works in the former language include commentaries on Apollonius' *Conics* and ibn al-Haytham's *Completion of the Conics* (Langermann, 1984).

A study of mathematics from Hebrew sources is an instructive exercise. Like Arab mathematics, of which it was an integral part at times, it went through different phases: translation, assimilation, synthesis, creation and transmission. Mathematical ideas and practices contained in Arabic texts were translated into Hebrew, and then assimilated by Jewish communities unfamiliar with Arabic sources. From their ranks came notable scholars who synthesized and developed the subject. Their writing became an important channel for the diffusion of ideas and texts before Europe discovered the original Greek texts.

Mathematics of the Maghreb

There is a tendency in studying Arab mathematics to confine one's attention to the activities around the Middle East and Spain, and ignore the work in North and North-west Africa, a large region referred to as the Maghreb, in which there occurred mathematical activity within the framework of the Islamic civilization. The starting point of this activity was the link between Andalusia in Spain and the Maghreb, where the close political, economic and cultural ties meant that it was difficult to separate the two regions of the Muslim West.

From the information available, the beginnings of mathematical activity began in Ifriqya (present-day Tunisia) around the end of the eighth century, when there is a record of a scholar known as Yahya al-Kharraz who wrote a book on metrology. For the next two centuries, the information about mathematical activity in the region is scarce although there is some evidence of knowledge moving westwards from Baghdad. Abu Sahl al-Qayarawani, who may have spent some time in Baghdad, is the first known mathematician from the Maghreb. He wrote a book entitled *Kitab fi al-hisab al-hindi* explaining Indian arithmetic. Around the same period, the ruler Ibrahim II (875–902) established a *Bait al-Hikma* (House of Wisdom) in Raqqada, modelled on the lines of the one in Baghdad, to promote the study of mathematics, astronomy and other subjects. This institution survived as a scientific centre until the establishment of the Fatimid Caliphate in Egypt. The patronage of learning was re-established during the reign of Fatimid Caliph al-Mu'izz and, although we know that the study of mathematics and astronomy flourished during this period, there is little information on the scholars and the type of work that they undertook.

There is considerably more information available on mathematical activity in the Maghreb during the Almohad period (twelfth and thirteenth centuries). Of the three mathematicians who have been identified and studied, Ibn Mun'im's (*d.* 1228) work is the most innovative. In his only extant work, *Fiqh al-hisab*, dealing with combinatory problems, he states the rule for determining all possible combinations of n colours p times and establishes, inductively, the resulting arithmetic triangle of the relationship:

$$^nC_p = {}^{n-1}C_{p-1} + {}^{n-2}C_{p-2} + \ldots + {}^{p-1}C_{p-1}$$

He applies similar formulae for permutations with and without repetitions using the Arabic alphabet for illustrative purposes. One of the more remarkable aspects of his work is the use of combinatorial reasoning, the earliest attempt in Arab mathematics to do so.

With the advent of the fourteenth century, the quantity of mathematical writings increased substantially, although many of them took the form of commentaries on and summaries of previous work. However, this century produced a mathematician of exceptional ability: Ibn al-Banna (*c.* 1256–1321) was born in Marrakesh (Morocco). He was a versatile scholar, having made a study of the Arabic language, the Qu'ran, astronomy, mathematics and medicine. His fame rests mainly on his work in mathematics. His best-known book, *Talkis a'mal al-hisab* ('Summary of Arithmetical Operations') is still extant. This was widely known in the Arab world because of its clarity and conciseness. He also wrote two other texts on calculation which have survived: *al-Qanun al-hisab* ('Manual of Mathematics') and *Raf'al-hijab* ('Lifting of the Veil'). Other texts include an introduction to Euclid, a book on algebra, a treatise on geometry and a popular astronomical almanac.

The Maghreb tradition carried on for a few centuries, but the quality of the work declined. Mathematicians were reduced to teaching or mundane activities such as advising on the determination of times for prayer, distributing inheritance or using astronomical instruments. Yet sparks of curiosity and inventiveness still occasionally appeared. Muhammad ibn Muhammad al-Fullani al-Kishnawi was a Fulani from northern Nigeria. He travelled to Egypt and, in 1732, he wrote a manuscript (in Arabic) on procedures for constructing magic squares up to order 11. As words of encouragement to the reader he writes, 'Do not give up, for that is ignorance and not according to the rules of this art . . . Like the lover, you cannot hope to achieve success without infinite perseverance.' He died in Cairo in 1741.

General methods for constructing magic squares first appeared in the Islamic world, including the Maghreb, during the ninth century. From the thirteenth century, magic and divinatory applications began to replace mathematical study. However, interest in the mathematics of construction survived, as shown by the work of Muhammad ibn Muhammad al-Fullani al-Kishnawi.

Bibliography

◇◇

In preparing this bibliography, I have been conscious of the fact that a number of the specialist references, particularly journal articles relating to Indian mathematics, are not easily accessible. There is no easy solution to this problem. For British readers in search of the more inaccessible references, I would recommend the following three sources, which I found useful:

1 The British Library, and in particular the India Office Library and Records.
2 The School of Oriental and African Studies, University of London.
3 The Inter-University Library Loans Scheme, which can quickly trace any reference to be found in a British university. Some of the larger public libraries can, given sufficient notice, acquire books and articles through their own Inter-Library Loan Scheme.

GENERAL WORKS

A. Al-Daffa and J. J. Stroyls (1984), *Studies in the Exact Sciences in Medieval Islam*, New York, John Wiley & Sons.

M. Ascher and R. Ascher (1981), *Code of the Quipu*, Ann Arbor, MI, University of Michigan Press.

A. K. Bag (1979), *Mathematics in Ancient and Medieval India*, Varanasi, Chaukhambha Orientalia.

E. T. Bell (1940), *Development of Mathematics*, New York, McGraw-Hill.

J. L. Berggren (1986), *Episodes in the Mathematics of Medieval Islam*, New York, Springer-Verlag.

J. D. Bernal (1969), *Science in History*, 4 volumes, London, Penguin.

M. Bernal (1987), *Black Athena*, London, Free Association Books.

D. M. Bose, S. N. Sen and B. V. Subbarayappa (eds) (1971), *A Concise History of Science in India*, New Delhi, Indian National Science Academy.

C. B. Boyer (1968), *A History of Mathematics*, New York, John Wiley (Reprinted by Princeton University Press, Princeton, NJ, 1985).

D. Chattopadhyaya (1986), *History of Science and Technology in Ancient India: The Beginnings*, Calcutta, Firma KLM PVT.

M. P. Closs (1986), *Native American Mathematics*, Austin, TX, University of Texas Press.

B. Datta and A. N. Singh (1962), *History of Hindu Mathematics*, 2 volumes, Bombay, Asia Publishing House.

W. M. Denevan (ed.) (1992), *The Native Population of the Americas in 1492*, 2nd Edition, Madison, WI, University of Wisconsin Press.

Dharampal (1971), *Indian Science and Technology in the Eighteenth Century*, New Delhi, Biblia Impex.

——— (1983), *The Beautiful Tree*, New Delhi, Biblia Impex.

M. Edwardes (1971), *East–West Passage*, London, Cassell.

R. Eglash (1997), 'When math worlds collide: intention and invention in ethnomathematics', *Science, Technology and Human Values*, **22**, (1), 79–97.

H. Eves (1983), *An Introduction to the History of Mathematics*, 5th Edition, Philadelphia, Saunders.

J. Fauvel and J. Gray (eds) (1987), *The History of Mathematics: A Reader*, London, Macmillan.

G. Flegg (1983), *Numbers: Their History and Meaning*, London, André Deutsch.

——— (ed.) (1989), *Numbers Through the Ages*, London, Macmillan, in association with the Open University.

C. C. Gillespie (ed.) (1969–), *Dictionary of Scientific Biography*, 15 volumes, New York, Charles Scribner's Sons.

R. J. Gillings (1972), *Mathematics in the Time of the Pharaohs*, Cambridge, MA, MIT Press.

R. Gurukkal (1992), *The Kerala Temple and the Early Medieval Agrarian System*, Sukapuram, Vallathol Vidyapeetham.

G. B. Halsted (1912), *On the Foundation and Technic of Arithmetic*, London, The Open Court Company.

T. L. Heath (1921), *A History of Greek Mathematics*, Oxford, Clarendon Press (Reprinted by Dover, New York, 1981).

M. G. S. Hodgson (1977), *The Venture of Islam: Conscience and History in a World Civilization*, Volume 1, Chicago, University of Chicago Press.

G. Ifrah (1985), *From One to Zero: A Universal History of Numbers*, New York, Viking Penguin.

G. G. Joseph, V. Reddy and M. Searle-Chatterjee (1990), 'Eurocentrism in the social sciences', *Race and Class*, **31**, (4), 1–26.

G. G. Joseph (1994a), 'The politics of anti-racist mathematics', *European Education Journal*, July, 67–74.

———— (1995a), 'Mathematics and Eurocentrism', in *Enduring Western Civilization*, S. Federici (ed.), Westport, CT, Praeger Publishers, 119–35.

———— (1995b), 'Cognitive encounters in India during the age of imperialism', *Race and Class*, **36**, (3), 39–56.

———— (1995c) (In Malayalam) 'Transmissions of mathematical knowledge during the medieval period', *Kerala Padanengal*, **5**, 126–140.

———— (1996a), 'Calculating people: origins of numeracy in India and the West', in *Challenging Ways of Knowing in English, Maths and Science*, D. Baker, J. Clay and C. Fox (eds), London, Falmer Press, 194–203.

———— (1997b), 'Foundations of Eurocentrism in mathematics', in *Ethnomathematics: Challenging Eurocentrism in Mathematics*, M. Frankenstein and A. Power (eds), Albany, NY, SUNY Press, 61–81.

———— (1997c), 'Mathematics', in *Encyclopaedia of the History of Science, Technology and Medicine in Non-Western Cultures*, H. Selin (ed.), Dordrecht, Kluwer, 604–10.

R. Kanigel (1991), *The Man Who Knew Infinity: A Life of the Genius Ramanujan*, New York, Charles Scribner's Sons.

B. Kemp (1991), *Ancient Egypt: Anatomy of a Civilization*, London, Routledge.

M. Kline (1953), *Mathematics in Western Culture*, New York, Oxford University Press (Reprinted by Penguin, London, 1972).

———— (1962), *Mathematics: A Cultural Approach*, Reading, MA, Addison-Wesley.

———— (1972), *Mathematical Thought from Ancient to Modern Times*, New York, Oxford University Press.

D. F. Lach (1965), *Asia in the Making of Europe*, 2 volumes, Chicago, Chicago University Press.

Lam Lay Yong and Ang Tian Se (1992), *Fleeting Footsteps: Tracing the Conception of Arithmetic and Algebra in Ancient China*, Singapore, World Scientific.

D. F. Lancy (1983), *Cross Studies in Cognition and Mathematics*, New York, Academic Press.

Li Yan and Du Shiran (1987), *Chinese Mathematics: A Concise History*, translated by J. N. Crossley and A. W.-C. Lun, Oxford, Clarendon Press.

U. Libbrecht (1973), *Chinese Mathematics in the Thirteenth Century*, Cambridge, MA, MIT Press.

K. Menninger (1969), *Number Words and Number Symbols: A Cultural History of Numbers*, Cambridge, MA, MIT Press.

H. O. Midonick (1965), *The Treasury of Mathematics*, 2 volumes, London, Peter Owen (Reprinted by Penguin, London, 1968).

Y. Mikami (1913), *The Development of Mathematics in China and Japan*, New York, Hafner (Reprinted by Chelsea, New York, 1974).

S. H. Nasr (1968), *Science and Civilization in Islam*, Cambridge, MA, Harvard University Press.

J. Needham (1954), *Science and Civilisation in China*, Vol. 1, Cambridge, Cambridge University Press.

———— (1959), *Science and Civilisation in China*, Vol. 3, Cambridge, Cambridge University Press, pp. 1–168.

O. Neugebauer (1962), *The Exact Sciences in Antiquity*, New York, Harper & Row (Reprinted by Dover, New York, 1969).

G. Robins and C. Shute (1989), *The Rhind Mathematical Papyrus*, London, British Museum Publications.

W. W. Rouse Ball (1908), *A Short Account of the History of Mathematics*, New York, Macmillan (Reprinted by Dover, New York, 1960).

E. Said (1978), *Orientalism*, New York and London, Vintage.

T. A. Sarasvati (1979), *Geometry in Ancient and Medieval India*, Delhi, Motilal Banarsidass.

G. Sarton (1927), *Introduction to the History of Science*, Vol. 1, Baltimore, MD, Williams & Williams Co.

D. E. Smith (1923/5), *History of Mathematics*, 2 volumes, Boston, MA, Ginn & Co. (Reprinted by Dover, New York, 1958).

———— and Y. Mikami (1914), *A History of Japanese Mathematics*, Chicago, Open Court.

D. J. Struik (1965), *A Concise History of Mathematics*, London, Bell.

I. Van Sertima (1983), *Blacks in Science*, New Brunswick, Transactions Books.

B. L. van der Waerden (1961), *Science Awakening*, New York, Oxford University Press.

———— (1983) *Geometry and Algebra in Ancient Civilizations*, Berlin, Springer-Verlag.

C. Zaslavsky (1973), *Africa Counts: Number and Pattern in African Culture*, Boston, MA, Prindle, Weber & Schmidt.

SPECIALIST REFERENCES

A. Aaboe (1954), 'Al-Kashi's iteration method for sin 1°', *Scripta Mathematica*, **20**, 24–9.

W. H. Abdi (1982), 'Some works of Indian mathematicians who wrote in Persian', *Ganita Bharati*, **4**, (2), 6–9.

José de Acosta (1596), *Historia Natural Moral de las Indias*, Madrid.

S. H. Alatas (1976), *Intellectuals in Developing Societies*, London, Frank Cass.

A. Alsdorf (1991), 'The *pratyayas*: Indian contribution to combinatorics', *Indian Journal of History of Science*, **26**, (1), 17–61.

G. E. Andrews (1979), 'An introduction to Ramanujan's "lost" notebook', *American Mathematics Monthly*, **86**, February, 89–108.

Ang Tian Se (1978), 'Chinese interest in right-angled triangles', *Historia Mathematica*, **5**, 253–66.

M. Ascher (1988), 'Graphs in cultures II. A study in ethnomathematics', *Archive for History of Exact Sciences*, **39**, 75–95.

———— and R. Ascher (1972), 'Numbers and relations from ancient Andean *quipus*', *Archive for History of Exact Sciences*, **8**, 288–320.

A. F. Aveni (1981), 'Tropical archeoastronomy', *Science*, **213**, 161–71.

A. K. Bag (1976), 'Madhava's sine and cosine series', *Indian Journal of History of Science*, **11**, (1), 54–7.

T. Baqir (1950), 'An important mathematical text from Tell Harmal', *Sumer*, **6**, 39–54.

———— (1951), 'Some more mathematical texts from Tell Harmal', *Sumer*, **7**, 28–45.

———— (1962), 'Foreword', *Sumer*, **18**, 5–14.

M. Bernal (1992), 'Animadversions on the origins of Western science', *ISIS*, **83**, 596–607.

———— (1994), 'Response to Palter', *History of Science*, **32**, 445–64.

J. Bogoshi, K. Naidoo and J. Webb (1987), 'The oldest mathematical artefact', *Mathematical Gazette*, **71**, (458), 294.

N. J. Bolton and D. N. Macleod (1977), 'The geometry of *Sriyantra*', *Religion*, **7**, (1), 66–75.

R. Bonala (1955), *Non-Euclidean Geometry*, New York, Dover.

C. B. Boyer (1944), 'Fundamental steps in the development of numeration', *ISIS*, **35**, 153–68.

M. Bruckheimer and Y. Salamon (1977), 'Some comments on R. J. Gillings' analysis of the $2/n$ table in the Rhind Papyrus', *Historia Mathematica*, **4**, 445–52.

E. M. Bruins (1955), 'Pythagorean triads in Babylonian mathematics: The errors in Plimpton 322', *Sumer*, **11**, 117–21.

A. B. Chace (1979), *The Rhind Mathematical Papyrus*, Reston, VA, National Council of Teachers of Mathematics.

———— L. S. Bull, H. P. Manning and R. C. Archibald (eds) (1927/9), *The Rhind Mathematical Papyrus*, 2 volumes, Buffalo, NY, Mathematical Association of America.

M. P. Closs (1992), 'I am a Kahal; my parents were scribes', *Research Reports on Ancient Maya Writing*, No. 39, Center for Maya Research, Washington, DC, USA.

———— (1995), 'Mathematicians and mathematical education in ancient Maya Society', Mimeo, privately circulated.

D. Crowe (1971), 'The geometry of African art: I. Bakuba Art', *Journal of Geometry*, **1**, 169–82.

———— (1975), 'The geometry of African art: II. A catalogue of Benin patterns', *Historia Mathematica*, **2**, 253–71.

———— (1982), 'The geometry of African art: III. The smoking pipes of Begho', in *The Geometric Vein*, the Coxeter Festschrift, C. Davis, B. Grunbaum and F. Sherk (eds), New York, Springer-Verlag, 177–89.

B. Datta (1929), 'The Bakhshali Manuscript', *Bulletin of the Calcutta Mathematical Society*, **21**, 1–60.

———— (1932), *The Science of the Sulbas: A Study in Early Hindu Geometry*, Calcutta, Calcutta University Press.

B. Davidson (1987), 'The ancient world and Africa', *Race and Class*, **29**, (2), 1–16.

C. L. Day (1967), *Quipus and Witches' Knots*, Lawrence, KS, University of Kansas Press.

A. Djebbar (1981), *Enseignement et recherche mathématique dans le Maghreb des XIIIe–XIVe siècles*, Paris, Publications Mathématiques d'Orsay.

———— (1985), *L'Analyse combinatoire au Maghreb: l'exemple de Ibn Munim (XIIe–XIIIe siècles)*, Paris, Publications Mathématiques d'Orsay.

———— (1990), *Mathématiques et mathématiciens du Maghreb, Médiéval (IX–XVI siècles)*, Contribution à l'étude des activités de l'Occident mussulman. Thèse de Doctorat, Université de Nantes.

———— (1997), 'Mathematics of Africa: The Maghreb', in *Encyclopaedia of the History of Science, Technology and Medicine in Non-Western Cultures*, H. Selin (ed.), Dordrecht, Kluwer, pp. 613–16.

R. Eglash (1995), 'Fractal geometry in African material culture. Symmetry: natural and artificial symmetry', *Culture and Science*, **6**, (1), 174–7.

———— (1997), 'When math worlds collide: Intention and invention in ethnomathematics', *Science, Technology and Human Values*, **22**, (1), 77–97.

H. Fahim (1982), *Indigenous Anthropology in Non-Western Countries*, North Carolina University Press.

D. H. Fowler (1987), *The Mathematics of Plato's Academy: A New Reconstruction*, Oxford, Clarendon Press.

J. Friberg (1981), 'Methods and traditions of Babylonian mathematics:

Plimpton 322, Pythagorean triples and the Babylonian triangle parameter equations', *Historia Mathematica*, **8**, 277–318.

S. Gandz (1938), 'The algebra of inheritance', *Osiris*, **5**, 319–91.

S. Ganguli (1929), 'Notes on Indian mathematics: A criticism of G. R. Kaye's interpretation', *ISIS*, **12**, 132–45.

P. Gerdes (1985), 'Three alternative methods of obtaining the ancient Egyptian formula for the area of a circle', *Historia Mathematica*, **12**, 261–8.

———— (1986), 'How to recognize hidden geometrical thinking: A contribution to the development of anthropological mathematics', *For the Learning of Mathematics*, **6**, (2), 10–17.

———— (1988a) 'On possible uses of traditional Angolan sand drawings in the mathematical classroom', *Educational Studies in Mathematics*, **19**, (1), 3–22.

———— (1988b), 'A widespread decorative motif and the Pythagorean theorem', *For the Learning of Mathematics*, **8**, (1), 35–9.

———— (1988c), 'On culture, geometrical thinking and mathematics education', *Educational Studies in Mathematics*, **19**, (3), 123–35.

———— (1995), *Ethnomathematics and Education in Africa*, Stockholm, Stockholm Universitet.

———— and J. Fauvel (1990), 'African slave and calculating prodigy: Bicentenary of the death of Thomas Fuller', *Historia Mathematica*, **17**, 141–51.

R. J. Gillings (1962), 'Problems 1 to 6 of the Rhind Mathematical Papyrus', *The Mathematics Teacher*, **55**, 61–9.

———— (1964), 'The volume of a truncated pyramid in ancient Egyptian papyri', *The Mathematics Teacher*, **57**, 552–5.

B. S. Gillon (1977), 'Introduction, translation and discussion of Chao Chung Ching's "Notes to the Diagrams of Short Legs and Long Legs and of Squares and Circles"', *Historia Mathematica*, **4**, 253–93.

S. R. K. Glanville (1927), 'The mathematical leather roll in the British Museum', *Journal of Egyptian Archaeology*, **13**, 232–8.

J. R. Gonzalez a Reguera (1980), 'Teoría fisiológica del placer est, tico', *I Conferencia Científica de Investigaciones Sobre Arte*, Havana, Instituto Superior de Arte.

B. Gunn and T. E. Peet (1929), 'Four geometrical problems from the Moscow mathematical papyrus', *Journal of Egyptian Archaeology*, **15**, 167–85.

R. C. Gupta (1975), 'Circumference of the *Jambudvipa* in Jaina cosmography', *Indian Journal of History of Science*, **10**, (1), 38–46.

———— (1977), 'Paramesvara's rule for the circumradius of a cyclic quadrilateral', *Historia Mathematica*, **4**, 67–74.

————— (1978), 'Indian values of the sinus totus', *Indian Journal of History of Science*, **13**, (2), 125–43.

T. A. Hayashi, T. Kusuba and M. Yano (1989), 'Indian values for π from Aryabhata's value', *Historia Scientiarum*, **37**, 1–16.

————— ————— ————— (1990), 'The correction of the Madhava series for the circumference of a circle', *Centaurus*, **33**, 149–74.

J. de Heinzelin (1962), 'Ishango', *Scientific American*, **206**, June, 105–16.

Herodotus (1984), *The Histories*, London, Penguin.

A. F. R. Hoernle (1888), 'The Bakhshali manuscript', *Indian Antiquary*, **17**, 33–48 and 275–9.

J. Hoyrup (1985), 'Varieties of mathematical discourse in pre-modern socio-cultural contexts: Mesopotamia, Greece and the Latin Middle Ages', *Science and Society*, **49**, (1), 4–41.

————— (1987), 'Algebra and naive geometry: An investigation of some basic aspects of Old Babylonian mathematical thought', *Filosofi og Videnskabsteori pa Roskilde Universitetscenteri*, Preprints and Reprints No. 2.

————— (1988), 'On parts of parts and ascending continued fractions', *Filosofi og Videnskabsteori pa Roskilde Universitetscenteri*, Preprints and Reprints No. 2.

————— (1994), 'Babylonian mathematics', in *Companion Encyclopedia of the History and Philosophy of the Mathematical Sciences*, I. Grattan-Guinness (ed.), London, Routledge, 21–29.

G. R. Hunter (1934), *The Script of Harappa and Mohenjodaro and its Connection with Other Scripts*, London, Kegan Paul, Trenchy, Trabner & Co.

L. C. Jain (1973), 'Set theory in the Jaina school of mathematics', *Indian Journal of History of Science*, **8**, (1/2), 1–27.

————— (1982), *Exact Sciences from Jaina Sources*, 2 volumes, Jaipur, Rajasthan Prakrit Bharti Sansthan.

P. Jeganathan (1993), *On the Structural Reading and the Evolution of the Indus Script*, Technical Report 220, Department of Statistics, University of Michigan, Ann Arbor.

W. T.. Johnson (1997), 'Mathematics in the Pacific', in *History of Science, Technology and Medicine in Non-Western Cultures*, H. Selin (ed.), Dordrecht, Kluwer, 657–9

G. G. Joseph (1994b), 'Tibetan astronomy and mathematics', in *Companion Encyclopedia of the History and Philosophy of the Mathematical Sciences*, I. Grattan-Guinness (ed.), London, Routledge, 131–6.

————— (1994c), 'Different ways of knowing: Contrasting styles of argument in Indian and Greek mathematical traditions', in *Mathematics, Education and*

Philosophy: An International Perspective, P. Ernest (ed.), London, Falmer Press, 194–204.

———— (1994d), 'The Arabs', in Nuffield Advanced Mathematics, *History of Mathematics*, London, Longman Education, 54–70.

———— (1996b), 'The geometry of Vedic altars', in *Nexus: Geometry and Architecture*, K. Williams (ed.), Fucecchio, Edizioni Dell'Erba, 97–113.

———— (1996c), 'Mathematics in ancient India. Harappa civilization: Mathematics in metrology', *Mathesis*, **12**, 236–83.

———— (1996d), '*Infinite series across cultures: A study of mathematical motivation*', University of Manchester, School of Economic Studies, Discussion Paper No. 9615.

———— (1997a), 'What is a square root? A study of geometrical representation in different mathematical traditions', *Mathematics in Schools*, **23**, (1), 4–9.

S. C. Kak (1989), 'Indus writing', *Mankind Quarterly*, **30**, 113–18.

———— (1990), 'Indus and Brahmi: Further connections', *Cryptologia*, **14**, (2), 169–83.

———— (1993), 'Astronomy of the Vedic altars', *Vistas in Astronomy*, **36**, 117–40.

D. S. Kasir (ed.) (1931), *The Algebra of Omar Khayyam*, New York, Teachers' College Press, Columbia University (Reprinted by College Press, New York, 1972).

V. J. Katz (1997), 'The transmission of mathematics from Islam to Europe', Abstracts of Papers Presented to the American Mathematical Society, **18**, 1, 5.

G. R. Kaye (1915), *Indian Mathematics*, Calcutta, Amd Simla.

———— (1933), 'The Bakhshali manuscript: A study in mediaeval mathematics', *Archaeological Survey of India*, New Imperial Series, **43**, Parts 1–3.

Y. W. Kim (1994), 'Korean mathematics', in *Companion Encyclopedia of the History and Philosophy of the Mathematical Sciences*, I. Grattan-Guinness (ed.), London, Routledge, 111–17.

A. P. Kulaichev (1984), '*Sriyantra* and its mathematical properties', *Indian Journal of History of Science*, **19**, 279–92.

R. P. Kulkarni (1971), 'Geometry as known to the people of Indus Civilisation', *Indian Journal of History of Science*, **13**, 117–24.

———— (1983), *Geometry According to Sulba Sutra*, Pune, Vaidika Samsodhana Mandala.

K. Kunjunni Raja (1963), 'Astronomy and mathematics in Kerala', *Adyar Library Bulletin*, **27**, 117–67.

I. Lakatos (1976), *Proofs and Refutations*, Cambridge, Cambridge University Press.

Lam Lay Yong (1970), 'The geometrical basis of the ancient Chinese square-root method', *ISIS*, **61**, 96–102.

——— (1974), 'Yang Hui's commentary on the *Ying Nu* chapter of the *Chiu Chang Suan Shu*', *Historia Mathematica*, **1**, 47–64.

——— (1977), *A Critical Study of the* Yang Hui Suan Fa, Singapore, University Press.

——— and Shen Kangsheng (1984), 'Right-angled triangles in ancient China', *Archive for History of Exact Sciences*, **30**, 87–112.

——— (1986), 'The conceptual origins of our numeral system and the symbolic form of algebra', *Archive for History of Exact Sciences*, **36**, 183–95.

——— (1987), 'Linkages: Exploring the similarities between the Chinese rod numeral system and our numeral system', *Archive for History of Exact Sciences*, **37**, 365–92.

Y. T. Langermann (1984), 'The mathematical writings of Maimonides', *Jewish Quarterly Review*, **75**, 57–65.

——— (1994), 'Mathematics in medieval Hebrew literature', in *Companion Encyclopedia of the History and Philosophy of the Mathematical Sciences*, I. Grattan-Guinness (ed.), London, Routledge, 137–42.

D. C. Laycock (1975), 'Observations on number systems and semantics', in *New Guinea Area Languages and Language Studies*, Volume 1, *Papua Languages and New Guinea Linguistic Scene*, S. A. Wurm (ed.), Canberra, Department of Linguistics, Research School of Pacific Studies, Australian National University.

G. A. Lean (1996), *Counting Systems of Papua New Guinea and Oceania*, 21 Volumes, Papua New Guinea, PNG Institute of Technology.

T. Levy (1996), 'Hebrew mathematics in the Middle Ages: An assessment', in *Tradition, Transmission, Transformation: Ancient Mathematics in Islamic and Occidental Cultures*, F. Jamil Ragep and Sally P. Rageb with Steven J. Livsey (eds), Leiden, E. J. Brill.

——— (1997), 'Mathematics of the Hebrew people', in *Encyclopaedia of the History of Science, Technology and Medicine in Non-Western Cultures*, H. Selin (ed.), Dordrecht, Kluwer, 632–4.

A. Lichtheim (1975), *Ancient Egyptian Literature*, Volume 1, Berkeley, University of California Press, 169–84.

L. L. Locke (1912), 'The ancient *quipu*, a Peruvian knot record', *American Anthropologist*, **14**, 325–52.

———— (1923), *The Ancient* Quipu *or Peruvian Knot Record*, New York, American Museum of Natural History.

B. Lumpkin (2000), 'The mathematical legacy of ancient Egypt: A response to Robert Palter', in *Debating Black Athena*, M. Bernal (ed.), Durham, NC, Duke University Press.

E. W. Mackie (1977), *The Megalithic Builders*, Oxford, Phaidon.

S. Madhavan (1991), 'Origin of *Katapyadi* system of numerals', *Sri Ravi Varma Samskrta Granthvali Journal*, **18**, (2), 35–48.

V. B. Mainikar (1984), 'Metrology in the Indus civilisation', in *Frontiers of the Indus Civilisation*, B. B. Lal and S. P. Gupta (eds), New Delhi, Motilal Banarsidass, 141–51.

A. Marshack (1972), *The Roots of Civilisation*, London, Weidenfeld & Nicolson.

J.-C. Martzloff (1994), 'Chinese mathematics', in *Companion Encyclopedia of the History and Philosophy of the Mathematical Sciences*, I. Grattan-Guinness (ed.), London, Routledge, 93–103.

M. J. Megitt (1958), 'Mae Enga time reckoning and calendar, New Guinea', *Man*, **58**, 74–7.

T. Murata (1975), 'Pour une interpretation de la destinée du Wasan – Aventure et mésaventure de ces mathématiques', in *Proceedings of the 14th International Congress of the History of Science*, Volume 2, Tokyo, 184–99.

———— (1980), 'Wallis' *Arithmetica infinitorum* and Takebe's *Tetsujutsu Sankei*: What underlies their similarities and dissimilarities?', *Historia Scientarium*, **19**, 77–100.

———— (1994), 'Indigenous Japanese mathematics, *Wasan*', in *Companion Encyclopedia of the History and Philosophy of the Mathematical Sciences*, I. Grattan-Guinness (ed.), London, Routledge, 104–10.

J. V. Murra (1968), 'An Aymara kingdom in 1567', *Ethnohistory*, **15**, 115–51.

O. Neugebauer (1935/7), *Mathematische Keilschrift-Texte Quellen und Studien*, 3 volumes, Berlin, Springer-Verlag.

———— and A. Sachs (1945), *Mathematical Cuneiform Texts*, New Haven, CT, Yale University Press.

H. Nissen, P. Damerow and R. Englund (1993), *Archaic Bookkeeping*, Chicago, Chicago University Press.

Y. Ohashi (1997), 'Astronomy in Tibet', in *Encyclopaedia of the History of Science, Technology and Medicine in Non-Western Cultures*, H. Selin (ed.), Dordrecht, Kluwer, 136–9.

Open University (1975), *Counting: I. Primitive and More Developed Counting Systems*, AM 289 N1, Milton Keynes, Open University Press.

———— (1976), *Written Numbers*, Vol. 3, 'History of mathematics: Counting, numerals and calculation', Milton Keynes, Open University Press.

R. Palter (1993), '*Black Athena*, Afro-centrism, and the history of science', *History of Science*, **31**, 227–87.

S. Parameswaran (1980), 'Kerala's contributions to mathematics and astronomy', *Journal of Kerala Studies*, **7**, 135–47.

———— (1983), 'Madhava of Sangamagramma', *Journal of Kerala Studies*, **10**, 185–217.

R. A. Parker (1972), *Demotic Mathematical Papyri*, Brown Egyptological Studies VII, Providence, RI.

T. E. Peet (1923), 'Arithmetic in the Middle Kingdom', *Journal of Egyptian Archaeology*, **9**, 91–5.

———— (1931), 'A problem in Egyptian geometry', *Journal of Egyptian Archaeology*, **17**, 100–106.

D. Pingree (1981), 'History of mathematical astronomy in India', in *Dictionary of Scientific Biography*, Vol. 15 (Suppl.), C. C. Gillespie (ed.), New York, Charles Scribner's Sons.

———— (1981), 'Power series in medieval Indian trigonometry', *Proceedings of the South Asian Seminar II*, University of Pennsylvania, 25–30.

Guaman Poma de Ayala (1936), *Nueva Corónica y Buen Gobierno*, Paris, Institute d'Éthnologie.

C. T. Rajagopal and K. Mukanda Marar (1944), 'On the quadrature of the circle', *Journal of the Royal Asiatic Society* (Bombay Branch), **20**, 65–82.

C. T. Rajagopal and M. S. Rangachari (1978), 'On an untapped source of medieval Keralese mathematics', *Archive for History of Exact Sciences*, **18**, 89–102.

———— ———— (1986), 'On medieval Keralese mathematics', *Archive for History of Exact Sciences*, **35**, 91–9.

———— and T. V. Vedamurthi Iyer (1952), 'On the Hindu proof of Gregory's series', *Scripta Mathematica*, **18**, 65–74.

———— and A. Venkataraman (1949), 'The sine and cosine power-series in Hindu mathematics', *Journal of the Royal Asiatic Society* (Bengal Branch), **25**, (1), 1–13.

P. Rajagopal (1991), 'The *Sthanga Sutra* programme in Indian mathematics', *Arhat Vacana*, **3**, (2), 1–8.

S. Ramanujan (1985), *Notebooks*, B. C. Berndt (ed.), New York, Springer-Verlag.

K. Ramasubramanian, M. D. Srinivas and M. S. Sriram (1994), Modification of the earlier Indian planetary theory by the Kerala astronomers (*c.* AD 1500) and the implied heliocentric picture of planetary motion, *Current Science*, **66**, (10), 784–90.

S. R. Rao (1973), *Lothal and the Indus Civilisation*, New York, Asia Publishing Co.

R. Rashed (1986), *Al-Tusi, Oeuvres mathématiques: Algèbre et géométrie au XII siècle*, Paris, Les Belles Lettres.

S. A. H. Rizvi (1983), 'On trisection of an angle leading to the derivation of a cubic equation and computation of value of sine', *Indian Journal of History of Science*, **19**, (1), 77–85.

C. S. Roero (1994), 'Egyptian mathematics', in *Companion Encyclopedia of the History and Philosophy of the Mathematical Sciences*, I. Grattan-Guinness (ed.), London, Routledge, 30–45.

F. Rosen (1831), *The Algebra of Muhamed ben Musa*, London, Oriental Translation Fund.

A. S. Saidan (1974), 'Arithmetic of Abul Wafa', *ISIS*, **65**, 367–75.

Sang-Woon Jeon (1974), *Science and Technology in Korea*, Cambridge, MA, MIT Press.

T. A. Sarasvati (1963), 'Development of mathematical series in India', *Bulletin of the National Institute of Sciences*, **21**, 320–43.

G. B. Sarfati (1968), *Mathematical Terminology in Hebrew Scientific Literature of the Middle Ages*, Jerusalem, Magnes Press.

K. V. Sarma (1972), *A History of the Kerala School of Hindu Astronomy*, Hoshiarpur, Vishveshvaranand Institute.

———— (1986), 'Some highlights of astronomy and mathematics in medieval India', *Sanskrit and World Culture*, **18**, 595–605.

———— and S. Hariharan (1991), 'Yuktibhasa of Jyesthadeva', *Indian Journal of History of Science*, **26**, 186–207.

R. M. Savory (1976), *Introduction to Islamic Civilisation*, Cambridge, Cambridge University Press.

G. Saxe (1982), 'Developing forms of arithmetical thought among the Oksapmin of Papua New Guinea', *Developmental Psychology*, **18**, (4), 583–94.

A. Sayili (1960), 'Thabit ibn Qurra's generalisation of the Pythagorean theorem', *ISIS*, **51**, 35–7.

A. Scharff (1922), 'Ein Rechnungsbuch des Königlichen Hofes aus der 13. Dynastie (Papyrus Boulaq Nr. 18)', *Zeitschrift für ägyptische Sprache und Altertumskunde*, **57**, 58–9.

A. Schmandt-Bessarat (1992), *Before Writing*, 2 volumes, Austin, TX, Texas University Press.

—————— (1994), 'Oneness, twoness, threeness', in *From Five Fingers to Infinity*, F. Swetz (ed.), New York, Open Court.

C. J. Scriba (1966), 'John Wallis's *Treatise on Angular Sections* and Thabit ibn Qurra's generalisations of the Pythagorean theorem', *ISIS*, **57**, 55–66.

A. Seidenberg (1962), 'The ritual origins of geometry', *Archive for History of Exact Sciences*, **1**, 488–522.

—————— (1983), 'The geometry of Vedic rituals', in *Agni, the Vedic Ritual of the Fire Altar*, F. Staal (ed.), Berkeley, CA, Asian Humanities Press, 95–126.

C. Selenius (1975), 'Rationale of the *Chakravala* process of Jayadeva and Bhaskara II', *Historia Mathematica*, **2**, 167–84.

Shigeru Jochi and B. Rosenfeld (1997), 'Seki Kowa', in *Encyclopaedia of the History of Science, Technology and Medicine in Non-Western Cultures*, H. Selin (ed.), Dordrecht, Kluwer, 890–92.

L. Shirley (1988), *Historical and Ethnomathematical Algorithms for Classroom Use*, Zaria, Nigeria, Ahamadu Bello University, Mimeo.

N. Singh (1984), 'Foundations of logic in ancient India: Linguistics and mathematics', in *Science and Technology in Indian Culture*, A. Rahman (ed.), New Delhi, National Institute of Science, Technology and Development Studies (NISTAD).

—————— (1987), *Jain Theory of Actual Infinities and Transfinite Infinities*, New Delhi, National Institute of Science, Technology and Development Studies (NISTAD).

P. Singh (1985), 'The so-called Fibonacci numbers in ancient and medieval India', *Historia Mathematica*, **12**, 229–44.

D. Singmaster (1994), 'Recreational mathematics', in *Companion Encyclopedia of the History and Philosophy of the Mathematical Sciences*, I. Grattan-Guinness (ed.), London, Routledge, 1568–75.

E. Sondheim and A. Rogerson (1981), *Numbers and Infinity*, Cambridge, Cambridge University Press.

H. J. Spinden (1924), 'The reduction of Mayan dates', Papers of the Peabody Museum of American Archaeology and Ethnology, Harvard University, Cambridge, MA.

F. Staal (1978), 'The ignorant Brahmin of the Agnicayana', *Annals of the Bhandarkar Oriental Research Institute*, Pune, Diamond Jubilee Number, 337–48.

—————— (1995), 'The Sanskrit of science', *Journal of Indian Philosophy*, **23**, 73–127.

I. Stewart (1981), *Concepts of Modern Mathematics*, London, Penguin.

B. V. Subbarayappa (1993), *Numerical System of the Indus Valley Civilisation: A Monograph on the Indus Script*, Bangalore, Indian Council of World Culture, Mimeo.

F. Swetz and T. I. Kao (1977), *Was Pythagoras Chinese? An Examination of Right Triangle Theory in Ancient China*, University Park, PA, Pennsylvania State University Press.

G.Thibaut (1875), 'On the Sulba-sutra', *Journal of the Asiatic Society of Bengal*, **44**, 227–75.

A. Thom (1967), *Megalithic Sites in Britain*, Oxford, Oxford University Press.

F. Thureau-Dangin (1938), *Textes mathématiques Babyloniens*, Leiden, Springer-Verlag.

UNESCO (1986), Meeting of Experts on Comparative Philosophical Studies on Changes in Relations between Science and Society, Working Document, New Delhi, National Institute of Science, Technology and Development Studies (NISTAD).

B. Krishna Tirthaji (1965), *Vedic Mathematics*, New Delhi, Motilal Banarsidass.

A. M. Tozzer (1941), *Land's Relacion de las Cosa de Yucatán. A Translation with Notes*, Papers of the Peabody Museum, Vol. 18. Cambridge, MA, Harvard University.

Garcilaso de la Vega (1966), *The Royal Commentaries of the Incas*, translated by H. V. Livermore, Austin, TX, University of Texas Press.

Wang Ling and J. Needham (1955), 'Horner's method in Chinese mathematics: Its origins in the root extraction procedures of the Han dynasty', *Toung Pao*, **43**, 345–88.

H. Wassen (1931), 'The ancient Peruvian abacus', *Comparative Ethnological Studies*, **9**, 191–205.

C. M. Whish (1835), 'On the Hindu quadrature and infinite series of the proportion of the circumference to the diameter exhibited in the four *Sastras*, the *Tantrasangraham*, *Yucti-Bhasa*, *Carana Padhati* and *Sadratnamala*', *Transactions of the Royal Asiatic Society of Great Britain and Ireland*, **3**, Part 3, 509–23.

K. Yabbuchi (1954), 'Indian and Arabian astronomy in China', in *Zinbun Kagaku Kenkyusyo*, Silver Jubilee Volume, Kyoto, Kyoto University Press.

Name Index

◇◇

Subject Index

◇◇◇

The Crest of the Peacock

contribution 303; *see also individual branches of mathematics and translation*
area, calculation of 82–4, 113, 115, 132, 157; Vedic 228, 230–33, 234–6
arithmetic: and algebra 249, 344–5; Arabs 309–10, 311, 316–23, 344–5; Babylonia 15; China 132; Egypt 63–76; India 218, 257, 269, 270
Asia, Central, 303, 304, 347
assumption, false 78–9, 158–9, 172, 211
Assyrian empire 17, 93, 94, 95, 379
astronomy 363: Arabs 221, 304, 305–6, 310–11, 344, 345, 347–8; Assyria 93, 94; Babylonia 11, 15, 17, 18, 93, 96, 265, 285; and calculus 298–9; China 11, 19, 131, 132, 136, 138, 179, (Indian influence) 19, 213; Greece 347; Hebrew 421; Hellenistic 11, 12, 18, 265; Inca 40–41; India 265–71, 275, 281, (influence abroad) 11, 213, 221, 304, 305–6, (influences on) 17, 18, 347–8, (Kerala) 219, 270–71, 287, 408, 410, 416, (Vedic) 18, 218, 226, 265, 398–9; Ishango Bone 26–7; Maya 15, 22, 54; Persia 15, 18, 303–4, 347–8; Seleucid 93, 96; Tibet 390–92; and trigonometry 281, 345; *see also* calendars
Atharvaveda 236–9, 244
Australia 41, 43
Aztecs 46–7

Babylonia 6, 91–124; and Alexandria 8, 312, 323, 377–84; algorithms 127, 129; approximation 126, 128; Arabs and 303, 305, 344–5; Egypt compared 101–2, 125–9; empiricism 8, 126, 127–8, 196, 345; and Greece 8, 9, 14–15, 16–17, 126, 196, 345; history 15, 91–5, 96, 135; and India 14–15,

17, 18, 180, 265; place-holder 101, 145; proof 126, 127–8; reciprocals 109, 128; sources 6, 95–6, (*see also* tablets); *see also under individual aspects of mathematics*
Baghdad 10, 11, 303, 305, 354
Bait al-Hikma 11, 303, 305
Bakhshali Manuscript 216, 218, 219, 241, 256–63, 272
binomial theorem 218
'body part' counting 359, 360
bones: Chinese oracle 130, 140–41; tallying 24–7
Buddhism 6n; China 19, 136, 212–13; India 18, 19, 218, 219, 221

calculus 20, 270, 298–300
calendars: African 24, 26–7; Arab 306, 309, 320; Babylonian 100; Chinese 136, 138, 178, 179, 207–8; Christian 312; duodecimal 44; Egyptian 59; Indian 397, 399; lunar 360, 361; Mayan 15, 22, 51–4, 55; Pacific 360; Tibetan 391
ceremonial dances 360
chakravala 279–80
China 2, 5, 12–13, 130–214, 385–94; assumption, false 158–9, 172, 211; background and sources 15, 130–40; colour symbols 31, 146; determinants 172, 176; dissection and reassembly principle 181–2, 188; divination 150; and Euclid 134, 139, 205, 214; European contacts 14–15, 20–21, 138, 191, 204–5, 212, 214; and India 14–15, 19, 22, 137, 139, 211, 212–13, 236, 265, (Kerala) 294, (zero) 145, 213; interpolation 134, 179; and Japan 137, 212; and Korea 137, 212; large numbers 132; magic squares 132, 134, 136, 148–56; numerology 149; oracle bones 130, 140–41; progressions 133, 158; proportions 158; rod numerals 15, 31, 48, 140–48, 176, 201; series 133, 138, 179, 296;

448

Micawber Books
Princeton, NJ
Wed 13 April 2005
#4.00 + .24 tax